ENVIRONMENTAL STATISTICS and DATA ANALYSIS

WAYNE R. OTT

Library of Congress Cataloging-in-Publication Data

Ott, Wayne
Environmental statistics and data analysis / Wayne R. Ott
p. cm.
Includes bibliographical references and index.
1. Environmental sciences—Statistical methods. I. Title.
GE45.S73088 1995
363.7′0072—dc 20 94–15435
ISBN 0–87371–848–8 (acid-free paper)

Direct all inquiries to CRC Press, Inc., 2000 Corporate Blvd., N.W., Boca Raton, Florida 33431.

Lewis Publishers is an imprint of CRC Press

International Standard Book Number 0–87371–848–8

Library of Congress Card Number 94–15435
Printed in the United States of America
2 3 4 5 6 7 8 9 0

To my father and mother,

Florian and Evelyn Ott

and my sister,

Le verne

Wayne R. Ott has been with the U.S. Environmental Protection Agency (EPA) and its predecessor agencies for over 28 years, conducting research on environmental statistics, environmental data analysis, environmental indices, environmental decision analysis, mathematical modeling of human exposure to environmental pollutants, Monte Carlo simulation, environmental field studies, measurement of human exposure, indoor air quality, stochastic modeling of environmental processes, human activity patterns, and quality assurance of environmental measurements.

Dr. Ott earned a B.A. degree from Claremont McKenna College in business economics; a B.S. degree in electrical engineering, an M.A. degree in communications, and an M.S. degree in engineering science from Stanford University; and received his Ph.D. in civil and environmental engineering from Stanford University. He is a member of the American Statistical Association, the International Society of Exposure Analysis, the Air and Waste Management Association, Kappa Mu Epsilon (mathematics honor society), Tau Beta Pi (engineering honor society), Sigma Xi (scientific research society), and Phi Beta Kappa (honorary member). He is a Commissioned Officer in the U.S. Public Health Service (PHS) with the rank of Captain.

In the 1970's, Dr. Ott developed mathematical techniques for analyzing water and air quality data, leading to the proposed structure of the Pollutant Standards Index (PSI), a nationally uniform air pollution index now used throughout the U.S. and in other countries. He received the Public Health Service Commendation medal in 1977 for his work in developing the PSI. In 1980, he received a competitive EPA Innovative Research Program grant award to become a visiting scholar at Stanford University's Department of Statistics in conjunction with the Societal Institute for the Mathematical Sciences (SIMS) research program. While at Stanford, he developed and wrote the computer source code for the Simulation of Human Air Pollutant Exposure (SHAPE) model, the first human activity pattern-exposure model.

He contributed to development of the Total Human Exposure concept and to EPA's Total Exposure Assessment Methodology (TEAM) field studies.

Dr. Ott has written or co-authored over 100 research reports, technical papers, and journal articles on statistical analysis of environmental data, mathematical modeling of human exposure, environmental indices, probabilistic concepts, indoor air quality, measurment methods, quality assurance, environmental data analysis, field study design, and other related topics.

Acknowledgments

I wish to acknowledge the skill, care, and tireless energy provided by the reviewers of this book, who were kind enough to examine the draft manuscript and recommend numerous changes and improvements over a period spanning a decade. Several chapters were reviewed by Steven Bayard, James Repace, Ralph Larsen, and Lynn Hildemann. All chapters were reviewed by David Mage, Lance Wallace, and William Sayers. I am extremely grateful for the important contribution they made to the completion of this book. I also appreciate the thoughtful suggestions by Paul Switzer on the concept of successive random dilutions applied to stream water quality. I am also grateful for the editorial assistance provided by Vivian Collier, Virginia Zapitz, and Le verne McClure.

NOTICE

Preface

Many random processes occur in nature. To make accurate predictions about the manner in which man's activities may alter these processes and affect their outcomes, it is necessary to construct models that faithfully represent reality, including its random components. In the environmental field, the purpose of most models is to make accurate predictions about the effect of environmental pollution control activities on environmental quality or on human exposure to pollutants. Although modeling the fate and transport of pollutants through the environment is well advanced, few models have been developed adequately to include the random, or *stochastic*, nature of environmental phenomena. Stochastic models treat the phenomenon being modeled probabilistically, thus including the random components in a statistical framework. Because random phenomena abound in the environment, stochastic modeling often is more important than other kinds of modeling. Despite the importance of stochastic models in environmental analyses and decision-making, few reference works are available covering these techniques and showing how to apply them to environmental problems. Of particular importance for concentrations measured in the environment are the right-skewed distributions, such as the lognormal, and techniques to take into account source controls, such as rollback models (see Chapter 9).

To help fill the need for a reference work on environmental statistics, this book seeks to develop a comprehensive and understandable framework for applying probabilistic techniques to environmental problems of all kinds. It includes statistical models for environmental decision-making, data analysis, and field survey design, along with the theoretical basis for each model wherever possible. A model that has a sound theoretical basis is more likely to make accurate predictions than one that does not. This book also includes a considerable body of original material, not previously published, on new theories and insights to help explain observed environmental phenomena and the control of these phenomena. The new theories are included to help guide data analysts and decision makers in applying statistical models to practical problems in the environment.

The book is intended to provide students, managers, researchers, field monitoring specialists, engineers, statisticians, data analysts, and environmental decision makers with both a reference source and a body of statistical procedures for analyzing environmental data and making environmental predictions. In its structure, this book includes full documentation of each probability model presented, the theoretical basis for its origin, and examples of its application to environmental problems. In addition, BASIC computer programs, which can be programmed readily on personal computers, are included in the book wherever possible. Typical uses of these probability models include analyzing environmental monitoring data, describing the frequency distribution of exposures of the population, deciding the degree to which environmental measurements comply with health-related standards, and predicting the effect of pollutant source reductions on environmental quality. The book also introduces several new techniques, not presented elsewhere, that are important for solving practical problems.

This book is dedicated to the concept that selection of a stochastic model should be based not merely on its presumed "good fit" to empirical data; rather, it should be consistent with the basic theory of the underlying physical laws responsible for generating the observed data. To assist in determining which models are theoretically appropriate for certain physical processes, the idealized physical conditions associated with each probability distribution are presented in detail. It may appear surprising to some readers, for example, to discover that commonly applied Gaussian diffusion plume models can arise naturally from probability theory and from fairly simple "random walk" Brownian motion examples.

In Chapter 8, the Theory of Successive Random Dilutions (SRD) is proposed to explain why lognormal appearing distributions occur so commonly in the diverse fields of indoor air quality, water quality, and geological systems. Hopefully, the concepts presented in this book will stimulate other investigators to consider why right-skewed distributions with a single mode occur so often in environmental pollution data. It is hoped that the SRD concepts presented in this book are fundamental to developing a general theory, and that others in the future will help extend these principles.

For more than three decades, papers have appeared in the literature on the effect of source reductions on environmental concentrations, but these predictions seem to be based more on hunches of the authors than on a proven theory or on experimental evidence. In Chapter 9, a Statistical Theory of Rollback (STR) is introduced for the first time to provide a basis for predicting the effect of changes in sources on observed environmental concentrations. The STR is derived from basic statistical concepts. Some experimental observations are included in Chapter 9 to illustrate how this theory might be applied in practice. It is hoped that this statistical theory will be applicable to a wide variety of environmental problems.

To ease presentation of the theories presented in this book, many practical examples and commonplace physical analogs are included, both to help show how a process works and to describe the physical laws responsible.

The objective of this volume is to communicate basic statistical theory to a broad environmental audience — chemists, engineers, data analysts, and managers — with as little abstract mathematical notation as possible, but without omitting important details and assumptions. It is assumed only that the reader has a basic knowledge of algebra and calculus. I hope that this book can serve as a reference source for those wishing to analyze, understand, and predict environmental quality variables in many practical settings. I also hope that this book will help provide a beginning toward establishing a comprehensive body of knowledge known as "Environmental Statistics."

Contents

ENVIRONMENTAL STATISTICS and DATA ANALYSIS

1 Random Processes

Random: A haphazard course — *at random:* without definite aim, direction, rule or method[1]

The concept of "randomness," as used in common English, is different from its meaning in statistics. To emphasize this difference, the word *stochastic* commonly is used in statistics for *random*, and a *stochastic process* is a *random process*. A stochastic process is one that includes any random components, and a process without random components is called *deterministic*. Because environmental phenomena nearly always include random components, the study of stochastic processes is essential for making valid environmental predictions.

To most of us, it is comforting to view the world we live in as consisting of many identifiable cause-effect relationships. A "cause-effect" relationship is characterized by the certain knowledge that, if a specified action takes place, a particular result always will occur, and there are no exceptions to this rule. Such a process is called deterministic, because the resulting outcome is determined completely by the specified cause, and the outcome can be predicted with certainty. Unfortunately, few components of our daily lives behave in this manner.

Consider the simple act of obtaining a glass of drinking water. Usually, one seeks a water faucet, and, after it is found, places an empty glass beneath the faucet and then turns the handle on the faucet. Turning the handle releases a piston inside the valve, allowing the water to flow. The process is a simple one: the act of turning the handle of the faucet (the "cause") brings about the desired event of water flowing (the "effect"), and soon the glass fills with water.

Like so many other events around us, this event is so familiar that we ordinarily take it for granted. If, before we operated the faucet, someone asked us, "What will happen if the handle of the faucet is turned?", we would be willing to predict, with considerable certainty, that "water will appear." If we had turned the handle and no water appeared, we probably would conclude that there is something wrong with the plumbing. Why do we feel so comfortable about making this simple cause-effect prediction? How did we arrive at this ability to predict a future event in reality?

In our mind, we possess a conceptual framework, or a "model," of this process. This model has been developed from two sources of information: (1) Our knowledge of the physical structure of faucets, valves, and water pipes and the manner in which they are assembled, and (2) Our historical experience with the behavior of other water faucets, and, perhaps, our experience with this particular faucet. The first source of knowledge comes from our understanding of the physical construction of the system and the basic principles that apply to water under pressure, valves that open and close, etc. For example, even if we had never seen a faucet or a valve before, we might be willing to predict, after the mechanism and attached pipe were described to us in detail, that turning the handle of the faucet would release the water. The second source of knowledge is derived from what we have learned from our experience with other, similar faucets. We reason thus: "Turning the faucet handle always has caused the water to flow in the past, so why shouldn't it do so in the future?" The first source of knowledge is theoretical and the second source is empirical (that is, based on actual observations). If only the second source of knowledge were available — say, 179 cases out of 179 tries in which the faucet handle is turned on and the water appears — we probably would be willing to predict (based on this information alone and with no knowledge of the internal workings of the system) that turning the handle the next time — for the 180*th* try — would allow us fill the glass with water.

These two independent sources of information — physical knowledge of the structure of a system and observational knowledge about its behavior — greatly strengthen our ability to make accurate predictions about the system's future behavior. From the first source of information, we can construct a *conceptual model* based on the internal workings of the system. From the second source of information, we can validate the conceptual model with real observations. The first source of information is theoretical in nature; the second one is empirical. A theoretical model validated by empirical observation usually provides a powerful tool for predicting future behavior of a system or process.

Unfortunately, the world about us does not always permit the luxury of obtaining both sources of information — theory and observation. Sometimes, our knowledge of the system's structure will be vague and uncertain. Sometimes, our observational information will be very limited. Despite our lack of information, it may be necessary to make a prediction about the future behavior of the system. Thus, a methodology that could help us analyze existing information about the system to improve the accuracy of our predictions about its future behavior would be extremely useful.

Consider the above example of the water faucet. Suppose little were known about its construction, attached pipes, and sources of water. Suppose that the faucet behaves erratically: when the handle is turned, sometimes the water flows and sometimes it does not, with no obvious pattern. With such uncertain behavior of the device, we probably would conclude that unknown factors (for example, clogged pipes, defective pumps, broken valves, inadequate water supplies, wells subject to rising and falling water tables) are affecting this system. The faucet may, in fact, be attached to a complex network of pipes, tanks, valves, filters, and other devices, some of which are faulty or controlled by outside forces. Because the arrival of water will depend on many unknown factors beyond the user's control, and because the outcome of each particular event is uncertain, the arrival of water from the faucet may behave as a stochastic process.

How can one make predictions about the behavior of such a process? The first step is to express the event of interest in some formal manner — such as a

"1" or "0" — denoting the presence or absence of water, or a quantitative measure (gm or m^3) denoting the amount of water arriving in a fixed time interval. Such a quantitative measure is called a *random variable*. A random variable is a function of other causative variables, some of which may or may not be known to the analyst. If all of the causative variables were known, and the cause-effect relationships were well-understood, then the process would be deterministic. In a deterministic process, there are no random variables; one can predict with certainty the rate at which the water will flow whenever the valve is opened by examining the status of all the other contributing variables.

In view of the uncertainty present in this system, how does one develop sufficient information to make a prediction? If we had no prior information at all — neither theoretical nor empirical — we might want to flip a coin, showing our total uncertainty, or lack of bias, about either possible outcome. Another approach is to conduct an *experiment*. Let K be a random variable denoting the arrival of water: if $K = 0$, water is absent, and if $K = 1$, water is present. Each turning of the faucet handle is viewed as a *trial* of the experiment, because we do not know beforehand whether or not the water will appear. Suppose that the faucet handle is tried 12 times, resulting in the following series of observations for K: $\{1,0,1,1,1,0,1,0,1,1,1,1\}$. Counting the ones indicates that the water flowed in 9 of the 12 cases, or 3/4 of the time. What should we predict for the 13*th* trial? The data we have collected suggest there may be a bias toward the presence of water, and our intuition tells us to predict a "success" on the 13*th* trial. Of course, this bias may have been merely the result of chance, and a different set of 12 trials might show a bias in the other direction. Each separate set of 12 trials is called a *realization* of this random process. If there are no dependencies between successive outcomes, and if the process does not change (i.e., remains "stationary") during the experiment, then the techniques for dealing with Bernoulli processes (Chapter 4) provide a formal methodology for modeling processes of this kind.

Suppose that we continue our experiment. The faucet is turned on 100 times, and we discover that the water appears in 75 of these trials. We wish to predict the outcome on the 101*st* trial. How much would we be willing to bet an opponent that the 101*st* trial will be successful? The information revealed from our experiment of the first 100 trials suggests a bias toward a successful outcome, and it is likely that an opponent, witnessing these outcomes, would not accept betting odds of 1:1. Rather, the bet might be set at odds of 3:1, the ratio of past successes to failures. The empirical information gained from our experiment has modified our future predictions about the behavior of this system, and we have created a model in our minds. If an additional number of trials now were undertaken, say 1,000, and if the basic process were to remain unchanged, we would not be surprised if water appeared in, say, 740 outcomes.

The problem becomes more difficult if we are asked to compare several experiments — say, two or more different groups of 1,000 observations from the faucet — to determine if a change has occurred in the basic process. Such comparison is a "trend" analysis, since it usually utilizes data from different time periods. Probabilistic concepts must be incorporated into such trend analyses to help assess whether a change is real or due to chance alone.

Our intuitive *model* of this process is derived purely from our empirical observations of its behavior. A model is an abstraction of reality allowing one to make predictions about the future behavior of reality. Obviously, our model will be more successful if it is based on a physical understanding of the characteristics of the process as well as its observed past behavior. The techniques described in this book provide formal procedures for constructing stochastic models of environmental processes, and these techniques are illustrated by applying them to examples of environmental problems.

Many of these techniques are new and have not been published elsewhere, while some rely on traditional approaches applied in the field of stochastic modeling. It is hoped that, by bridging many fields, and by presenting several new theories, each technique will present something new that will provide the reader with new insight or present a practical tool that the reader will find useful.

STOCHASTIC PROCESSES IN THE ENVIRONMENT

The process described above is an extremely simple one. The environmental variables actually observed are the consequence of thousands of events, some of which may be poorly defined or imperfectly understood. For example, the concentration of a pesticide observed in a stream results from the combined influence of many complex factors, such as the amount of pesticide applied to crops in the area, the amount of pesticide deposited on the soil, irrigation, rainfall, seepage into the soil, the contours of the surrounding terrain, porosity of the soil, mixing and dilution as the pesticide travels to the stream, flow rates of adjoining tributaries, chemical reactions of the pesticide, and many other factors. These factors will change with time, and the quantity of pesticide observed in the stream also varies with time. Similarly, the concentrations of an air pollutant observed in a city often are influenced by hundreds or thousands of sources in the area, atmospheric variables (wind speed and direction, temperature, and atmospheric stability), mechanical mixing and dilution, chemical reactions in the atmosphere, interaction with physical surfaces or biological systems, and other phenomena. Even more complex are the factors that affect pollutants as they move through the food chain — from sources to soils, to plants, to animals, and to man — ultimately becoming deposited in human tissue or in body fluids. Despite the complexity of environmental phenomena, many of these processes share certain traits in common, and it is possible to model them stochastically. There is a growing awareness within the environmental community of the stochastic nature of environmental problems.

Ward and Loftis[2] note that, with the passage of the Clean Water Act (Public Law 92-500), water quality management expanded both its programs (permits and planning) and the money devoted to wastewater treatment plants. The data collected at fixed water quality monitoring stations* assumed a new role: to identify waters in violation of standards and to evaluate the effectiveness of expenditures of the taxpayers' money. They conclude that water quality monitoring was expected to serve as a "feedback loop" by which to evaluate the effec-

*A fixed water quality monitoring station usually consists of a set of sampling points (stations) at which samples are taken (usually "grab" samples) approximately once per month.[2,3]

tiveness of regulatory programs. Unfortunately, unless the stochastic properties of these data are taken into account, the inherent randomness of these data will conceal real changes in environmental conditions:

> When data, collected to check only if a sample meets a standard, are used to evaluate management's success, the stochastic variation in the data often completely masks any improvement in controlling society's impact on water quality. Since the data cannot show management's effectiveness, the conclusion is that fixed station data are useless. However, the data are not useless: they are simply being asked to provide information they cannot show, without further statistical analysis.[2]

Standards in air and water pollution control usually are long-term goals that are well-suited to statistical formulation. Previous environmental standards often have been deterministic, however, perhaps because it was believed that probabilistic forms would complicate enforcement activities. Practically, it is impossible to design a regulatory program that can guarantee that any reasonable standard *never* will be violated, and there is a growing awareness that probabilistic concepts should be an integral part of the standard setting process. Ewing[4] states that water quality standards should be formulated in a probabilistic manner:

> The establishment of state water quality standards for both interstate and intrastate streams has recently been accomplished. In practically every case, DO [dissolved oxygen] requirements have been set without any reference to the probability of these levels being exceeded. It would seem that the state-of-the-art is rapidly advancing to the point, however, where the probabilistic concept should be recognized more specifically in the statement of the water quality standards themselves.

Drechsler and Nemetz[5] believe that incorporation of probabilistic concepts places greater demands on the design of the monitoring program but will yield a more efficient and effective water pollution control program:

> We recommend that, where appropriate, standards be altered to reflect the probability of occurrence of pollution events. This will require a greater degree of information concerning both the distribution of pollutant discharges and biological damage functions.... These measures will help overcome some of the significant weaknesses in the current regulatory system for the control of water pollution and will be more efficient and effective in the protection of both corporate and social interests.

In the air pollution field, significant progress has been made toward incorporating probabilistic concepts into the basic form of the standards. In 1979, for example, the Environmental Protection Agency (EPA) revised its National Ambient Air Quality Standard (NAAQS) for ozone from a deterministic form to a probabilistic form. The probabilistic form was based on the "expected number of days" in which the measured ozone level exceeds a certain hourly average value. The NAAQS apply uniformly on a nationwide basis, and compliance is evaluated using data collected at fixed air monitoring stations* operated in most U.S. cities. The Clean Air Act (Public Law 91-604) requires that the NAAQS be evaluated and revised from time to time, and it is anticipated that future NAAQS will be formulated on a similar statistical framework. As indicated by Curran,[6] probabilistic air quality standards have important advantages for accommodating missing data and handling rare events occurring in an unusual year:

> While the initial short-term NAAQS promulgated by EPA in 1971 were stated as levels not to be exceeded more than once per year, the revised ozone standard, which was promulgated in 1979, stated that the expected annual exceedance rate should not be greater than one. From an air quality data analysis viewpoint, this change is intuitively appealing in that it incorporates an adjustment for incomplete sampling and provides a framework for quantifying the impact of an unusual year, particularly in the development of design values. From a statistical viewpoint, this change represents a transition from the simple use of observed measurements to the use of estimates of underlying parameters.

Statistical techniques for judging compliance with these probabilistic standards are presented in Chapters 4 and 5. It appears likely that there will be steady progress toward incorporating probabilistic concepts into future environmental standards.

STRUCTURE OF BOOK

The present chapter discusses the need for probabilistic concepts in the environmental sciences and introduces common terms in statistics (e.g., deterministic and stochastic processes, models, random variables, experiments, trials, realizations). Chapter 2 briefly presents and reviews the theory of probability, including such concepts as the union and joint occurrence of events, conditional probability, and Bayes' Theorem. It includes a bibliography of books written on probability, statistics, and data analysis over four decades. Chapter 3 introduces the formal concept of probability models, both discrete and continuous, discussing measures of central tendency, dispersion, skewness, and kurtosis. It also discusses data analysis, histograms, probability plotting, and fitting probability models to observations, including goodness-of-fit tests. Chapter 4 presents Bernoulli processes, discussing the theoretical conditions required for the binomial distribution, and illustrating the binomial model by applying it to cigarette smoking activities. The binomial probability model then is applied to the interpretation of air and water quality standards, discussing how one might han-

*Fixed air monitoring stations generally consist of a group of instruments that sample air continuously, housed in a single structure or building in the city, generating 1-hour or 24-hour concentration averages. Large cities may have 10 or more of these fixed stations.

dle environmental situations that do not meet the independence and stationarity conditions of the model. In Chapter 5, the theoretical conditions required for a Poisson process are presented, and several similar environmental applications of the Poisson probability model are discussed. It is shown that some interpretations of air quality standards can be handled more effectively by modeling them as Poisson rather than Bernoulli processes. Chapter 6 is intended as a transition from discrete probability models to continuous probability models. Using fairly simple examples, it shows how the symmetrical normal distribution arises naturally from diffusion processes, satisfying the diffusion equation, and it develops the theoretical basis for the Gaussian plume model. Chapter 7 presents the normal distribution in detail, showing an important practical use of the model: calculation of confidence intervals in random sampling surveys. Another class of physical processes — those involving dilution — gives rise to right-skewed distributions. Chapter 8 introduces, in detail, the Theory of Successive Random Dilutions to help explain why distributions that are approximately lognormal (i.e., the distribution of the logarithm of concentration is approximately normal) can arise from such a great variety of environmental phenomena. Chapter 8 is intended as a transition from symmetrical distributions to right-skewed concentration distributions, and Chapter 9 formally presents the lognormal distribution. Four different methods for estimating the parameters of the lognormal distribution from data are presented. Equations are given to assist the analyst in converting from the "normal parameters" to the "geometric parameters" to the "arithmetic parameters," or to any combination of these. Chapter 9 also introduces the Statistical Theory of Rollback, which provides statistical principles for predicting pollutant concentrations after controlling the sources of pollution. The objective is to make predictions about the distribution of concentrations after a source is controlled using information about the distribution before it is controlled. Although statistical rollback theory is general and applies to any distribution, Chapter 9 illustrates statistical rollback applied to the lognormal distribution. It illustrates statistical rollback approaches using exposure field measurements on a U.S. arterial highway at two time periods separated by 11 years. Chapter 9 brings together many of the principles and concepts presented earlier in the book, introducing techniques that draw upon principles presented in earlier chapters.

REFERENCES

1. Merriam Webster's Collegiate Dictionary, Tenth Edition (Springfield, MA: Merriam-Webster, Incorporated, 1993).
2. Ward, Robert C. and Jim C. Loftis, "Incorporating the Stochastic Nature of Water Quality Into Management," *J. Water Poll. Control Feder.* 55(4):408–414 (April 1983).
3. Ott, Wayne, *Environmental Indices: Theory and Practice* (Ann Arbor, MI: Ann Arbor Science Publishers, 1978).
4. Ewing, B.B., "Probabilistic Consideration in Water Quality Management," *Water Wastes Eng.* 7:50 (1970).

5. Drechsler, Herbert D. and Peter N. Nemetz, "The Effect of Some Basic Statistical and Biological Principles on Water Pollution Control," *Water Resour. Bull.* 14(5):1094-1104 (October 1978).
6. Curran, T.C., "Transition to Statistical Ambient Air Quality Standards," *Environmetrics 81: Selected Papers* (Philadelphia, PA: Society for Industrial and Applied Mathematics, 1981).

2 Theory of Probability

Over the last four decades, numerous introductory books and detailed texts have been published on the theory of probability, statistics, and data analysis, many of which are listed in the bibliography at the end of this chapter.[1–116] Simplified basic presentations of probability theory are available, such as Freedman et al.,[13] Freund,[15] Kattsoff and Simone,[27] and elementary textbooks such as Levin and Rubin,[30] McClave and Dietrich,[33] Triola,[46] Weiss and Hassett,[50] and Witte.[51] Comprehensive basic treatments of probability theory and statistics can be found in textbooks by Feller;[10,11] Lindgren;[32] Mood, Graybill, and Boes;[35] Parzen;[38] and Snedecor and Cochran.[41] Some books — such as Olkin, Gleser, and Derman;[37] Kendall and Stuart;[28] and Stuart and Ord[43,44] — give advanced presentations of probability theory and models. Books on *mathematical statistics*, such as Freund,[14] Hoel,[23] and Larsen,[29] emphasize the application of these methods to real-world problems. Some textbooks emphasize the analysis and interpretation of data, such as Anderson;[1] Berenson, Levine, and Rindskopf;[3] Brant;[5] Glenberg;[17] Hamburg;[20] Taylor;[45] and Tukey.[49] Some books deal with specific statistical topics, such as *extreme value theory* (Castillo;[53] Kinnison[54]) or *simulation* (Banks and Carson;[55] Bratley et al.[56]).

Textbooks also can be found on the application of probability theory to more specialized fields, such as business and management,[57–70] the social sciences,[71–74] and engineering and the physical sciences.[75–91] *Engineering statistics* is a sufficiently well-developed field that a number of comprehensive textbooks are available. Some engineering statistics books specialize in more narrow engineering fields, such as the application of statistics to civil engineering problems by Benjamin and Cornell.[75]

A number of books have been written on the application of probability theory to the biological and health sciences,[92–101] or *biostatistics*, such as Altman,[92] Daniel,[94] Glantz,[95] Kuzma,[96] Milton,[97] and Remington and Schork.[99] Specialized books dealing with stochastic models of ecological systems, such as Pielou,[98] also have been written. Many of the textbooks on stochastic processes[101–115] are relevant to environmental modeling and ecological systems.

Despite the large number of books available on general statistical methods,[1–56] stochastic processes,[102–115] applications of statistics to business problems,[57–70] the social sciences,[71–74] engineering,[75–91] and the health sciences,[92–101] there are very few books on statistical methods in the environmental sciences.

Existing texts on stochastic processes tend to be theoretical in nature—focusing on topics such as Brownian motion—with few practical applications of interest to environmental professionals. Hopefully, the present book will help fill the need for a practical reference on statistical methods in the environment and will provide useful tools for persons evaluating environmental quality, assessing trends, and making environmental decisions and predictions.

That there are so few books available on environmental statistics is surprising when one considers the vast quantities of data collected by environmental agencies, universities, and consulting firms on air pollutants, water pollutants, pesticides, ground water contaminants, and many other topics. A book by Gilbert[116] deals with the statistics of environmental monitoring and sampling. Because available texts provide excellent introductory treatments of probability theory, the present chapter is intended as a brief review, and the interested reader may wish to seek other textbooks on probability to study this topic in greater depth.

PROBABILITY CONCEPTS

Most of us, if we have ever gambled, have developed subjective notions about probability. If each of several events is equally likely to occur, then we assign to each the same probability. If a coin is equally likely to land on its "head" as on its "tail," then we customarily assign the probability of either a "head" or a "tail" to be $p = 1/2$. Similarly, a fair six-sided die has a probability $p = 1/6$ of landing on any one side. In general, if n events are equally likely to occur, then the probability of any one event is $p = 1/n$.

What do these assigned probabilities mean? Perhaps the most intuitively useful concept of probability is that of an asymptotic ratio. If the coin were flipped 10,000 times, we might guess that the total number of "heads" appearing will be very close to 5,000, making the proportion of flips resulting in "heads" very close to 1/2. Bulmer[7] reports an experiment conducted by Kerrich in Denmark during the second World War in which a coin actually was tossed 10,000 times, yielding 5067 heads. If m denotes the number of "heads" in n coin flips, then the ratio $m/n = 0.5067$ for this experiment. One can imagine the experiment continuing for even a larger number of trials, say $n = 1$ million, or $n = 1$ trillion. As the number of trials increases, it is reasonable to believe that the proportion m/n will become increasingly close to 1/2, and the probability $p = 1/2$ can be viewed as the limiting value of this ratio as the trials continue indefinitely.

Instead of the tedious chore of flipping a real coin 10,000 times, today's computers can be programmed to generate pseudorandom numbers and to accomplish essentially the same result by computer simulation. To simulate Kerrich's coin-flipping experiment, a computer program was written in Microsoft BASIC* to run on an IBM* personal computer (Figure 2.1). This simple program reads in the desired number of coin flips as N (line #130). For each of N coin flips in the computation loop (lines #150 to #220), it generates a uniformly distributed random variable between 0 and 1 using the BASIC function RND in line #160 (see Problem 1). If the resulting value is equal to or greater than 0.5,

*Microsoft is a registered trademark of Microsoft Corporation, and IBM is a registered trademark of International Business Machines.

```
100 REM PROGRAM TO FLIP A COIN AND COUNT THE NUMBER OF HEADS
110 SUM = 0!
120 CLS : PRINT : PRINT
130 INPUT "No. of Coin Flips Desired"; N
140 PRINT : PRINT "Count         Heads       No. of Heads      Prop. of Heads"
150 FOR I = 1 TO N
160 R = RND
170 H = 0
180 IF R >= .5 THEN H = 1
190 IF H = 1 THEN SUM = SUM + 1
200 PROP = SUM / I
210 PRINT I, H, SUM, PROP
220 NEXT I
230 PRINT : PRINT
240 PRINT "Final Values: N ="; N, SUM; : PRINT "Heads"
250 END
```

Figure 2.1. BASIC computer program to generate any number of simulated flips of a fair coin and to compute the proportion that comes out "heads."

it assumes a "head" has appeared and sets H = 1 (line #180); otherwise, H = 0. SUM counts the total number of heads that appear, since it is incremented by 1 whenever H = 1 (line #190). The counter I keeps track of the number of trials, beginning at I = 1 and ending at I = N, and the proportion of the trials resulting in "heads" is printed out on each coin flip (line #210).

This computer program was run to generate the outcomes of a series of successive coin flips. On the first 68 trials, the proportion m/n fluctuated above and below 1/2 several times, becoming more stable as n increased (Figure 2.2). The first coin flip gave a "head," making the ratio $m/n = 1$. A "tail" appeared on the second flip, giving a ratio of 0.5; on the third trial, a "head" made the ratio $m/n = 2/3 = 0.66667$. After 10 coin flips, there were 6 "heads," giving a

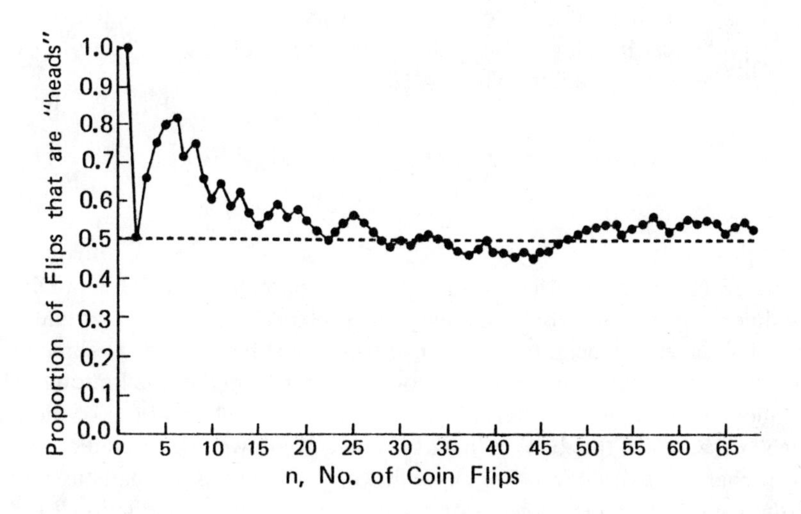

Figure 2.2. Proportion of "heads" appearing in a series of n flips of a fair coin, from computer simulation.

Table 2.1. Results of Computer Simulation of a Coin-Flipping Experiment

	Case 1	Case 2	Case 3	Case 4	Case 5
No. of flips n	10	100	1,000	10,000	20,000
No. of "heads" m	6	55	504	5,008	9,998
No. of "tails" n–m	4	45	496	4,992	10,002
Error[a]	+10%	+5%	+ 0.4%	+ 0.08%	– 0.01%

[a]Percent deviation from one-half of the proportion of flips that are "heads."

ratio of 0.6. These results represent a particular realization of this stochastic process. Using a different random number starting point (i.e., the "seed" of the random number generator; see Problem 1) would create a different realization, but the series would have the same statistical properties.

If the experiment is continued, the proportion m/n wanders above and below 1/2, sometimes spending long periods above or below 1/2. The results for 5 cases ranging from n = 10 to n = 10,000 show that the percentage deviation from 1/2 steadily declines as n increases, ranging from +10% for n = 10 to – 0.01% for n = 20,000 (Table 2.1). Actually, the percentage deviations from 1/2 for n = 10,000 and n = 20,000 in this realization are unusually small, and use of a different random number seed could make these deviations much greater (Problem 1). When this program was run to n = 100,000, a total of 50,118 "heads" appeared giving a deviation of 0.118%, which is larger than the result for n = 10,000 or n = 20,000. However, such fluctuations are within the normal statistical range expected for such a process. One's intuition suggests that the fluctuations would continue to decrease if the experiment were continued indefinitely, and the proportion of "heads" would cluster more and more closely to a limiting value which would be very near, if not exactly, one-half. This hypothetical limiting value is the probability p of a "head" in a fair coin, or $p = P\{\text{"head"}\}$. For a fair coin, $p = 1/2$:

$$\lim_{n \to \infty} \left(\frac{m}{n} \right) = p = 1/2$$

Notice that the probability is an asymptotic parameter of the coin-flipping experiment which one would see only if the experiment were continued forever. In a real experiment, the number of trials must be curtailed at some point, and the resulting ratio will deviate in some manner from the true limiting value. The above experiment is a Bernoulli process (Chapter 4), and use of the techniques presented in this book (Chapters 4 and 7) allows one to predict the statistical characteristics of m/n for any value of n. An important statistic describing the "spread" of these deviations, called the "standard deviation," can be shown to decrease as $1/\sqrt{n}$ for this process. Understanding such processes provides analytical insight into many environmental problems.

PROBABILITY LAWS

A Venn diagram is a graph allowing one conceptually to represent the occurence of various events of interest. Suppose A denotes the occurrence of a particular event, and $\bar{A}$ denotes the complementary event, its nonoccurrence. Then a Venn diagram representing the *sample space* (all possible events) would contain all the elements in A (circle with diagonal lines in Figure 2.3) and all the elements in $\bar{A}$ (remaining shaded area surrounding the circle in Figure 2.3). If $P\{A\}$ denotes the probability of the event A occurring and $P\{\bar{A}\}$ denotes the probability of the event not occurring, then:

$$P\{A\} = 1 - P\{\bar{A}\}$$

The entire sample space in Figure 2.3 is then rectangular box containing both A and $\bar{A}$, and is represented by I.

If A and B are two *independent* events, then the probability that both will happen, the *joint probability*—also known as the *intersection* A∩B (or just AB) of the two events—is the product of their respective probabilities:

$$P\{A \cap B\} = P\{AB\} = P\{A\}P\{B\}$$

In a Venn diagram, the intersection of independent events is depicted by the shaded area in which elements contained in circles A and B overlap (Figure 2.4).

If A and B are *mutually exclusive* events, then either A or B may occur, but not both; that is, $P\{AB\} = 0$. Then the probability that either one or the other of these events will take place is the sum of their probabilities:

$$P\{A \text{ or } B\} = P\{A\} + P\{B\}$$

A Venn diagram for mutually exclusive events shows two regions which do not intersect (Figure 2.5).

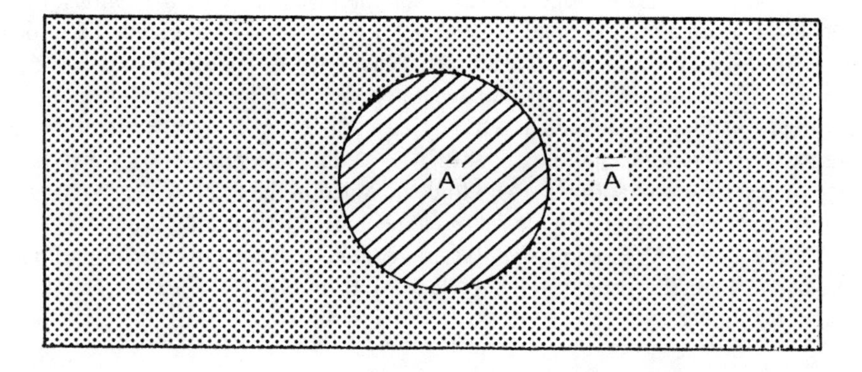

Figure 2.3. Venn diagram showing the sample space and the events A and $\bar{A}$.

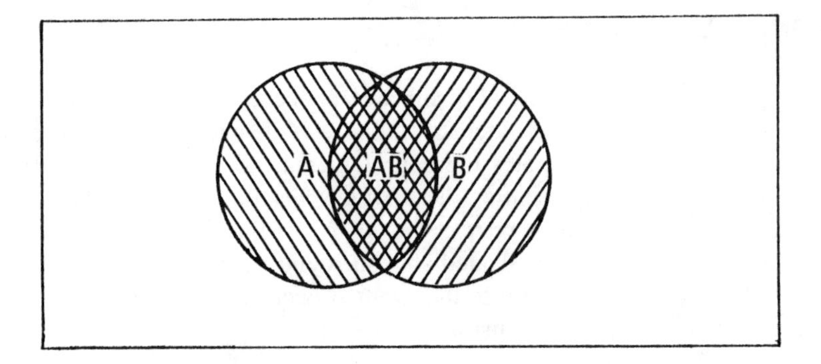

Figure 2.4. Venn diagram showing two events that are not mutually exclusive.

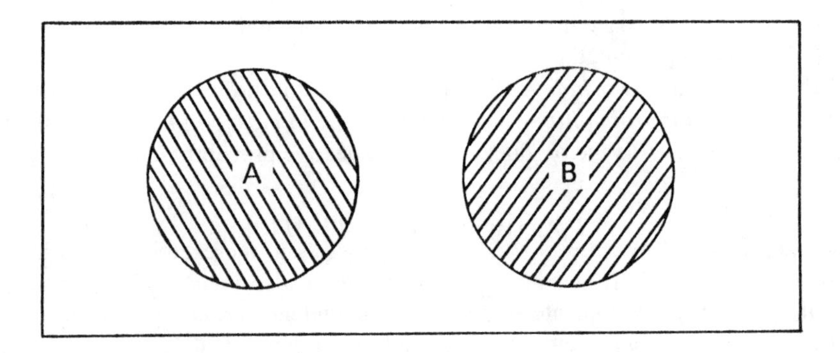

Figure 2.5. Venn diagram for two mutually exclusive events.

If the events A and B are independent but not mutually exclusive—that is, $P\{AB\} \neq 0$—then the probability of at least one of these events occurring is given by the sum of the individual probabilities less their joint probability:

$$P\{\text{A and/or B}\} = P\{A + B\} = P\{A\} + P\{B\} - P\{AB\}$$

This expression (and the one above) gives the probability of the *union* of two events, also sometimes denoted as $P\{A + B\} = P\{A \cup B\}$. Notice that the probability of the joint occurrence is subtracted just once in the above expression. In the Venn diagram (Figure 2.4), the event A consists of all of the elements comprising A, including elements in AB. Similarly, the event B consists of all the elements in B, including the elements in AB. Adding the probabilities $P\{A\}$ and $P\{B\}$ causes the probability $P\{AB\}$ to be included in the total twice, so the probability corresponding to a single set of elements $P\{AB\}$ must be subtracted once.

If three events A, B, and C are considered, and these events are not mutually exclusive, then the probability $P\{\text{A and/or B and/or C}\} = P\{A + B + C\}$ is given as follows:

$$P\{A + B + C\} = P\{A\} + P\{B\} + P\{C\} - P\{AB\} - P\{AC\} - P\{BC\} + P\{ABC\}$$

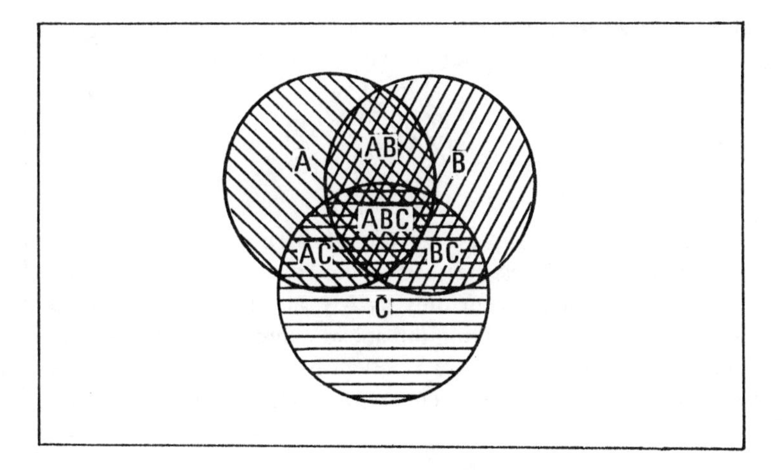

Figure 2.6. Venn diagram for three events A, B, and C that are not mutually exclusive.

To explain how this expression is developed, note, from the Venn diagram (Figure 2.6), that the probabilities for the joint occurrences AB, BC, and AC must be subtracted from the total because they are included twice by adding the probabilities associated with A, B, and C, just as in the example for A and B above (Figure 2.4). However, the elements in A also include the overlapping elements in ABC, and the same is true for the elements in B and C. Thus, inclusion of the elements in A, B, and C includes the elements ABC three times. Therefore, the sum $P\{A\} + P\{B\} + P\{C\}$ includes $3P\{ABC\}$. Similarly, the joint occurrences AB, BC, and AC each include the elements in ABC, thereby contributing the elements in ABC three times. Subtracting the joint probabilities $P\{AB\}$, $P\{BC\}$, and $P\{AB\}$ removes $3P\{ABC\}$, thereby completely removing the probability associated with the joint occurrence ABC. To once again include the probability for the missing intersection ABC, we must add $P\{ABC\}$ back into the total, giving the above expression.

A simpler way to view $P\{A + B + C\}$ is to note that it is logically equivalent to the probability that "at least one or more of the three events A, B, C happens." If the events are independent, this is the same as one minus the joint probability that none of these events happen. That is,

$$\begin{aligned} P[A+B+C\} &= 1 - P\{\bar{A}\}P\{\bar{B}\}P\{\bar{C}\} \\ &= 1 - [(1 - P\{A\})(1 - P\{B\})(1 - P\{C\})] \\ &= P\{A\} + P\{B\} + P\{C\} - P\{AB\} - P\{BC\} - P\{AC\} + P\{ABC\} \end{aligned}$$

CONDITIONAL PROBABILITY AND BAYES' THEOREM

Often, events are not independent, and the occurrence of one alters the probability of the second. We may be interested in the *conditional probability* of the second, given that the first one has occurred. The conditional probability of event B, given that A has occurred, is written as $P\{B|A\}$. In general,

$$P[B|A] = \frac{P\{AB\}}{P\{A\}}$$

In the Venn diagram for two events (Figure 2.4), once it is known that A has occurred, the set A replaces the set I as the sample space. Because A already has occurred, B then can occur only as AB, and the elements representing B without AB are excluded. Solving the above expression for $P\{AB\}$, the probability of the joint occurrence AB is given by the product of the conditional probability of A, given that B has occurred, and the probability of A:

$$P\{AB\} = P\{B|A\}P\{A\}$$

Example 1

The following example illustrates these concepts. In a suburban neighborhood in the U.S., 25% of the homes have slab foundations, and 75% do not. Previous studies have shown that 80% of the homes with slab foundations in this region have indoor radon problems due to intrusion of radon gas from the soil beneath. Of those homes without slab floors, 10% have indoor radon problems due to the remaining indoor sources of radon, such as stone construction materials.

The following events are possible in this example:

A = event "house has a slab foundation"

$\overline{A}$ = event "house does not have a slab foundation"

B = event "house has an indoor radon problem"

$\overline{B}$ = event "house does not have an indoor radon problem"

Representing the percentages as probabilities, this problem is summarized formally as follows:

$$P\{A\} = 0.25 \qquad P\{\overline{A}\} = 0.75$$
$$P\{B|A\} = 0.80 \qquad P\{\overline{B}|A\} = 0.20$$
$$P\{B|\overline{A}\} = 0.10 \qquad P\{\overline{B}|\overline{A}\} = 0.90$$

Then, the probability of finding a house with a slab foundation (A) and a radon problem (B) is given by the product of the conditional probability that a house has a radon problem, given that it has a slab foundation, and the probability that it has a slab foundation:

$$P\{B|A\}P\{A\} = P\{AB\} = (0.80)(0.25) = 0.20$$

Similarly, the probability that a house chosen at random has a radon problem and does not have a slab foundation is given by the product of the conditional probability of a radon problem, given that the house does not have a slab foundation, and the probability that the house does not have a slab foundation:

$$P\{B|\bar{A}\}P\{\bar{A}\} = P\{\bar{A}B\} = (0.10)(0.75) = 0.075$$

Because the occurrence of an indoor radon problem in a house that has a slab foundation and the occurrence of a radon problem in a house that does not are mutually exclusive events, it follows that the probability of an indoor radon problem, irrespective of the type of foundation, is the union of the two events, or the sum of the joint probabilities of the two possible cases:

$$\begin{aligned} P\{B\} &= P\{B|A\}P\{A\} + P\{B|\bar{A}\}P\{\bar{A}\} \\ &= 0.20 + 0.075 = 0.275 \end{aligned}$$

Thus, 27.5% of the homes will have indoor radon problems. (The result of multiplying the probability $p = 0.275$ by 100% gives an *expected percentage* of 27.5%, as discussed in Chapter 4.)

Bayes' Theorem

A useful law called *Bayes' Theorem* can be developed from the concept of conditional probability. In the conditional probability expression $P\{B|A\}$, B is called the *posterior event* and A is called the *prior event.* Thus, $P\{A\}$ is known as the *prior probability*, and $P\{B\}$ is known as the *posterior probability.*

Suppose the problem above is reversed in the following manner. We discover a house that has a radon problem, and we wish to determine the probability that it has a slab foundation. This problem considers the following question: if posterior event B is known to have occurred, then what is the probability that prior event A has occurred? Thus, we seek the probability of A, given that B has occurred, or $P\{A|B\}$.

Following reasoning similar to that presented above,

$$P\{A|B\} = \frac{P\{AB\}}{P\{B\}}$$

The basis for this relationship is the same as for $P\{B|A\}$. Because A is known to have occurred, the set B in the Venn diagram (Figure 2.4) replaces I as the sample space. Now A can occur only as AB, and the elements representing A without AB are excluded.

It was shown above that $P\{AB\} = P\{B|A\}P\{A\}$; substituting this expression into the above equation gives the following result:

$$P\{A|B\} = \frac{P\{B|A\}P\{A\}}{P\{B\}}$$

Because $P\{B\}$ is the sum of the probabilities of two mutually exclusive events, $P\{B\} = P\{AB\} + P\{\bar{A}B\} = P\{B|A\}P\{A\} + P\{B|\bar{A}\}P\{\bar{A}\}$. Substituting this expression into the above equation for $P\{B\}$ gives the following "general form" of this relationship, which is known as *Bayes' Theorem:*

$$P\{A|B\} = \frac{P\{B|A\}}{P\{B|A\}P\{A\} + P\{B|\bar{A}\}P\{\bar{A}\}} P\{A\}$$

Dividing both the numerator and denominator by $P\{B|A\}$, this equation for Bayes' Theorem also can be written in the following "simplified form":

$$P\{A|B\} = \frac{1}{1 + \dfrac{P\{B|\bar{A}\}P\{\bar{A}\}}{P\{B|A\}P\{A\}}}$$

In our problem, all the probabilities required in the right side of this expression are known; substituting them gives the following:

$$P\{A|B\} = \frac{1}{1 + \dfrac{(0.10)(0.75)}{(0.80)(0.25)}} = \frac{1}{1 + 0.375} = 0.7273$$

Thus, rounding off this result, we can expect approximately 73% of the homes reporting radon problems to have slab foundations. Notice that this result is not immediately apparent from the original facts given for this problem.

It is instructive to review the steps that led to this result. Initially, the prior probability of a slab foundation was $P\{A\} = 0.25$. Once a radon problem was found in a particular house, it was possible to "update" the original probability for that house, and the probability that the house had a slab foundation became $P\{A|B\} = P\{A'\} = 0.7273$ based on the new information. Here $P\{A'\}$ is the updated, or *posterior*, probability.

Example 2

Suppose a carbon monoxide (CO) measuring badge has been developed that is imperfect, with random error that has been studied and is well understood. Originally, the design called for a monitor that would give the user a positive reading whenever the integrated CO concentration exceeded some public health criterion (e.g., 72 ppm-hr in 8 hours). Experiments conducted on the badge show that the monitor correctly indicated a violation of the health criterion in 95% of the cases in which a true violation occurred. The monitor also occasionally gives false positive readings: 10% of the time it reports a CO problem when none really exists.

Suppose it is known the CO problems occur in approximately 10% of the high-rise buildings in a given city. If this badge is deployed in a field study, and a positive reading is obtained for a given building, what is the probability that a real CO problem exists there?

Let A denote the event "a real CO problem exists" (i.e., a real violation of the criterion value) and B denote the event "the monitor reports a problem." This example can be summarized by the following events and probabilities:

A = event "a real CO problem in this building"

B = event "positive reading on the monitor"

$P\{A\} = 0.10$ $\quad P\{\bar{A}\} = 0.90$

$P\{B|A\} = 0.95$ $\quad P\{\bar{B}|A\} = 0.05$ (false negative)

$P\{B|\bar{A}\} = 0.10$ (false positive) $\quad P\{\bar{B}|\bar{A}\} = 0.90$

Suppose the monitor reports a positive reading in a particular building. What is the probability that a real CO problem exists, taking into account the known error of the instrument? Formally, we seek the conditional probability that a real CO problem exists, given that the monitor reports such a problem, or the prior conditional probability $P\{A|B\}$. Using the simplified expression for Bayes' Theorem,

$$P\{A|B\} = \frac{1}{1 + \dfrac{P\{B|\bar{A}\}P\{\bar{A}\}}{P\{B|A\}P\{A\}}}$$

$$P\{A|B\} = \frac{1}{1 + \dfrac{(0.10)(0.90)}{(0.95)(0.10)}} = \frac{1}{1 + 0.9474} = 0.5135$$

Thus, the observed positive reading, along with Bayes' Theorem, has changed the probability of a real CO problem in this building from the original estimate of $P\{A\} = 0.10$ to a new value of $P\{A|B\} = P\{A'\} = 0.5135$.

Suppose readings are taken in 1,000 buildings as part of an indoor air quality survey. How many correct and incorrect readings can we expect? Consider first the cases in which a true CO problem exists. The probability of obtaining a positive reading and a true problem is given by $P\{B|A\}P\{A\} = (0.95)(0.10) = 0.095$, and the expected number of readings in this category will be $(0.095)(1{,}000) = 95$. Similarly, the probability of a negative reading and a true CO problem is given by $P\{\bar{B}|A\}P\{A\} = (0.05)(0.10) = 0.005$, giving an expected number of false negative readings of $(0.005)(1{,}000) = 5$. The probability of a false positive reading with no CO problem is given by $P\{B|\bar{A}\}P\{\bar{A}\} = (0.10)(0.90) = 0.09$, and the expected number of cases will be $(0.09)(1{,}000) = 90$. Finally, the probability of a negative reading and no CO problem is given by $P\{\bar{B}|\bar{A}\}P\{\bar{A}\} = (0.90)(0.90) = 0.81$, and the expected number of cases in this category will be $(0.81)(1{,}000) = 810$. These results are summarized in

Table 2.2. Summary of Expected Outcomes of 1,000 Readings from a Building Survey Using a CO Monitoring Badge with Known Error

	No. of Positive Readings	No. of Negative Readings
Correct Result:	95	810
Incorrect Result:	90	5
Total:	185	815

Table 2.2. Notice that the proportion of the 185 positive readings in which a true CO problem was present is 95/185 = 0.5135, the same result as was obtained from Bayes' Theorem.

Conditional probabilities can be represented graphically by a tree diagram, as suggested by Raiffa.[67] Each node in the tree depicts a trial, and the lines emanating from the node are the trial's possible outcomes. In this example, there are four possible outcomes from each CO sample, and the conditional probabilities are listed on the four lines at the ends of the tree (Figure 2.7). The Bayesian probability $P\{A|B\} = 0.5135$ results when the total number of positive readings (95 + 90 = 185) becomes the new sample space after the positive reading is obtained.

The proportion of the negative readings with an incorrect result is much smaller than the proportion of positive readings with an incorrect result; of 815 negative readings, only 5, or 0.6%, were wrong. Of course, a false negative may be more important than a false positive from a health standpoint, since it hides the existence of a pollution problem (Problem 7). The large proportion of false positive readings in this example occurs because so many buildings (90%) do not have a CO problem, but the instrument had a finite probability of reporting a problem when none existed. If the monitor's tendency to report false positives were corrected; that is, if $P\{B|\bar{A}\} = 0$, then $P\{A|B\} = 1/(1 + 0) = 1$, and a positive reading would denote the presence of a CO problem with certainty.

Considerable uncertainty remains after the first positive sample, because the probability is close to one-half, the same as for a coin flipping experiment. A good approach for reducing this uncertainty is to take another sample. The pos-

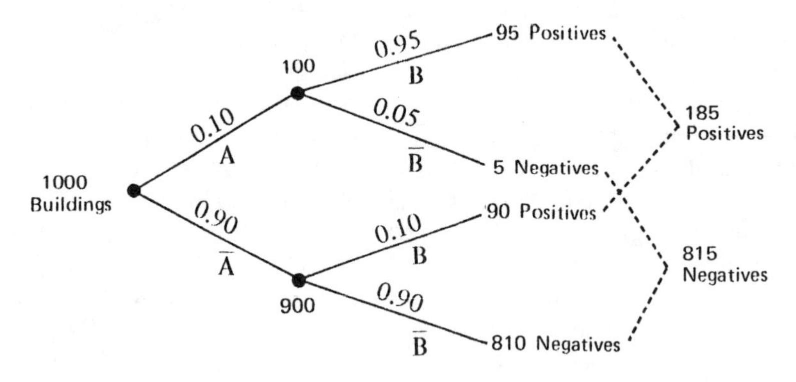

Figure 2.7. Probability tree for indoor air quality survey of CO problems in 1,000 buildings (Example 2).

terior probability of 0.5135 then will become the prior probability; that is, $P\{A'\} = 0.5135$ and $P\{\overline{A'}\} = 1 - 0.5135 = 0.4865$. Suppose the second sample gives a positive reading; what will be the probability that a real CO problem exists? Using the simplified form of Bayes' Theorem again and substituting $P\{A\}$ for $P\{A'\}$,

$$P\{A|B\} = \frac{1}{1+\dfrac{(0.10)(0.4865)}{(0.95)(0.5135)}} = 0.9093$$

Now the analyst can state that a CO problem exists with probability $p = 0.9093$, or a *confidence level* of more than 90%. Suppose that a confidence level of at least 95% is desired. If a third sample yields a positive reading, will the desired level of confidence be met? Once again, the posterior probability 0.9093 becomes the prior probability:

$$P\{B|A\} = \frac{1}{1+\dfrac{(0.10)(0.0907)}{(0.95)(0.9093)}} = 0.9896$$

Thus, we can now state with 98.96% certainty that a true CO problem exists in the building. Even though the measuring system has considerable error, it was possible, with only three readings, to exceed the 95% level of confidence. These steps show how a probabilistic analysis can be undertaken hand-in-hand with a monitoring survey to help determine the fewest readings needed to achieve a desired level of confidence.

SUMMARY

This chapter has introduced the concept of probability as an asymptotic ratio. If a fair coin is tossed indefinitely, the proportion of "heads" appearing eventually will approach the probability $p = 1/2$. A Venn diagram is a graphical approach for representing various events of interest within the sample space. If two events are independent, the probability of their intersection, or joint occurrence, is given by the product of their individual probabilities. The probability of the union of two events—the probability that one or the other (or both) occurs—is given by the sum of their individual probabilities less their joint probability. For mutually exclusive events, the joint probability is zero, so the union's probability then is the sum of the two probabilities.

Conditional probability is a useful concept that considers the probability of a second event, given that the first one has occurred. Bayes' Theorem considers the problem in reverse: "If the second event is known to have occurred, what is the probability that the first one occurred?" Using the Theorem, one can recalculate (i.e., "update") the probability that the original event occurred each time a new sample is taken and its outcome is known. Environmental measuring systems usually introduce error; nevertheless, the data from these systems must be used to draw inferences about environmental problems. Interpreting such data is equivalent to observing an outcome (response from the measuring system) and seeking to determine the probability of the original event (absence or presence of pollution). Bayesian approaches can be used to design efficient field data collection strategies that produce the desired level of confidence with a minimum of samples.

PROBLEMS

1. Copy the computer program given in Figure 2.1 and change the random seed. Run the program for selected values of n: (a) 10, (b) 100, (c) 1,000, (d) 10,000, (e) 20,000, and (f) 50,000. Discuss the deviations from one-half of the proportion of the outcomes resulting in "heads." Examine several sequences of 100 successive outcomes, and discuss the pattern of the fluctuations. [Note: In Microsoft BASIC, the statement RANDOMIZE(j) reseeds the random number generator RND with seed j.]
2. Change the computer program in Figure 2.1 so that a biased coin is used in the simulation with $P\{\text{"head"}\} = p = 0.35$. Run this program for several large values of n and discuss the fluctuations about the limiting value.
3. Two manufacturing plants located in a valley operate the same chemical process on dates scattered throughout the year. Whenever the process operates, emissions of toxic air pollutants are released into the air. Out of 260 weekdays per year, Plant A always operates 150 weekdays, and Plant B always operates 30 weekdays. If the two plants' operations are independent of each other and are not related to the day of the week or the season, what is the probability that toxic pollutants are released into the valley on a given weekday? [Answer: $P\{A + B\} = P\{A\} + P\{B\} - P\{AB\} = 0.5769 + 0.1154 - 0.0666 = 0.6257$] What is the probability that emissions will be released on at least one day in a given week? [Answer: $1 - (1 - 0.6257)^5 = 0.9927$]
4. In Problem 3, suppose that emissions are observed on a given date; what is the probability that they were caused by Plant A but not Plant B? [Answer: $P\{A\bar{B}|A + B\} = (P\{A\} - P\{AB\})/(P\{A\} + P\{B\} - P\{AB\}) = 0.8156$]
5. Consider the independent events A, B, C, and D, none of which are mutually exclusive. Using the simplified approach presented in this chapter, develop an expression for $P\{A + B + C + D\}$. [Answer: $P\{A\} + P\{B\} + P\{C\} + P\{D\} - P\{AB\} - P\{AC\} - P\{AD\} - P\{BC\} - P\{BD\} - P\{CD\} + P\{ABC\} + P\{ABD\} + P\{ACD\} + P\{BCD\} - P\{ABCD\}$]
6. Consider the radon example presented in this chapter. Suppose a house does not have a radon problem; what is the probability that it has a slab foundation? [$P\{A|\bar{B}\} = 0.06896$]
7. In the CO building survey, suppose a negative result is obtained on the first sample; what is the probability that a CO problem exists in this building? [Answer: $P\{A|\bar{B}\} = 1/(1 + P\{\bar{B}|\bar{A}\}P\{\bar{A}\}/P\{\bar{B}|A\}P\{A\}) = 0.00613$] Suppose a second sample is taken after the first negative, and a positive reading is obtained. What is the probability that a CO problem exists? Does this result

meet the 95% confidence level that the building is free of CO problems? [Answer: 0.05535; No]

8. Write a BASIC computer program to apply Bayes' Theorem to a sampling problem. The program should read in the prior and conditional probabilities. It should then request the outcome of each trial and update the prior probability whenever an outcome is known. Apply this program to the CO building survey example. Compute the prior probabilities for all possible outcomes of four successive samples.

REFERENCES

1. Anderson, Alan J.B., *Interpreting Data: A First Course in Statistics* (New York, NY: Chapman and Hall, 1988).
2. Barlow, Roger, *Statistics* (New York, NY: John Wiley & Sons, 1989).
3. Berenson, Mark L., David M. Levine, and David Rindskopf, *Applied Statistics: A First Course* (Englewood Cliffs, NJ: Prentice-Hall, 1988).
4. Berry, Donald A., and Bernard W. Lindgren, *Statistics: Theory and Methods* (Pacific Grove, CA: Brooks/Cole Publishing Co., 1990).
5. Brant, Siegund, *Statistical and Computational Methods in Data Analysis* (New York: Elsevier/North-Holland, 1983).
6. Breiman, Leo, *Probability* (Philadelphia, PA: Society for Industrial and Applied Mathematics [SIAM], 1992).
7. Bulmer, M.G., *Principles of Statistics* (Cambridge, MA: M.I.T. Press, 1967).
8. Chatfield, Christopher, *Statistics for Technology: A Course in Applied Statistics* (New York, NY: Chapman and Hall, 1983).
9. Dixon, Wilrid J., and Frank J. Massey, Jr., *Introduction to Statistical Analysis* (New York, NY: McGraw-Hill, 1983).
10. Feller, William, *An Introduction to Probability Theory and Its Applications,* Volume I (New York, NY: John Wiley & Sons, 1968).
11. Feller, William, *An Introduction to Probability Theory and Its Applications*, Volume II (New York, NY: John Wiley & Sons, 1971).
12. Fraser, D.A.A., *Statistics: An Introduction* (New York, NY: John Wiley & Sons, 1958).
13. Freedman, David, Robert Pisani, Roger Purves, and Ani Adhikari, *Statistics*, 2nd Edition (New York, NY: W.W. Norton & Co., 1991).
14. Freund, John E., *Mathematical Statistics* (Englewood Cliffs, NJ: Prentice-Hall, 1992).
15. Freund, John E., *Statistics: A First Course* (Englewood Cliffs, NJ: Prentice-Hall, 1976).
16. Freund, John E., and Gary A. Simon, *Modern Elementary Statistics*, 8th Edition (Englewood Cliffs, NJ: Prentice-Hall, 1992).
17. Glenberg, Arthur M., *Learning from Data: An Introduction to Statistical Reasoning* (Orlando, FL: Harcourt Brace Jovanovich, 1978).
18. Gnedenko, B.V., *The Theory of Probability and the Elements of Statistics* (New York, NY: Chelsea, 1989).
19. Goodman, Roe, *Introduction to Stochastic Models* (Menlo Park, CA: Benjamin/Cummings, 1988).

20. Hamburg, Morris, *Statistical Analysis for Decision Making* (Orlando, FL: Harcourt Brace Jovanovich, 1991).
21. Harp, William L., *Statistics* (Orlando, FL: Holt, Rinehart, and Winston, 1988).
22. Hodges, J.L., Jr., and E.L. Lehmann, *Basic Concepts of Probability and Statistics* (San Francisco: Holden-Day, 1970).
23. Hoel, Paul G., *Introduction to Mathematical Statistics* (New York: John Wiley & Sons, 1971).
24. Hoel, Paul G., Sidney C. Port, and Charles J. Stone, *Introduction to Probability Theory* (Boston, MA: Houghton Mifflin, 1971).
25. Johnson, Richard A., and Gourik K. Bhattacharya, *Statistics: Principles and Methods*, 2nd Edition (New York, NY: John Wiley & Sons, 1992).
26. Johnson, Robert, *Elementary Statistics* (Boston, MA, PWS-Kent Publishing Co., 1992).
27. Kattsoff, Louis O., and Albert J. Simone, *Foundations of Contemporary Mathematics* (New York, NY: McGraw-Hill Book Co., 1967).
28. Kendall, Sir Maurice, and Alan Stuart, *The Advanced Theory of Statistics* (New York, NY: Macmillan Publishing Co., 1977).
29. Larsen, Richard J., *An Introduction to Mathematical Statistics and Its Application* (Englewood Cliffs, NJ: Prentice-Hall, 1986).
30. Levin, Richard I., and David S. Rubin, *Applied Elementary Statistics* (Englewood Cliffs, NJ: Prentice-Hall, 1980).
31. Kirk, Roger E., *Statistics: An Introduction*, 3rd Edition (New York, NY: Macmillan, 1976).
32. Lindgren, Bernard W., *Statistical Theory*, 3rd Edition (New York, NY: Macmillan, 1976).
33. McClave, James T., and Frank H. Dietrich, II, *A First Course in Statistics* (New York, NY: Macmillan, 1992).
34. Mendenhall, William, Dennis D. Wackerly, and Richard L. Scheaffer, *Mathematical Statistics* (Boston, MA: PWS-Kent, 1990).
35. Mood, Alexander M., Franklin A. Graybill, and Duane C. Boes, *Introduction to the Theory of Statistics*, 3rd Edition (New York, NY: McGraw-Hill, 1974).
36. Neter, John, William Wasserman, and E.A. Whitmore, *Applied Statistics*, 4th Edition (Boston, MA: Allyn and Bacon, 1992).
37. Olkin, Ingram, Leon J. Gleser, and Cyrus Derman, *Probability Models and Applications* (New York, NY: Macmillan Publishing Co., 1980).
38. Parzen, Emanuel, *Modern Probability Theory and Its Applications* (New York, NY: John Wiley & Sons, 1992).
39. Ross, Sheldon M. *Introduction to Probability Models*, 4th Edition (Boston, MA: Academic Press, 1989).
40. Ross, Sheldon M., *A First Course in Probability*, 3rd Edition (New York, NY: Macmillan, 1988).
41. Snedecor, George W. and William G. Cochran, *Statistical Methods* (Ames, IA: The Iowa State University Press, 1989).
42. Spence, Janet T., John W. Cotton, Benton J. Underwood, and Carl P. Duncan, *Elementary Statistics* (Englewood Cliffs, New Jersey: Prentice-Hall, 1992).

43. Stuart, Alan, and J. Keith Ord, *Kendall's Advanced Theory of Statistics: Volume 1. Distribution Theory*, 5th Edition (New York, NY: Oxford University Press, 1987).
44. Stuart, Alan, and J. Keith Ord, *Kendall's Advanced Theory of Statistics: Volume 2. Classical Inference and Relationship*, 5th Edition (New York, NY: Oxford University Press, 1987).
45. Taylor, John, *Statistical Techniques for Data Analysis* (Chelsea, MI: Lewis Publishers, 1990).
46. Triola, Mario, *Elementary Statistics*, 5th Edition (Reading, MA: Addison-Wesley, 1992).
47. Thomas, John B., *An Introduction to Applied Probability and Random Processes* (Malabar, FL: Robert E. Krieger, 1981).
48. Thomas, John B., *Introduction to Probability* (New York, NY: Springer-Verlag, 1986).
49. Tukey, John W., *Exploratory Data Analysis* (Reading, MA: Addison-Wesley, 1977).
50. Weiss, Neil and Matthew Hassett, *Introductory Statistics*, 3rd Edition (Reading, MA: Addison-Wesley, 1991).
51. Witte, Robert S., *Statistics* (Orlando, FL: Harcourt Brace Jovanovich, 1993).
52. Wonnacott, Thomas H., and Ronald J. Wonnacott, *Introductory Statistics*, 5th Edition (New York, NY: John Wiley & Sons, 1990).
53. Castillo, Enrique, *Extreme Value Theory in Engineering* (San Diego, CA: Academic Press, 1988)
54. Kinnison, R.R., *Applied Extreme Value Statistics* (New York, NY: Macmillan, 1984).
55. Banks, Jerry, and John S. Carson, II, *Discrete-Event System Simulation* (Prentice-Hall: Englewood Cliffs, NJ, 1984).
56. Bratley, Paul, Bennett L. Fox, and Linus E. Schrage, *A Guide to Simulation*, 2nd Edition (New York, NY: Springer-Verlag, 1987).
57. Aczel, Amir D., *Complete Business Statistics* (Boston, MA: Irwin, 1989).
58. Freund, John E., and Frank J. Williams, *Elementary Business Statistics* (Englewood Cliffs, NJ: Prentice-Hall, 1972).
59. Haccon, Patsy and Evert Meelis, *Statistical Analysis of Behavioral Data: An Approach Based on Time-Structured Models* (New York, NY: Oxford University Press, 1992).
60. Hall, Owen P., Jr., and Harvey E. Adelman, *Business Statistics* (Boston, MA: Richard D. Irwin, 1991).
61. Harnett, Donald L., and Ashok K. Soni, *Statistical Methods for Business and Economics* (Reading, MA: Addison-Wesley Publishing Co., 1991).
62. Levin, Richard F., *Statistics for Management* (Englewood Cliffs, NJ: Prentice-Hall, 1984).
63. Lupin, Lawrence L., *Statistics for Modern Business Decisions* (Orlando, FL: Harcourt Brace Jovanovich, 1990).
64. Mason, Robert D., and Douglas A. Lind, *Statistical Techniques in Business and Economics* (Boston, MA: Irwin, 1993).
65. McClave, James T., and P. George Benson, *A First Course in Business Statistics*, 5th Edition (New York, NY: Macmillan, 1992).

66. Mendenhall, William, and Robert J. Beaver, *A Course in Business Statistics* (Boston, MA: PWS-Kent, 1992).
67. Raiffa, Howard, *Decision Analysis: Introductory Lectures on Choices Uncertainty* (Reading, MA: Addison-Wesley, 1968).
68. Sincich, Terry, *A Course in Modern Business Statistics* (Riverside, NJ: Dellen, 1991).
69. Sincich, Terry, *Business Statistics by Example* (New York, NY: Macmillan, 1992).
70. Webster, Allen, *Applied Statistics for Business and Economics* (Boston, MA: Irwin, 1992).
71. Arney, William Ray, *Understanding Statistics in the Social Sciences* (New York: W.H. Freeman and Company, 1990).
72. Bakeman, Roger, *Understanding Social Science Statistics: A Spreadsheet Approach* (Hillsdale, NJ: Lawrence Erlbaum Associates, 1992).
73. Gravetter, Frederick J., and Larry B. Wallnau, *Statistics for the Behavioral Sciences: A First Course for Students of Psychology and Education* (St. Paul, MN: West Publishing Co., 1992).
74. McNemar, Quinn, *Psychological Statistics* (New York, NY: John Wiley & Sons, 1962).
75. Benjamin, Jack R., and C. Allen Cornell, *Probability, Statistics, and Decision for Civil Engineers* (New York, NY: McGraw-Hill Book Co., 1970).
76. Bethea, Robert M., and Russell Rhinehart, *Applied Engineering Statistics* (New York, NY: Marcel Deckker, 1991)
77. Bowker, Albert H., and Gerald J. Lieberman, *Engineering Statistics*, 2nd Edition (Englewood Cliffs, NJ: Prentice-Hall, 1959).
78. Brownlee, K.A., *Statistical Theory and Methodology in Science and Engineering*, 2nd Edition (Malabar, FL: Robert G. Krieger Publishing Co., 1984).
79. Devore, Jay L., *Probability and Statistics for Engineering and the Sciences,* 3rd Edition (Pacific Grove, CA: Brooks/Cole Publishing Co., 1990).
80. Dougherty, Edward R., *Probability and Statistics for the Engineering, Computing, and Physical Sciences* (Englewood Cliffs, NJ: Prentice-Hall, 1990).
81. Hahn, Gerald J., and Samuel S. Shapiro, *Statistical Models in Engineering* (New York, NY: John Wiley & Sons, 1967).
82. Hald, A., *Statistical Theory with Engineering Applications* (New York, NY: John Wiley & Sons, 1952).
83. Hamming, Richard W., *The Art of Probability for Scientists and Engineers* (Redwood City, CA: Addison-Wesley Publishing Co., 1991).
84. Hogg, Robert V., and Johannes Ledolter, *Applied Statistics for Engineers and Physical Scientists*, 2nd Edition (Boston, MA: PWS-Kent Publishing Co., 1990).
85. Hillier, Frederick S., and Gerald J. Lieberman, *Operations Research* (San Francisco, CA: Holden-Day, Inc., 1974).
86. Lapin, Lawrence L., *Probability and Statistics for Modern Engineering*, 2nd Edition (Boston, MA: PWS-Kent Publishing Co, 1990).

87. Mendenhall, William, and Terry Sincich, *Statistics for Engineering and the Sciences* (New York, NY: Macmillan, 1992).
88. Miller, Irwin, John E. Freund, and Richard A. Johnson, *Probability and Statistics for Engineers*, 4th Edition (Englewood Cliffs, NJ: Prentice-Hall, 1990).
89. Milton, J.S., and Jesse C. Arnold, *Introduction to Probability and Statistics: Principles and Applications for Engineering and the Computing Sciences*, 2nd Edition (New York, NY: McGraw-Hill, 1990).
90. Wadsworth, Harrison M., *Handbook of Statistical Methods for Engineers and Scientists* (New York, NY: McGraw-Hill, 1990).
91. Walpole, Ronald E., and Raymond H. Myers, *Probability and Statistics for Engineers and Scientists*, 4th Edition (New York, NY: Macmillan, 1989).
92. Altman, Douglas G., *Practical Statistics for Medical Research* (New York, NY: Chapman & Hall, 1991).
93. Brown, Byron Wm., Jr., and Myles Hollander, *Statistics: A Biomedical Introduction* (New York, NY: John Wiley & Sons, 1977).
94. Daniel, Wayne W., *Biostatistics: A Foundation for Analysis in the Health Sciences*, 5th Edition (New York, NY: John Wiley & Sons, 1991).
95. Glantz, Stanton A., *Primer of Biostatistics*, 3rd Edition (New York, NY: McGraw-Hill, 1992).
96. Kuzma, Jan W., *Basic Statistics for the Health Sciences* (Palo Alto, CA: Mayfield Publishing Co., 1992).
97. Milton, J. Susan, *Statistical Methods in the Biological and Health Sciences,* 2nd Edition (New York, NY: McGraw-Hill, 1992).
98. Pielou, E. C., *An Introduction to Mathematical Ecology* (New York, NY: John Wiley & Sons, 1969).
99. Remington, Richard D., and M. Anthony Schork, *Statistics with Applications to Biological and Health Sciences* (Englewood Cliffs, NJ: Prentice-Hall, 1970).
100. Sokal, Robert R., and F. James Rohif, *Introduction to Biostatistics* (New York, NY: W.F. Freeman and Company, 1987).
101. Zar, Jerrold, H., *Biostatistical Analysis*, 2nd Edition (Englewood Cliffs, NJ: Prentice-Hall, 1984).
102. Bailey, Norman T. J., *The Elements of Stochastic Processes* (New York, NY: John Wiley & Sons, 1964).
103. Bhattacharya, Rabi N., and Edward C. Waymire, *Stochastic Processes with Applications* (New York, NY: John Wiley & Sons, 1990).
104. Cinlar, Erhan, *Introduction to Stochastic Processes* (Englewood Cliffs, NJ: Prentice-Hall, 1975).
105. Goodman, Roe, *Introduction to Stochastic Models* (Menlo Park, CA: Benjamin/Cummings Publishing Co., 1988).
106. Harrison, J. Michael, *Brownian Motion and Stochastic Flow Systems* (Malabar, FL: Robert E. Krieger Publishing Co., 1990).
107. Hoel, Paul G., Sidney C. Port, and Charles J. Stone, *Introduction to Stochastic Processes* (Boston, MA: Houghton Mifflin Co., 1972).
108. Kannan, D., *An Introduction to Stochastic Processes* (New York, NY: Elsevier/North Holland, 1979).

109. Karlin, Samuel, and Howard M. Taylor, *A First Course in Stochastic Processes* (New York, NY: Academic Press, 1975).
110. Melsa, James L., and Andrew P. Sage, *An Introduction to Probability and Stochastic Processes* (Englewood Cliffs, NJ: Prentice-Hall, 1973).
111. Papoulis, Athanasios, *Probability, Random Variables, and Stochastic Processes* (New York, NY: McGraw-Hill Book Co., 1991).
112. Parzen, Emanuel, *Stochastic Processes* (San Francisco, CA: Holden-Day, 1967).
113. Solomon, Frederick, *Probability and Stochastic Processes* (Englewood Cliffs, NJ: Prentice-Hall, 1987).
114. Stirzaker, D.R., and G.R. Grimmett, *Probability and Random Processes* (New York, NY: Oxford University Press, 1992).
115. Tedorovic, Petar, *An Introduction to Stochastic Processes and Their Applications* (New York, NY: Springer-Verlag, 1992).
116. Gilbert, Richard O., *Statistical Methods for Environmental Pollution Monitoring* (New York, NY: Van Nostrand Reinhold, 1987).

3 Probability Models

A "model" is an abstraction designed to represent some real phenomenon. Usually, a model is a formal conceptual framework for facilitating understanding or helping one to predict future events. Some models—such as spheres depicting the placement of atoms in a molecule, mechanical analogs of physical processes, sculptures of buildings, or scale prototypes of aircraft—do not embody equations. Equations simply are rules expressing supposed relationships among the variables of a model. A *probability model* consists of equations expressing the probabilities of all possible outcomes in the sample space:

> A probability model is a set of rules describing the probabilities of all possible outcomes in the sample space.

DISCRETE PROBABILITY MODELS

Discrete random variables are those which can take on only a finite number of values. For example, if K is a random variable describing the possible outcomes from the throw of a six-sided die, then K can take on any of six possible integer values in the sample space $\{1,2,3,4,5,6\}$. If the die is fair, then each face of the die is equally likely to appear, and we can write the following general expression for the probability of each possible outcome:

$$P_K\{k\} = P\{K = k\} = p = \frac{1}{6} \quad \text{for } k = 1, 2, 3, 4, 5, 6$$

Here, the lower case k denotes the numerical values that the random variable K takes on. $P_K\{k\}$ is a briefer way to write $P\{K = k\}$, and the above expression is known as the Probability Mass Function (PMF) of the random variable K. The PMF describes the probabilities of all possible outcomes in the sample space of a *discrete* random variable, and the sum of these probabilities is unity:

$$\sum_{k=1}^{6} P_K\{k\} = \frac{1}{6}+\frac{1}{6}+\frac{1}{6}+\frac{1}{6}+\frac{1}{6}+\frac{1}{6} = 1$$

The probabilities generated by this model can be represented by vertical bars

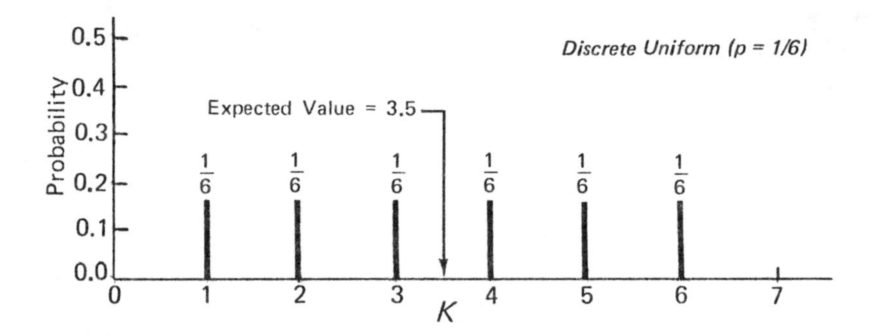

Figure 3.1. Discrete uniform probability distribution for the outcomes of the throw of a fair, six-sided die.

(Figure 3.1), and the result is the *probability distribution* of the discrete random variable K. Because all the probabilities are equal, K is said to have a *uniform distribution.* In this book, the discrete uniform distribution is represented by the notation $\mathbf{U}(p) = \mathbf{U}(1/6)$.

It is useful to describe the characteristics of probability distributions using several standardized parameters. One of these parameters is the *mean* or, more formally, the *expected value* $E[K]$. The expected value is computed by adding together the individual values of the random variable, weighted by the probability of each value's occurrence:

$$E[K] = \sum_{k=1}^{6} kP\{K = k\} = (1)\frac{1}{6} + (2)\frac{1}{6} + (3)\frac{1}{6} + (4)\frac{1}{6} + (5)\frac{1}{6} + (6)\frac{1}{6} = \frac{21}{6} = 3\frac{1}{2}$$

Geometric Distribution

As another example, consider a pollutant emission control device installed on a new motor vehicle. Suppose that the device has a probability of $p = 0.9$ of working effectively until the end of any year, given that it worked satisfactorily at the beginning of the year, and a probability of $q = 1 - p = 0.1$ of failing. If failures in successive years are independent, then the probability that it survives the first year and fails during the second year will be $pq = (0.9)(0.1) = 0.09$. Similarly, the probability that the device survives the first two years will be $p^2 = (0.9)(0.9) = 0.81$, and the probability that it then fails in the third year will be $p^2q = (0.9)^2(0.1) = 0.081$. The probability that it fails in the fourth year is $p^3q = (0.9)^3(0.1) = 0.073$.

Let K be a random variable denoting the year in which the control device fails. By examining the pattern of these probabilities as K increases, we inductively arrive at the following general expression for the probability that the device fails in any year k:

$$P\{\text{failure in year } k\} = p^{k-1}q$$

The individual probabilities form the series q, pq, p^2q, ... , and, if we add all the terms of the series, we obtain:

$$(1+p+p^2+p^3+\ldots)q = q\sum_{k=1}^{\infty} p^{k-1} = q\left(\frac{1}{1-p}\right) = 1$$

The sum of the infinite series p^{k-1} is $1/(1-p)$, which readily can be seen by dividing the quantity $1-p$ into 1. This result satisfies a necessary condition of all probability distributions: the sum of the probabilities for all outcomes in the sample space is one.

The individual probabilities $P_K\{k\} = qp^{k-1}$ comprise the *geometric* probability distribution (Figure 3.2), which is represented in this book as **GM**(p) = **GM**(0.9). Notice that it was possible to develop this distribution merely from the laws of probability (Chapter 2) and from our physical understanding of the problem. The only assumptions required were that the conditional probability that a device survives to the end of the year, given that it worked at the beginning of the year, remained constant at $p = 0.9$, and that failures on successive years were independent.

The above process has the characteristic, called a Markov property, that the probability on any trial depends *only* on the outcome of the previous trial and on the conditional probability p. Such a process sometimes is called "memoryless", because one does not need to know the past history of outcomes (prior to the previous outcome) to compute the probability of the next outcome. For real motor vehicle control devices, the probability of survival is likely to vary with the age of the device. Similar models can be developed for the birth and death of biological organisms, including humans. The present example is a very simple birth-death process. Birth-death processes are treated in greater detail by Pielou[1] and are applied to human populations by Keyfitz,[2-4] Keyfitz and Flieger,[5] Barlow and Saboia,[6] and others.

In this example, suppose we have a fleet of 100 new vehicles with control devices that all work initially. We can expect $100p = (100)(0.9) = 90$ to be

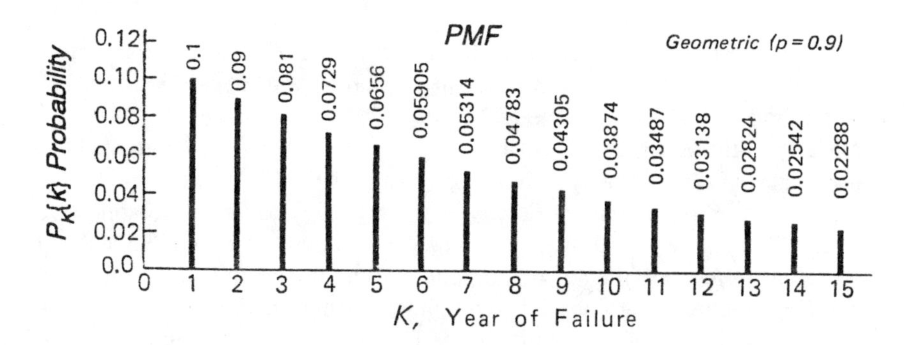

Figure 3.2. PMF of the geometric probability distribution **GM**(0.9), giving the probability that a motor vehicle emission control dvice fails in year K.

working at the end of the first year. Of these, (90)(0.9) = 81 will survive through the second year, and we can expect 90 – 81 = 9 failures in year 2. Or, using the geometric probability distribution **GM**(0.9) for failure of a device in a given year, we obtain the same result: $100pq$ = (100)(0.9)(0.1) = 9 failures in year $K = 2$.

Multiplying the total number of vehicles by the probability of an event, as we have done here, yields the "expected number of vehicles" in this category. Similarly, multiplying 100% by the individual probabilities yields the "expected percentage" in each category. For example, it is expected that 10% of the original population will fail in the first year, 9% in the second, 8.1% on the third, 7.29% in the fourth, and so on.

Expected numbers and expected percentages should not be confused with the expected value of a distribution, which is based on the entire distribution, and not just a particular outcome. For the geometric probability distribution, the expected value $E[K]$ is computed as follows:

$$E[K] = kP_K\{k\} = \sum_{k=1}^{\infty} kp^{k-1}q = q(1+2p+3p^2+\ldots) = \frac{q}{(1-p)^2} = \frac{1}{1-p}$$

For this example, the expected value will be $E[K] = 1/(1 - 0.9) = 10$ years, which can be interpreted as the "average life expectancy" of the control device.

Suppose we are interested not just in the number of vehicles failing in any year but in the total number of vehicles that have failed by the end of some year k. One approach is to add up the expected percentages until one arrives at the year of interest. A more formal approach is to utilize the *cumulative distribution function* (CDF) for the random variable K, defined as follows:

$$F_K(k) = P\{K \le k\} = \sum_{i=1}^{k} P_K\{i\}$$

For this example, the probability that a vehicle will fail on or before the fourth year is given by:

$$P\{K \le 4\} = F_K(4) = P\{K=1\} + P\{K=2\} + P\{K=3\} + P\{K=4\}$$

$$= 0.10 + 0.09 + 0.081 + 0.0729 = 0.3439$$

Thus, over one-third of the vehicles will fail within the first four years. Similarly, $F_K(7) = 0.5217$ (see Figure 3.3), so more than 50% of the vehicles are expected to fail within the first 7 years.

Notice that the vertical scales of Figures 3.2 and 3.3 differ, although the horizontal scales are the same. The probability bars of the PMF in Figure 3.2 approach zero as K increases indefinitely, while the CDF in Figure 3.3 approaches one. Although the probability bar of $F_K(1) = 0.1$ begins in Figure 3.3 at $K = 1$ and extends to just less than $K = 2$, it is not correct to conclude that values between $K = 1$ and $K = 2$ are possible. Only integer values of K can occur, and all other values occur with probability zero. The CDF in this example gives the probability that an emission control device failed on or before the end of year K. If a census of the vehicle population were taken at the end of each year, then

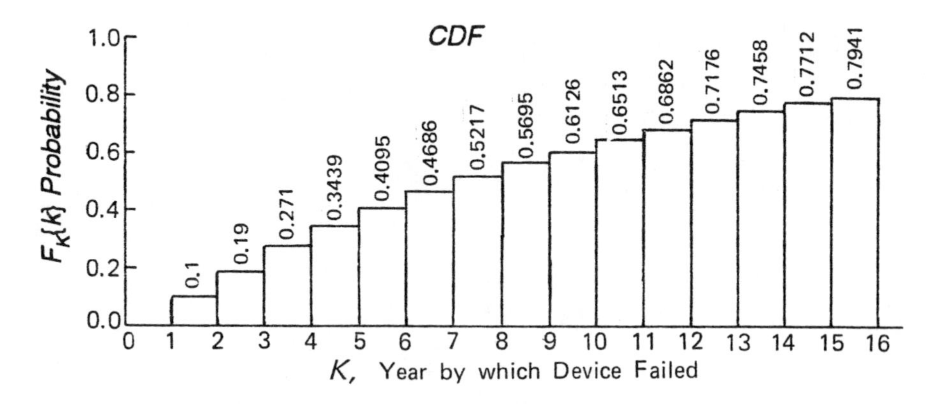

Figure 3.3. CDF of the geometric probability distribution **GM**(0.9), giving the probability that a motor vehicle emission control device fails on or before year *K*.

the CDF could be interpreted as the proportion of the vehicles experiencing a device failure up to (and including) the census date.

The CDF also is useful for determining the probability that a device fails *after* some specified year, since:

$$P\{K > k\} = 1 - P\{k \leq K\} = 1 - F_K(k)$$

Using the CDF, the probability that a device fails when a vehicle is five or more years old (i.e., more than four years old) is given by $P\{K > 4\} = 1 - F_K(4) = 1 - 0.3439 = 0.6561$. The CDF also is useful for obtaining probabilities associated with ranges of the random variable. Since $F_K(k) = P\{K \leq k\}$ and $F_K(j) = P\{K \leq j\}$, then, if $k > j$,

$$P\{j < K \leq k\} = F_K(k) - F_K(j)$$

Thus, the probability that an emission control device fails in any of years 2, 3, or 4 is given by:

$$P\{1 < K \leq 4\} = F_K(4) - F_K(1) = 0.3439 - 0.1 = 0.2439$$

The CDF of a discrete random variable resembles a step function (Figure 3.3), because $F_K(k)$ increases in incremental steps corresponding to the individual probabilities as k increases. In the limit, as k approaches infinity, $F_K(k)$ approaches 1.

CONTINUOUS RANDOM VARIABLES

Many random variables of interest are not limited to a finite number of values but can take on numerous possible values within some specified range. For example, an infinite number of different decimals are possible between 0 and 1.

Such *continuous random variables* can be handled by using the CDF to specify probabilities associated with particular ranges.

Let X be a continuous random variable. Then the CDF gives the probability that X is less than or equal to some value x:

$$P\{X \le x\} = F_X(x)$$

Thus, the probability that X exceeds some value x is given by,

$$P\{X > x\} = 1 - F_X(x)$$

Therefore, the probability that X lies between values a and b, where $b > a$, is the difference of the CDFs at these two points:

$$P\{a < X \le b\} = F_X(b) - F_X(a)$$

For a continuous random variable, the CDF has the following properties:

$$F_X(x) \ge 0 \text{ for all } x$$

$$F_X(b) \ge F_X(a) \text{ if } b > a$$

$$\lim_{x \to -\infty} \{F_X(x)\} = F_X(-\infty) = 0$$

$$\lim_{x \to \infty} \{F_X(x)\} = F_X(\infty) = 1$$

For a discrete random variable, the probability $P\{K = k\}$ was defined as the probability associated with the point k. For a continuous random variable, such a straightforward definition is not suitable. Instead, we define the *probability density function* (PDF) as the derivative of the CDF:

$$f_X(x) = \lim_{\Delta x \to 0} \left\{ \frac{P\{x < X \le x + \Delta x\}}{\Delta x} \right\} = \frac{d}{dx} F_X(x) = \frac{d}{dx} P\{X \le x\}$$

The term "density" is used because, although $f_X(x)$ is not a probability, $f_X(x)\Delta x$ is approximately equal to $P\{x < X < x + \Delta x\}$ if Δx is sufficiently small. Solving the above equation for $F_X(x)$, integrating, and using u as a dummy variable, we obtain,

$$P\{X \le x\} = F_X(x) = \int_{-\infty}^{x} f_X(u)du$$

Similarly,

$$P\{X > x\} = 1 - F_X(x) = \int_{x}^{\infty} f_X(u)du$$

Thus, the probability associated with a particular range of the random variable X is given by,

$$P\{a < X \le b\} = F_X(b) - F_X(a) = \int_{-\infty}^{b} f_X(u)du - \int_{-\infty}^{a} f_X(u)du = \int_{a}^{b} f_X(u)du$$

Unlike the discrete case, probabilities are defined over ranges only, and probabilities technically do not exist for any particular point within the range. From a mathematical standpoint, therefore, $P\{a < X\} = P\{a \le X\}$, and the following cases all are equivalent:

$$P\{a < X < b\} = P\{a \le X < b\} = P\{a < X \le b\} = P\{a \le X \le b\}$$

These mathematical requirements cause no difficulty because the CDF expresses the probability over specified ranges of the random variable, and one usually is interested in the probability associated with particular ranges of this variable. The following discussion introduces two examples of probability models of continuous random variables: the continuous uniform distribution and the exponential distribution.

Uniform Distribution

Consider a random variable that ranges between a and b, taking on any of the infinite number of possible decimal values within this range with equal probability. The PDF for this random variable is a rectangular function bounded by a and b, in which $b > a$ (Figure 3.4). The PDF and CDF, respectively, for this distribution, which is called the *uniform distribution*, are as follows:

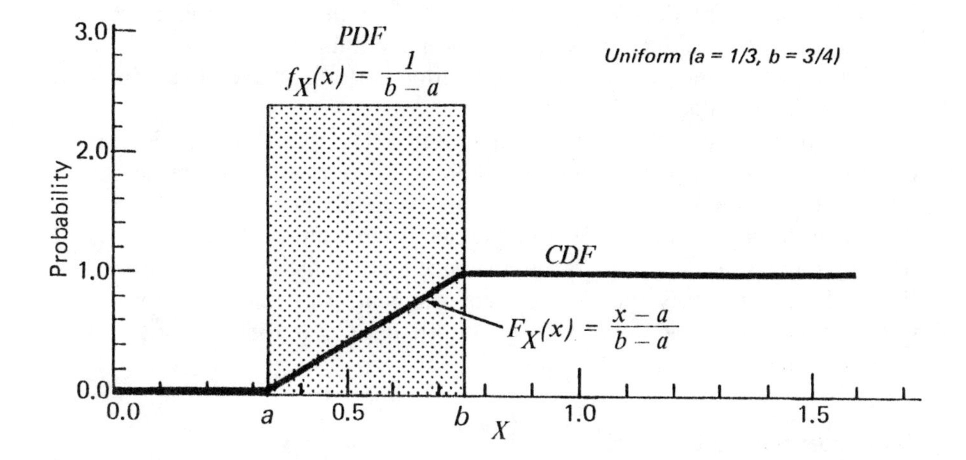

Figure 3.4. Probability density function (PDF) and cumulative distribution function (CDF) for a continuous random variable that has a uniform distribution **U**(a,b).

$$f_X(x) = \frac{1}{b-a} \quad \text{for } b > a$$

$$F_X(x) = \frac{x-a}{b-a}$$

In this book, the uniform distribution for a continuous random variable is represented by **U**(a,b). This notation should not be confused with the discrete uniform distribution **U**(p) introduced earlier, which has just one parameter p.

Suppose, for example, that $a = 1/3$ and $b = 3/4$. Then $f_X(x) = 1/(3/4 - 1/3) = 12/5 = 2.4$, and the CDF is $F_X(x) = 2.4x - 0.8$. Notice that, although the PDF over this range is greater than 1, the area under the rectangle (computed from the CDF; i.e., the integral over the sample space {1/3, 3/4}), still is unity:

$$F_X(3/4) - F_X(1/3) = \int_{1/3}^{3/4} f_X(x)dx = \Big[_{1/3}^{3/4} \; 2.4x - 0.8 = 1$$

The PDF and CDF for this example are plotted in Figure 3.4, and the area (shaded) of the rectangle also can be obtained by multiplying the height of the rectangle by its width (12/5)(3/4 – 1/3) = 1.

The expected value $E[X]$ of a continuous random variable is calculated in a manner analogous to that for a discrete random variable. Instead of forming the probability-weighted sum, we integrate the PDF, weighted by x, over the full range of the random variable:

$$E[X] = \int_a^b f_X(x)xdx = \int_a^b \frac{x}{b-a}dx = \Big[_a^b \frac{x^2}{2(b-a)} = \frac{b^2 - a^2}{2(b-a)} = \frac{1}{2}(b+a)$$

For the example above, $E[X] = 1/2(3/4 + 1/3) = 1/2$, the midpoint of the sample space.

The uniform distribution has many useful applications. The random number generator in the computer coin-flipping experiment (Chapter 2) simulates a uniformly distributed random variable **U**$(0,1)$. Such a random variable, in which $a = 0$ and $b = 1$, is said to have a "unit uniform distribution." As described in the following section, the unit uniform distribution performs a useful role in computer simulation models and similar experiments.

The uniform distribution can arise naturally in certain processes. For example, an experimenter pouring liquids without any knowledge of the quantities measured over some specified range can inadvertently generate concentrations that are uniformly distributed over that range. If the dilution process is repeated in many successive stages, the resulting concentration distributions will be similar to those commonly found in the environment. This reasoning forms the basis for the Theory of Successive Random Dilutions introduced later in this book (Chapter 8).

Computer Simulation

The uniform distribution is useful for generating continuous random variables with particular distributions specified by the user. If $F_X(x)$ is an analytical expression involving x, often it is possible to solve this expression for x as a function of $F_X(x)$, providing the inverse transformation of the CDF: $x = F_X^{-1}(x)$. If the expression cannot be solved analytically for x, nevertheless it may be possible to obtain an algorithm on the computer that approximates $F_X^{-1}(x)$. For any value of $F_X(x)$ between 0 and 1, this algorithm can generate the corresponding value of x using the inverse transformation $x = F_X^{-1}(x)$.

Suppose a continuous random variable X is known to have a particular probability distribution. If the CDF $F_X(x)$ for this distribution is made to vary as a unit uniform distribution **U**(0,1), then the inverse transformation of the CDF, $x = F_X^{-1}(x)$, will have the desired distribution. In other words, if U is uniformly distributed as **U**(0,1), then the value of X for which $F_X(x) = U$ has the desired distribution.

This approach, called the Inverse Transformation Method by Shedler,[7] provides an effective technique for generating random variables on large or small computers. Examples of this approach are given in the next section and in Chapter 8. When the random variables generated by this approach are used as inputs to mathematical models of environmental or other processes, the approach is called *computer simulation*, or sometimes *Monte Carlo simulation.*

Exponential Distribution

Another important probability model for continuous random variables is the exponential distribution, which has the following PDF:

$$f_X(x) = \frac{1}{\theta} e^{-x/\theta} \quad \text{for } x \geq 0, \ \theta > 0$$

The PDF of the exponential distribution begins at the origin, where $f_X(0) = 1/\theta$, and decays asymptotically toward zero as x increases without limit. For fixed increments along the abscissa, the height of the curve will be a fixed percentage of its previous height, a basic property of exponential functions. The exponential distribution for a continuous random variable therefore is analogous to the geometric distribution for a discrete random variable (see Problem 5).

The CDF for the exponential distribution is obtained by integrating the PDF directly:

$$F_X(x) = \int_0^x \frac{1}{\theta} e^{-u/\theta} du = \Big[_0^x \ e^{-u/\theta} = 1 - e^{-x/\theta}$$

The CDF begins at $F_X(0) = 0$ at the origin and asymptotically approaches 1 as x approaches infinity. When plotted, the CDF appears as an upside-down decaying exponential function (Figures 3.5 and 5.2).

If objects or particles are known to arrive at an observation point in a steady stream at a constant average rate $\lambda = 1/\theta$, and if arrivals are independent and sufficiently rare, then the exponential distribution is the correct probability distribution for the elapsed times between arrivals. The PDF of the elapsed time between the arrival of successive particles will have a single-parameter exponential **E**$(1/\lambda)$ distribution (Figure 5.2). The theoretical basis for this result is presented in this book's development of Poisson processes (Chapter 5).

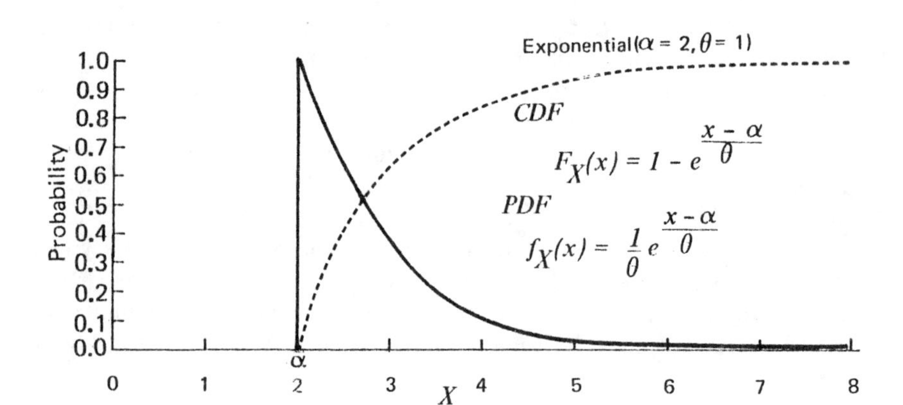

Figure 3.5. PDF and CDF of the two-parameter exponential probability model.

The expected value $E[X]$ of this distribution is computed in the same manner as for the uniform distribution:

$$E(X) = \int_0^\infty x f_X(x)dx = \int_0^\infty \frac{1}{\theta} x e^{-x/\theta} dx = \Big[_0^\infty \theta e^{-x/\theta}(-x/\theta - 1) = \theta$$

In this book, the single-parameter exponential probability model is represented as **E**(θ).

The parameter θ allows scaling of the random variable, but the single-parameter exponential model has limited flexibility. In many applications, it may be useful to include a second parameter α, which allows the distribution to be translated horizontally along the X-axis, thus serving as a location parameter. The PDF and CDF for the two-parameter exponential model are as follows:

$$f_X(x) = \frac{1}{\theta} e^{-\frac{x-a}{\theta}} \quad \text{for } x \geq \alpha,\ \theta > 0,\ -\infty < \alpha < \infty$$

$$F_X(x) = 1 - e^{-\frac{x-a}{\theta}}$$

The two-parameter exponential model is represented as **E**(α,θ). The single-parameter exponential model is a special case of the two-parameter exponential model in which α = 0. The two-parameter exponential model (Figure 3.5) has the same shape as the single-parameter exponential model, except that the PDF and CDF are displaced horizontally and begin at $x = \alpha$. Both forms of the exponential model have important environmental applications.

MOMENTS, EXPECTED VALUE, AND CENTRAL TENDENCY

The above discussion introduced the concept of the expected value by calculating $E[X]$ for both discrete (geometric) and continuous (uniform and exponential) probability models. The expected value $E[X]$ of a distribution is an important measure of its "central tendency." The expected value also is called the *average*, the *arithmetic mean*, or simply the *mean*. To summarize, the expected value is calculated as follows for probability models of both discrete and continuous random variables:

$$E[X] = \sum_{\text{all } k} KP_K\{k\}$$ if K is a discrete random variable with PMF $P_K\{k\}$

$$E[K] = \int_{-\infty}^{\infty} xf_X(x)dx$$ if X is a continuous random variable with PDF $f_X(x)$

Because of its importance, the expected value often is represented as $E[X] = \mu$.

Computation of expected values can be viewed as the result of a more general mathematical operation. If $g(k)$ is any integer mathematical function or $g(x)$ is any continuous mathematical function, then the following expressions are the *expectation operators* for discrete and continuous random variables:

$$E[g(k)] = \sum_{\text{all } k} g(k)P_K\{k\}$$ if K is a discrete random variable with PMF $P_K\{k\}$

$$E[g(x)] = \int_{-\infty}^{\infty} g(x)f_X(x)$$ if X is a continuous random variable with PDF $f_X(x)$

If, for the continuous case, we set $g(x) = x$, then the expected value $\mu = E[X]$ results, which also is known as the first *moment about the origin.* If $g(x) = x^2$, then the result gives the second moment of the random variable X about the origin. Similarly, if $g(x) = x^3$, the third moment about the origin is obtained. In general, the jth moment about the origin, in which j is a positive integer, is obtained as follows:

$$E[X^j] = \int_{-\infty}^{\infty} x^j f_X(x)dx$$

As a consequence of the mathematical form of the expectation operator, several important rules apply to the expected value:

1. The expected value of a constant a equals the constant a itself:

$$E[a] = a$$

2. The expected value of a constant a times a random variable equals the constant a times the expected value of the random variable itself:

$$E[aX] = aE[X]$$

3. The expected value of the sum of n independent random variables X_1, X_2, ... , X_n is the sum of the expected values of the individual expected values:

$$E[X_1 + X_2 + \ldots X_n] = E[X_1] + E[X_2] + \ldots + E[X_n]$$

4. Combining 2 and 3, we obtain the following general result in which a, b, and c are constants.

$$E[aX_1 + bX_2 + cX_3] = aE[X_1] + bE[X_2] + cE[X_3]$$

Suppose we set $g(x)$ in the expectation operator for a continuous random variable to an arbitrary weighting function of the form $g(x) = e^{-sx}$, in which s is a parameter. Then the expected value of this weighting function will be computed as:

$$M[X] = E[e^{-sx}] = \int_0^\infty e^{-sx} f_X(x)dx$$

This particular weighting function is of special importance, since this result is called the *moment generating function*, and it is often useful for computing the moments of a given probability model. For some probability models, the moments are more easily obtained using the expectation operator $E[X]$ than the moment generating function $M[X]$. For others, the reverse is true. Once an analytical expression for the moment generating function has been obtained, the jth moment of the probability model can be found by taking the jth derivative of $M[X]$ with respect to s and then setting $s = 0$:

$$E[X^j] = \frac{d^j\{M[0]\}}{ds^j}$$

Other measures of central tendency are the *median* and the *mode*. The median is the value of the random variable $X = x_{50}$ for which the CDF is 0.5. That is,

$$F_X(x_{50}) = 0.5$$

If CDF values are expressed as decimals, the corresponding values of the random variable sometimes are called "fractiles" or "quantiles." If CDF values are expressed as percentages, the corresponding values sometimes are called "percentiles." Thus, the median is the value of X at the 50-percentile of the CDF. This is the value of X at which 50% of the area under the PDF is enclosed.

For the single-parameter exponential probability model, for example, we compute x_{50} by setting the CDF to 0.5:

$$F_X(x_{50}) = 1 - e^{-x/\theta} = 0.5$$

Solving this equation for x gives $x_{50} = -\theta\ln(0.5) = 0.69314\theta$.

The mode x_m of a probability model is the value of X at which the PDF has a maximum. For a right-skewed distribution with a single mode (called "unimodal"), the mode, median, and mean do not usually coincide (Figure 3.6).

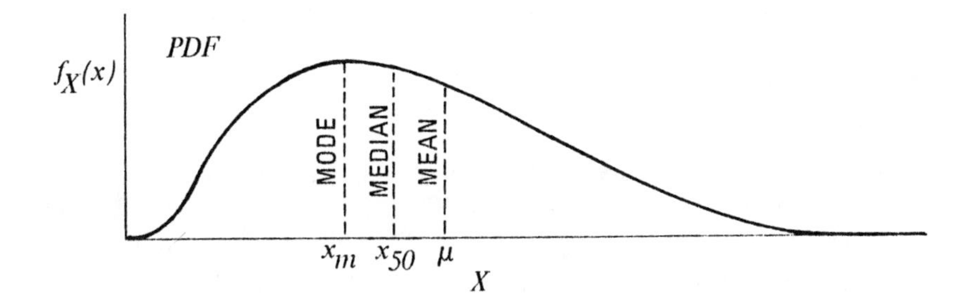

Figure 3.6. Typical locations of the mode, median, and mean for a right-skewed probability distribution.

VARIANCE, KURTOSIS, AND SKEWNESS

A variety of additional parameters are useful for describing the properties of probability models. Some of these parameters can be calculated from the moments of the model. Suppose we choose the function $g(x) = (x - \mu)^j$ in the expectation operator. Then the expectation $E[(X - \mu)^j]$ is known as the jth *moment about the mean* of the random variable X:

$$E\left[(X-\mu)^j\right] = \int_{-\infty}^{\infty}(x-\mu)^j f_X(x)dx$$

For a discrete random variable K, the jth moment about the mean is calculated in an analogous manner:

$$E\left[(K-\mu)^j\right] = \sum_{\text{all } k}(k-\mu)^j P_K\{k\}$$

Because of its importance in probability theory, the second moment about the mean is given a special name, the "variance" or simply $Var(X)$:

$$Var(X) = E\left[(X-\mu)^2\right] = \int_{-\infty}^{\infty}(x-\mu)^2 f_X(x)$$

If the squared term inside the parentheses above is expanded, we obtain the following "simplified form" of the variance expression by substituting $E[X] = \mu$:

$$\begin{aligned} Var(X) &= E[X-\mu]^2 = E[X^2 - 2\mu E[X] + \mu^2] \\ &= E[X^2] - 2\mu E[X] + \mu^2 = E[X^2] - \mu^2 \end{aligned}$$

The simplified form of the variance expression often is very useful, because it expresses the variance in terms of the moments about the origin, which already may be available to the analyst.

In this book, we shall use the notation μ_j to denote the jth moment about the origin, while m_j denotes the jth moment about the mean. That is,

$$\mu_j = E[X^j]$$

$$m_j = E[(X - \mu)^j]$$

Thus, μ_1, or simply "μ", denotes the first moment about the origin (the arithmetic mean), and m_2 denotes the variance, *Var(X)*. The notation σ^2 also commonly is used for the variance of a probability distribution; that is $m_2 = \sigma^2$. The square root of the variance is known as the *standard deviation*, or $\sigma = \sqrt{m_2}$.

As a consequence of the mathematical form of the variance, the following general rules apply to variances:

1. The variance of a constant a is zero:

$$Var(a) = 0$$

2. The variance of a constant times a random variable equals the square of the constant times the variance of the random variable:

$$Var(aX) = a^2 Var(X)$$

3. The variance of the sum of n independent random variables X_1, X_2, ... , X_n equals the sum of the individual variances:

$$Var(X_1 + X_2 + \ldots + X_n) = Var(X_1) + Var(X_2) + \ldots + Var(X_n)$$

4. Combining Rules 2 and 3, we obtain the following general result, in which *a, b,* and *c* are constants:

$$Var(aX_1 + bX_2 + cX_3) = a^2 Var(X_1) + b^2 Var(X_2) + c^2 Var(X_3)$$

These rules for variances are used later in this book. Another important concept is the *covariance* between random variables, $Cov(X_1,X_2)$. The covariance expresses the degree of association between the two random variances X_1 and X_2. If X_1 and X_2 are independent, then their covariance is zero: $Cov(X_1,X_2) = 0$.

To obtain the variance of the exponential probability model using the simplified form of the variance expression, for example, we first compute μ_2, the second moment about the origin:

$$\mu_2 = E[X^2] = \int_0^\infty \frac{x^2}{\theta} e^{-x/\theta} dx = \left[x^2 e^{-x/\theta}\right]_0^\infty + \left[2\theta^2 e^{-x/\theta}(-x/\theta - 1)\right]_0^\infty = 2\theta^2$$

Then we substitute this result into the simplified expression for *Var(X)*, noting that the mean of the exponential model is $\mu = \theta$:

$$Var(X) = m_2 = E[X^2] - \{E[X]\}^2 = \mu_2 - \mu^2 = 2\theta^2 - \theta^2 = \theta^2$$

From this result, the standard deviation of the exponential probability model is $\sigma = \theta$, which is equal to its mean.

The higher moments about the mean (for example, $m_3, m_4, \ldots$) can be expressed in terms of moments about the origin using algebraic expansions similar to the simplified variance equation. Using the same approach, the first four moments about the mean can be summarized as follows:

$$m_1 = 0$$

$$m_2 = \mu_2 - \mu_1^2$$

$$m_3 = \mu_3 - 3\mu_1\mu_2 + 2\mu_1^3$$

$$m_4 = \mu_4 - 4\mu_1\mu_3 + 6\mu_1^2\mu_2 - 3\mu_1^4$$

Equations for the higher moments are found in the book by Kendall and Stuart.[8]

Two other parameters that are useful for describing the shape of a probability distribution are the *coefficient of skewness* $\sqrt{\beta_1}$ and the *coefficient of kurtosis* β_2, which both can be written in terms of the moments about the mean:

$$\sqrt{\beta_1} = \frac{m_3}{m_2^{1.5}}$$

$$\beta_2 = \frac{m_4}{m_2^{2}}$$

The coefficient of skewness gives a measure of the relative skewness of a distribution, or its skewness normalized by its spread. It is possible, therefore, to compare the symmetry (lack of skewness) of two probability distributions whose scales differ. For symmetrical distributions, $\sqrt{\beta_1} = 0$ (see Figure 3.7). For distributions that have tails extending to the right, $\sqrt{\beta_1}$ is positive, and the result is called "right-skewed." For distributions that have tails extending to the left, $\sqrt{\beta_1}$ is negative, and the result is called "left-skewed."

For an exponentially distributed random variable, it previously was shown that the first and second moments about the origin are given by $\mu_1 = \theta$ and $\mu_2 = 2\theta^2$.

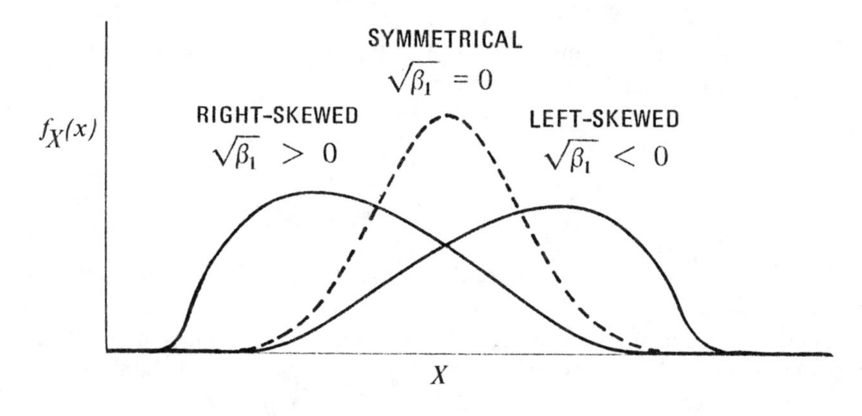

Figure 3.7. The coefficient of skewness indicates the direction of the preponderant tail of a distribution.

To compute m_3 for the exponential probability model, we first compute the third moment about the origin μ_3:

$$\mu_3 = E[X^3] = \int_0^\infty x^3 \frac{e^{-x/\theta}}{\theta} dx = 6\theta^3$$

Then we compute the third moment about the mean m_3 using the simplified equation for m_3 given above:

$$m_3 = \mathrm{E}[X-\mu]^3 = 6\theta^3 - 3\theta(2\theta^2) + 2\theta^3 = 2\theta^3$$

Since $m_2 = \theta^2$ for an exponentially distributed random variable, we compute the coefficient of skewness $\sqrt{\beta_1}$ as follows:

$$\sqrt{\beta_1} = \frac{m_3}{m_2^{3/2}} = \frac{2\theta^3}{(\theta^2)^{3/2}} = 2$$

The coefficient of kurtosis provides a standardized measure of a distribution's "peakedness"—the relationship of its "height" to its "width." Distributions with large values of β_2 have PDFs that are tall and narrow, while distributions with small values of β_2 have PDFs that appear short and squat. For the exponential probability model, we compute the coefficient of kurtosis by first computing the fourth moment about the origin μ_4:

$$\mu_4 = E[X^4] = \int_0^\infty x \frac{e^{-x/\theta}}{\theta} dx = 24\theta^4$$

Substituting this result and the other values for moments about the origin—$\mu_1 = \theta$, $\mu_2 = \theta^2$, and $\mu_3 = 6\theta^3$—for the exponential probability model into the above expression for m_4, we obtain:

$$m_4 = 24\theta^2 - 4\theta(6\theta)^3 + 6\theta^2(2\theta^2) - 3\theta^4 = 9\theta^4$$

Substituting this result into the expression of β_2 gives the following coefficient of kurtosis for the exponential probability model:

$$\beta_2 = \frac{m_4}{m_2^2} = \frac{9\theta^4}{(\theta^2)^2} = 9$$

Tables 3.1, 3.2, and 3.3 summarize three of the probability models presented thus far in this chapter. Each table contains the PDF and CDF, along with four parameters: the mean and variance and the coefficients of skewness and kurtosis. These tables are intended to provide a convenient reference. In this book, a similar table is given each time a new probability model is introduced.

Although it is not shown in the tables, another common parameter is the *coefficient of variation v*, which is the ratio of the standard deviation to the mean.

Table 3.1. The Geometric Distribution **GM**(p)

Probability Mass Function:

$$P_k\{k\} = (1-p)p^{k-1} = qp^{k-1} \qquad \begin{cases} k = 1, 2, \ldots \\ 0 < p < 1 \end{cases}$$

Expected Value:

$$E[K] = \frac{1}{1-p} = \frac{1}{q}$$

Variance:

$$Var(K) = \frac{p}{(1-p)^2} = \frac{p}{q^2}$$

Coefficient of Skewness:

$$\sqrt{\beta_1} = \frac{1+p}{\sqrt{p}}$$

Coefficient of Kurtosis:

$$\beta_2 = \frac{(1-p)^2 + 9p}{p}$$

Table 3.2. The Uniform Distribution **U**(a,b)

Probability Density Function:

$$f_X(x) = \begin{cases} \dfrac{1}{b-a}, & a \le x \le b \\ 0, & \text{elsewhere} \end{cases}$$

Expected Value:

$$E[K] = \frac{a+b}{2}$$

Variance:

$$Var(X) = \frac{(b-a)^2}{12}$$

Coefficient of Skewness:

$$\sqrt{\beta_1} = 0$$

Coefficient of Kurtosis:

$$\beta_2 = 1.8$$

Table 3.3. The Exponential Distribution $\mathbf{E}(\theta)$

Probability Density Function:

$$f_X(x) = \begin{cases} \frac{1}{\theta} e^{-x/\theta}, & x \geq 0,\ \theta > 0 \\ 0, & \text{elsewhere} \end{cases}$$

Expected Value:

$$E[X] = \theta$$

Variance:

$$Var(X) = \theta^2$$

Coefficient of Skewness:

$$\sqrt{\beta_1} = 2.0$$

Coefficient of Kurtosis:

$$\beta_2 = 9.0$$

The coefficient of variation for the geometric probability model is $v = \sqrt{p}$. For the uniform distribution, $v = (b - a)/[\sqrt{3}(b + a)]$. For the exponential distribution, $v = 1$.

ANALYSIS OF OBSERVED DATA

Usually, environmental data come to us without information about the processes responsible for generating them. The distributional form of the data also usually is unknown. To help data users interpret the data, it is important to present the data in a simplified, understandable form.

Computing Statistics from Data

One approach is to summarize the data using standardized descriptive statistics. A "statistic" is any value that is calculated from the data. A useful statistic is the arithmetic mean, which provides a measure of the central tendency of the observations. Let $x_1, x_2, \ldots, x_n$ denote the raw observations of some environmental variable. The arithmetic mean $\bar{x}$ is computed by summing the n observations and dividing by n:

$$\bar{x} = \frac{1}{n}\sum_{i=1}^{n} x_i$$

The arithmetic mean also is known as the first moment of the data about the origin, denoted as $\bar{x}_1 = \bar{x}$, which is analogous to the first moment about the origin $\mu_1 = E[X]$ for the probability models. All the moments of the observations are computed in a manner similar to that used to calculate the moments of probability models, except that the PMF or PDF in the expectation operator is replaced by $1/n$, and the actual observed values are used in place of $\bar{x}$:

$$\bar{x}_j = \frac{1}{n}\sum_{i=1}^{n} x_i^j$$

Thus, the second moment of the observations about the origin is computed by setting $j = 2$ in the above equation for $\bar{x}_j$:

$$\bar{x}_2 = \frac{1}{n}\sum_{i=1}^{n} x_i^2$$

For probability models, moments about the mean $m_1, m_2, \ldots, m_j$ were defined earlier. For observed data, analogous moments about the mean exist, and we shall denote them as $s_1, s_2, \ldots, s_j$, which are calculated by the following general relationship:

$$s_j = \frac{1}{n}\sum_{i=1}^{n} (x_i - \bar{x})^j$$

Of particular importance is the second moment about the mean, which also is known as the variance s_2:

$$s_2 = \frac{1}{n}\sum_{i=1}^{n} (x_i - \bar{x})^2$$

Calculating the mean $\bar{x}$ required n separate data points, and calculating the variance s_2 used the mean and the same n separate data points. Suppose we had calculated $\bar{x}$ and had begun calculating s_2 when we lost the last data point x_n. What would happen? We can rewrite the above equation for s_2 as consisting of two parts, one based on the first $n - 1$ values and the other based on the missing nth value:

$$s_2 = \frac{1}{n}\sum_{i=1}^{n-1} (x_i - \bar{x})^2 + \frac{1}{n}(x_n - \bar{x})^2$$

Similarly, the mean can be written as the sum of two components, one based on $n - 1$ data points and another based on the last point x_n:

$$\bar{x} = \frac{1}{n}\sum_{i=1}^{n} x_i = \frac{1}{n}\sum_{i=1}^{n-1} x_i + \frac{1}{n} x_n$$

Solving this equation for x_n gives:

$$x_n = n\bar{x} - \sum_{i=1}^{n-1} x_i$$

Substituting this result for x_n into the previous equation for s_2 gives the following result:

$$s_2 = \frac{1}{n}\sum_{i=1}^{n-1}(x_i - \bar{x})^2 + \frac{1}{n}\left[(n-1)\bar{x} - \sum_{i=1}^{n-1} x_i\right]^2$$

Notice that only the first $n - 1$ data points, n, and $\bar{x}$ appear in the right-hand side of this equation. This result shows that, by using this expression, s_2 can be computed from only $n - 1$ data points, and x_n actually is not needed. This last data point x_n is redundant, because the necessary information is contained in $\bar{x}$. Thus, although computation of s_2 ordinarily involves n data points, only $n - 1$ separate pieces of information are used, and s_2 is said to have $n - 1$ *degrees of freedom.* The concept of degrees of freedom is important in statistics.

An example illustrates this concept. Suppose we have five observations: $x_1 = 5$, $x_2 = 3$, $x_3 = 4$, $x_4 = 7$, and $x_5 = 6$. The observed mean will be given by $\bar{x} = (5 + 3 + 4 + 7 + 6)/5 = 25/5 = 5$. The observed variance s_2 will be computed as the sum of the following terms divided by 5:

$$\begin{aligned}
(x_1 - \bar{x})^2 &= (5-5)^2 = (0)^2 = 0\\
(x_2 - \bar{x})^2 &= (3-5)^2 = (-2)^2 = 4\\
(x_3 - \bar{x})^2 &= (4-5)^2 = (-1)^2 = 1\\
(x_4 - \bar{x})^2 &= (7-5)^2 = (2)^2 = 4\\
(x_5 - \bar{x})^2 &= (6-5)^2 = (1)^2 = 1\\
&\text{Sum} \quad 10
\end{aligned}$$

Thus, the variance will be $s_2 = 10/5 = 2$.

Suppose, after calculating the observed mean, we attempted to calculate the observed variance without the fifth data point x_5. Using the expression above for s_2 that is based on $n - 1$ data points, we obtain the same value for s_2 using only the first four data points:

$$s_2 = \frac{1}{5}(0+4+1+4) + \frac{1}{5}[(4)(5) - (5+3+4+7)]^2 = \frac{10}{5} = 2$$

Similarly, it is possible to discover the value of the missing data point using just the mean and the first four data points:

$$x_5 = 5\bar{x} - \sum_{i=1}^{4} x_i = (5)(5) - (5+3+4+7) = 25 - 19 = 6$$

There is nothing special about the last data point. This same approach would work if any of the five data points were deleted. This example shows that only four of the five observations are needed to calculate s_2 after the mean has been calculated, and the fifth observation adds no new information.

Because $n - 1$ separate pieces of information are used to calculate s_2 but n appears in the denominator, it is called a "biased" estimate of the variance. An "unbiased" estimate of the variance, denoted as s^2, is obtained by dividing the second moment about the mean by $n - 1$ instead of n:

$$s^2 = \frac{1}{n-1}\sum_{i=1}^{n}(x_i - \bar{x})^2$$

The square root of the variance s is known as the standard deviation of the observations. Because s^2 and s_2 are related to each other as follows, the biased and unbiased estimate of the variance are nearly equal as n becomes increasingly large, because the quantity $n/(n - 1)$ gradually approaches unity:

$$s^2 = s_2\left(\frac{n}{n-1}\right)$$

The quantity s^2 always is greater than s_2, but it decreases rapidly as n increases (Table 3.4). For n greater than 21, s^2 always is less than 5% larger than s_2. For n = 50 observations, the difference between s^2 and s_2 is only 2%, and the difference is only 1% for 100 observations.

With actual observations, s^2 is preferred to s_2 as an estimate of the variance. It is easy to obtain s^2 from s_2 using the above formula. Calculating s_2 from data sometimes is more convenient if the following expression derived by expanding s_2, is used:

$$s_2 = \frac{1}{n}\sum_{i=1}^{n}(x_i - \bar{x})^2 = \frac{1}{n}\sum_{i=1}^{n}(x_1^2 - 2\bar{x}x_i + \bar{x}^2) =$$

$$\frac{1}{n}\sum_{i=1}^{n}x_i^2 - 2\bar{x}\frac{1}{n}\sum_{i=1}^{n}x_i + \bar{x}^2 = \frac{1}{n}\sum_{i=1}^{n}x_i^2 - \bar{x}^2 = \bar{x}_2 - \bar{x}^2$$

This simplified expression for the biased estimate of the variance of observations is analogous to the simplified expression for the variance of a probability model discussed above, $Var(X) = \mu_2 - \mu^2 = E[X^2] - \mu^2$. If we now substitute $s_2 = s^2$

Table 3.4. Comparison of s^2 and s_2 for Selected Values of n

n	Deviation, %	n	Deviation, %
5	25	55	1.8
10	11	60	1.7
15	7.1	65	1.6
20	5.2	70	1.5
25	4.2	75	1.4
30	3.5	80	1.3
35	2.9	85	1.2
40	2.6	90	1.1
45	2.3	95	1.1
50	2.0	100	1.0

$(n-1)/n$ into the above expression and solve for s^2, the following simplified expression for the unbiased estimate of the variance results:

$$s^2 = \frac{1}{n-1}\sum_{i=1}^{n} x_i^2 - \frac{n}{n-1}\bar{x}^2 = \frac{1}{n-1}\sum_{i=1}^{n} x_i^2 - \frac{1}{n(n-1)}\left(\sum_{i=1}^{n} x_i\right)^2$$

These simplified expressions for the biased and unbiased estimates of the variance have an important computational advantage: one can proceed through a list of observations directly without first calculating the mean. Thus, these expressions often are called "one-pass" formulas. In the "two-pass" formulas, one proceeds through all the observations to calculate $\bar{x}$ on the first pass; then one proceeds through the observations again to calculate $(x_i - \bar{x})$ for each observation i on the second pass. Noting that $\bar{x} = \bar{x}_1$, similar one-pass formulas exist for the higher moments of the observations about the mean:

$$s_3 = \bar{x}_3 - 3\bar{x}\bar{x}_2 + 2\bar{x}^3$$

$$s_4 = \bar{x}_4 - 4\bar{x}\bar{x}_3 + 6\bar{x}^2\bar{x}_2 - 3\bar{x}^4$$

The one-pass formulas often are useful in certain computer programs and in hand calculations.

The values for s_2, s_3, and s_4 computed from these expressions can be used to compute the coefficient of skewness $\sqrt{b_1}$ and the coefficient of kurtosis b_2 for the observations:

$$\sqrt{b_1} = \frac{s_3}{s_2^{1.5}}$$

$$b_2 = \frac{s_4}{s_2^{2}}$$

The statistics $\sqrt{b_1}$ and b_2 computed from the observations are analogous to the parameters $\sqrt{\beta_1}$ and β_2 for the probability models. If the observations are known to arise from a particular probability distribution, then $\sqrt{b_1}$ will be an estimate of $\sqrt{\beta_1}$, and b_2 will be an estimate of β_2.

Since s_3 is developed from the cube of the difference between each observation and the mean, $(x_i - \bar{x})^3$, a single unusual observation (i.e., an "outlier") can exert considerable influence on s_3. Similarly, s_4 is developed from this same difference raised to the fourth power, $(x_i - \bar{x})^4$, so a single outlier exerts very great influence on s_4. Consequently, the statistics $\sqrt{b_1}$ and b_2 are very sensitive to occasional extreme values found among the observations. If the magnitude of just one or two observations can affect the value of a statistic, then it is not considered a very *robust* statistic. Robust statistics are those which are relatively insensitive to the occurrence of a few outliers among the data.

An example illustrates the computation of $\bar{x}$, s^2, b_1, and b_2 by these methods. Suppose one obtains 24 observations from a particular environmental process (Table 3.5). These 24 observations might represent the concentrations of a metal observed in a stream on 24 different dates: x_1 = 1.983 μg/m^3, x_2 = 0.433 μg/m^3, … , x_{24} = 5.748 μg/m^3. (As discussed later in this chapter, these ob-

Table 3.5. Example Illustrating Computation of Mean and Variance of 24 Observations of a Pollutant Concentration, μg/m³

i	x_i	$x_i - \bar{x}$	$(x_i - \bar{x})^2$	x_i^2
1	1.983	–0.111	0.012	3.932
2	0.433	–1.661	2.760	0.187
3	1.992	–0.102	0.010	3.968
4	3.969	1.875	3.515	15.753
5	2.666	0.572	0.327	7.108
6	3.706	1.612	2.598	13.734
7	0.052	–2.042	4.171	0.003
8	4.764	2.670	7.127	22.696
9	0.353	–1.741	3.032	0.125
10	0.288	–1.806	3.263	0.083
11	2.408	0.314	0.098	5.798
12	1.358	–0.736	0.542	1.844
13	3.550	1.456	2.119	12.603
14	0.224	–1.870	3.498	0.050
15	0.212	–1.882	3.543	0.045
16	1.787	–0.307	0.094	3.193
17	4.352	2.258	5.097	18.940
18	0.298	–1.796	3.227	0.089
19	2.109	0.015	0.000	4.448
20	0.553	–1.541	2.376	0.306
21	0.635	–1.459	2.130	0.403
22	0.491	–1.603	2.570	0.241
23	6.332	4.238	17.958	40.094
24	5.748	3.654	13.350	33.040
Sum:	50.263	0.007	83.417	188.683

servations actually were generated by computer simulation from a known distribution.) The arithmetic mean of these observations is computed by dividing their sum (bottom of the column marked "x_i") by $n = 24$:

$$\bar{x} = 50.263/24 = 2.094\ \mu g/m^3$$

To compute the mean $\bar{x}$, it was necessary to use each observation on the first pass through the data. The next column marked "$x_i - \bar{x}$" gives the difference between each observation and the mean and is the result of a second pass through the data. Its sum, 0.007 $\mu g/m^3$ is of no particular significance, but it shows that the numerical sum of these deviations gives little information about the spread of the observations, because positive and negative deviations from the mean tend to cancel each other. If these deviations are squared and added, however, the result can be used to compute either the biased estimate of the variance s_2 or the unbiased estimate s^2, each of which provides a measure of the spread, or dispersion, of the observations:

$$s_2 = 83.417 / 24 = 3.476\ \mu g/m^3 \qquad s^2 = 83.417/23 = 3.627\ (\mu g/m^3)^2$$

The unbiased estimate of the standard deviation will be $s = \sqrt{3.627} = 1.904$ $\mu g/m^3$.

The above calculations required two passes through the data and illustrate computation of the variance by the two-pass method. In the single-pass method, we compute the sum of each observation and the sum of the square of each observation (last column marked "x_i^2") on our first pass through the observations:

$$s^2 = \frac{1}{n-1}\sum_{i=1}^{n} x_i^2 - \frac{n}{n-1}\bar{x}^2 = \frac{1}{23}(188.683) - \frac{24}{23}(2.0943)^2 = 3.627\ (\mu g/m^3)^2$$

As anticipated, this one-pass approach gives the same result as the two-pass approach for the unbiased estimate of the standard deviation, or $s = \sqrt{3.627} = 1.904$ $\mu g/m^3$. Although the one-pass approach often is easier to apply than the two-pass approach, notice that the sum of the square of the observations, $\Sigma x_i^2 = 188.683$, is somewhat large. In some situations, this sum can become extremely large and may cause numerical overflows on computers or hand calculators, so careful programming may be needed.

If we compute the sum of the observations raised to the third and fourth powers, Σx_i^3 - 859.5 and $\Sigma x_i^4 = 4{,}317.6$, and divide by $n = 24$, the third and fourth moments of the observations about the origin are obtained:

$$\bar{x}_3 = \frac{859.5}{24} = 35.81 \qquad \bar{x}_4 = \frac{4317.6}{24} = 179.9$$

Applying the one-pass formulas for the third and fourth moments about the mean, and noting that $\bar{x}_2 = 188.683/24 = 7.862$ $\mu g/m^3$, we obtain:

$$s_3 = \bar{x}_3 - 3\bar{x}\bar{x}_2 + 2\bar{x}^3 = 35.8 - 3(2.094)(7.862) + 2(2.094)^3 = 4.79$$

$$s_4 = \bar{x}_4 - 4\bar{x}\bar{x}_3 + 6\bar{x}^2\bar{x}_2 - 3\bar{x}^4 = 179.9 - 4(2.094)(35.81)$$
$$+ 6(2.094)^2(7.862) - 3(2.094)^4 = 29.1$$

Finally, the coefficients of skewness $\sqrt{b_1}$ and kurtosis b_2 for these observations are computed as follows:

$$\sqrt{b_1} = \frac{s_3}{s_2^{1.5}} = \frac{4.79}{(3.476)^{1.5}} = 0.739$$

$$b_2 = \frac{s_4}{s_2^{2}} = \frac{29.1}{(3.476)^{2}} = 2.41$$

These statistics are dimensionless. Computation of these statistics by the two-pass formulas will yield the same answers.

Another statistic sometimes used to describe data is the *geometric mean* $\bar{x}_g$, which is the nth root of the product of the n observations:

$$\bar{x}_g = \left(\prod_{i=1}^{n} x_i\right)^{1/n}$$

Like the uppercase Greek letter *sigma*, which denotes addition, the uppercase Greek letter *pi* denotes multiplication. If logarithms of both sides of this equation are taken, the right-hand side of the equation becomes a summation:

$$\log \bar{x}_g = \frac{1}{n}\sum_{i=1}^{n} \log x_i$$

The right-hand side actually is the arithmetic mean of the logarithms of the observations. If any observation is zero, the geometric mean is undefined. For the example given in Table 3.5, the product of the 24 observations is $x_1 x_2 \ldots x_{24} = 28.56$, and the geometric mean is given by $(28.56)^{1/24} = 1.150$ μg/m³.

These statistics provide a standardized way to describe the characteristics of the observations in a parsimonious manner. The arithmetic and geometric means give indications of the central tendency of the observations. The standard deviation (biased or unbiased) gives an indication of the dispersion of the observations. Another measure of the "spread" of the observations is the range r, which is the difference between the highest and lowest observations. For this example, the range is $r = 6.332 - 0.052 = 6.280$ μg/m³. The coefficient of skewness gives an indication of the direction of the tail of the distribution of observations, if any. The coefficient of kurtosis gives an indication of the peakedness of the distribution of observations.

If only these statistics were available for the data, then one could obtain some sense of the characteristics and values of the observations by examining the value of each statistic. However, these "point estimates" seldom provide sufficient information about details of the distribution—whether the observations are equally spaced or clustered in certain locations—over the range of the data. To provide the analyst with a more detailed "feel" about the manner in which the observations are distributed, graphical techniques can be utilized, as discussed in the following section.

Histograms and Frequency Plots

One of the simplest possible graphical techniques is a "linear scale plot" which displays the location of every observation. That is, the location of each data point relative to the origin is linearly proportional to the value of the data point. When the 24 observations given in this example (Table 3.5) are shown on a linear scale plot (Figure 3.8a), a dense cluster appears in the range of about $x = 0$ to $x = 0.7$, with the other values unequally scattered over the remaining portions of the scale. If more than 24 observations were considered, the linear scale plot soon would become crowded, and so many observations would overlap that we might be unable to determine where the greatest number of observations occurred. Thus, the linear scale plot is useful only when the data set contains a small number of observations.

An improvement on the linear scale plot is the *histogram*. Here the same linear scale is used, but it is divided into equally spaced intervals. For each interval, a vertical bar is plotted whose height is proportional to the number of observations falling into the interval. If, for the 24 observations, we choose an interval width of $\Delta x = 0.5$ μg/m³, then 8 observations will fall into the interval $0 \le x_i < 0.5$, and

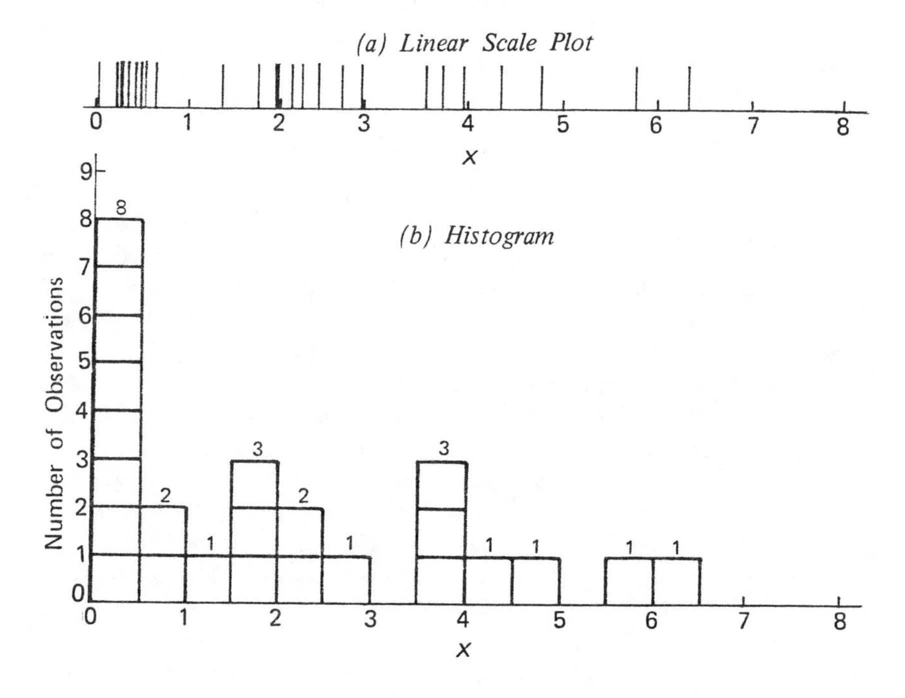

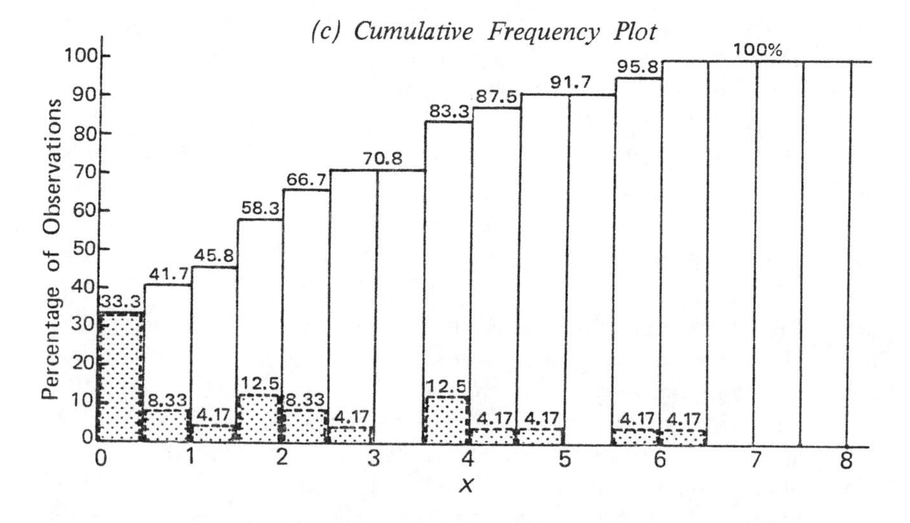

Figure 3.8. Graphical representation of 24 observations from example plotted on (a) linear scale plot, (b) histogram, and (c) individual and cumulative frequency plots.

a vertical bar 8 units in height will represent this cluster (Figure 3.8b). Notice that the general locations of the clusters of the data points can be more easily seen in the histogram than in the linear scale plot.

Construction of a histogram can be expressed more formally as follows. If m denotes the number of intervals, then the histogram is constructed by sorting each data point x_i into the jth interval such that:

$$(j-1)\Delta x \leq x_1 < j\Delta x \quad \text{for } j = 1, 2, \ldots, m$$

Choosing the size of Δx usually is a matter of judgment. If Δx is too small, there will be too many bars on the histogram, and the figure will appear unnecessarily complex and busy. If Δx is too large, the figure will not show sufficient detail about the locations of the clusters, if any, and it will be difficult to get a feel for the distribution of the data.

It is easy to divide the individual counts within each bar of the histogram by n and replot the graph. The result will give the *proportion* of the total number of observations found in each interval, or an *individual frequency* histogram. If each frequency is multiplied by 100, the results can be expressed as percentages (Figure 3.8c, shaded bars). If the individual frequencies are added from left to right and the result plotted, the *cumulative frequency* graph is obtained. For example, the individual frequency of the 8 observations in the interval between 0 and 0.5 $\mu g/m^3$ is (8/24)(100) = 33.33%, so the first cumulative frequency bar is shown as 33.3%. The relative frequency of the 2 observations in the interval between 0.5 and 1.0 $\mu g/m^3$ is (2/24)(100) = 8.33%, so the first cumulative frequency bar is the sum of the two individual frequency bars: 33.33 + 8.33 = 41.66 = 41.7%. The incremental steps comprising the cumulative frequency plot gradually approach 100% when all the 24 observations have been included (i.e., $x_i >$ 6.5).

Each of the individual drawings in Figure 3.8 conveys the same information, but is displayed differently. Actually, each successive drawing may be considered to be derived from the previous one. The middle drawing (Figure 3.8b) is constructed from the top one (Figure 3.8a). Similarly, the bottom drawing (Figure 3.8c) is constructed from the middle one (Figure 3.8b).

Another important graphical technique for displaying observed data is to use probability plotting paper. A variety of probability papers are available (normal, lognormal, extreme value, etc.) from private companies such as Keuffel & Esser Co. and the Technical and Engineering Aids for Management (TEAM) Co.[9] These papers usually can be obtained at drafting supply houses or through mail order. Personal computers can make these plots using programs such as SigmaPlot™ Scientific Graphing Software by Jandel Scientific Corp., San Rafael, CA.

When one is confronted by new measurements obtained from some process, it seldom is evident which probability paper is the best one to choose. One approach, usually rather tedious, is to try various probability papers to see which one gives the "best picture" of the observations. A better approach is to rely on any information we may have about the process itself. If there is a history of past observations from this same process, or from a similar class of processes, or if something is known about the physical laws responsible for generating the observations, then it may be possible to select in advance the most appropriate paper for plotting the observations. If the observations consist of pollutant concentrations measured in the environment, then a good rule is to begin with logarithmic-probability paper (Chapter 9). On this paper, observations whose logarithms are normally distributed (Chapter 7) will plot as a straight line. As indicated in this book (Chapter 8), concentrations of chemicals found in environmental media (air, water, soil, food, etc.) usually have undergone dilution in a particular manner that causes them to be approximately lognormally distributed.

In the present example giving 24 observations, we actually know that the distribution of the observations is exponential, because they were generated artificially by computer simulation (see next section). However, we will plot the observations on logarithmic-probability paper anyway to see how they look.

If there are only a few observations, as in the present case, the data do not have to be grouped, and the individual frequency for each data point can be plotted. Individual frequency plotting is preferred to grouped frequency plotting, because it gives improved resolution and detail. However, the individual frequency plots are more sensitive to unusual observations and outliers. Thus, grouped data plotting is a more robust analytical technique than individual frequency plotting and is preferred if there are serious anomalies among the observations.

To make an individual frequency plot, we first arrange the observations by sorting them from lowest value to highest value (Table 3.6). Table 3.6 contains the same 24 observations given in Table 3.5, except that it begins with the lowest value, 0.052 μg/m^3, and it ends with the highest value, 6.332 μg/m^3. Next, we compute the plotting position f_i for each observation by using the following formula:

$$f_i = \sum_{i=1}^{n} \frac{i}{n+1} \quad \text{for } i = 1, 2, \ldots, n$$

Table 3.6. 24 Observations from Previous Example Sorted In Increasing Order for Plotting on Probability Paper

i	x_i	Plotting Position, f_i%
1	0.052	4
2	0.212	8
3	0.224	12
4	0.288	16
5	0.298	20
6	0.353	24
7	0.433	28
8	0.491	32
9	0.553	36
10	0.635	40
11	1.358	44
12	1.787	48
13	1.983	52
14	1.992	56
15	2.109	60
16	2.408	64
17	2.666	68
18	3.550	72
19	3.706	76
20	3.969	80
21	4.352	84
22	4.764	88
23	5.748	92
24	6.332	96

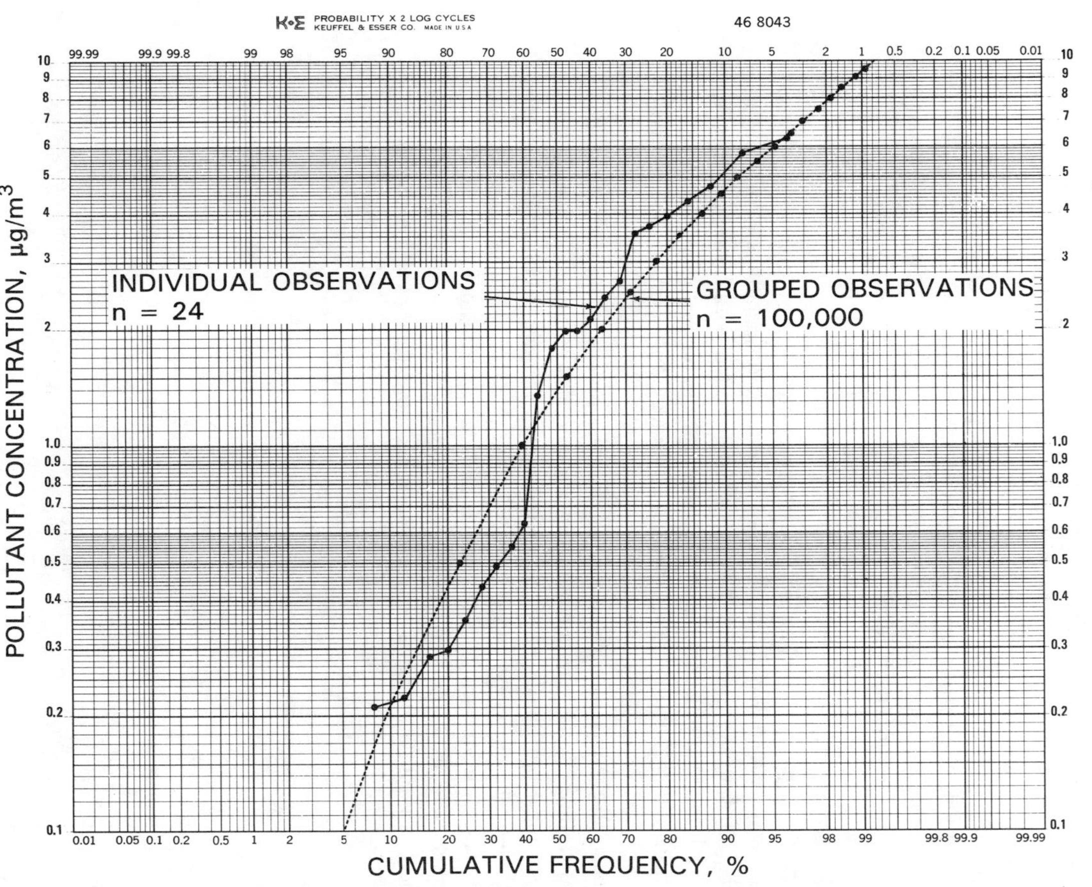

Figure 3.9. Logarithmic-probability plot of example giving 24 observations, along with grouped data plot from same (exponential) distribution from computer simulation, plotted on Keuffel & Esser 2-cycle logarithmic-probability paper no. 46 8043.

For the present example, $f_i = i/(24 + i) = i/25 = 0.05i$. When expressed in percentages, the plotting positions for these observations become 4%, 8%, 12%, etc., which form a series (see Table 3.6). Notice that the last term in the series will be 96% rather than 100%, and the last observation, 6.332 μg/m³, will be plotted at a frequency of $f_{24} = 0.96$. This happens because the plotting position formula can never become 100%, a desirable feature since the probability paper never reaches 100% due to the long tail of the lognormal distribution. This plotting position formula also has other desirable properties, and it is recommended by Mage[10] over the one used by Larsen[11] for analyzing air quality data. It also is the formula preferred in most texts on probability and statistics.

Finally, the individual observations are plotted on logarithmic-probability paper (see dots connected by solid line in Figure 3.9). Because the lowest value, 0.052 μg/m³, is off the scale of this two-cycle logarithmic-probability paper, we must begin with the second observation, $x_2 = 0.212$ μg/m³, which is plotted at a cumulative frequency of 8%. The last observation, $x_{24} = 6.332$ μg/m³, is plotted at a cumulative frequency of 96%.

As discussed in the next section, this same data generating process was allowed to continue to generate observations on the computer after the first 24 were obtained, giving a total of 100,000 observations. With so many data points, an individual frequency plot was impossible, and the observations were grouped using an interval width of $\Delta x = 0.5$ μg/m³ and plotted on a histogram (Figure 3.10). Then the resulting cumulative frequencies from the histogram were plotted on logarithmic-probability paper (see dots connected by dotted line in Figure 3.9).

Because so many observations are used in the grouped data plot, the dotted line drawn through the points exhibits gentle curvature, unlike the jagged solid line resulting from the 24-observation individual frequency plot (Figure 3.9). Nevertheless, the closeness and similarity of the two lines is surprising, because one line results from only 24 observations and the other results from 10,000 observations, both from the same stochastic process. This example shows that a great deal of information about a process can be obtained from relatively few observations, a basic principle in sampling theory. Both sets of observations were

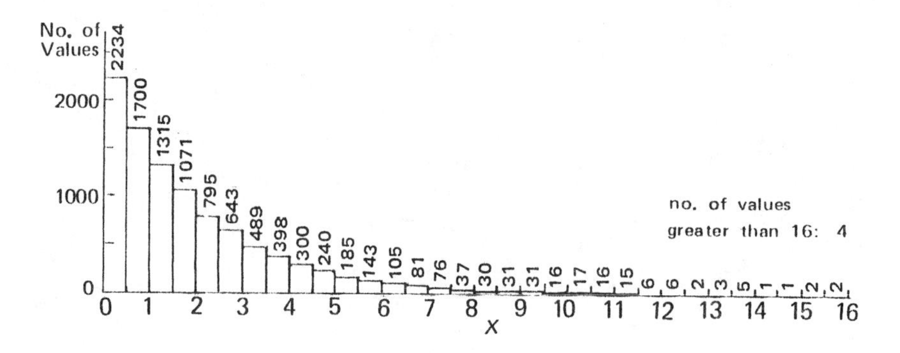

Figure 3.10. Histogram of 10,000 observations generated by computer simulation experiment from a single-parameter exponential distribution with $\theta = 2$.

generated from an exponential distribution, with $\theta = 2$. Thus, this result also illustrates that distributions other than the lognormal can be plotted on logarithmic-probability paper and that useful comparisons can be made between different curves depicting widely different sample sizes. This discussion also illustrates how to plot observations on logarithmic-probability paper. The same general methodology can be followed in plotting observations on other types of probability papers.

Fitting Probability Models to Environmental Data

In the example discussed above, we began by examining the data as though they arose from a real environmental process. Actually, however, we already know the stochastic process responsible for generating the data, since the values were generated on a personal computer using the Inverse Transformation Method for generating random variables (see earlier section on Computer Simulation, page 36). Thus, the distributional form—the probability model and its parameters—was known to the investigator, and was under his control. The distribution was a single-parameter exponential **E**(2).

To generate an exponentially distributed random variable using the Inverse Transformation Method, one begins with the CDF of the single-parameter exponential probability model: $F_X(x) = 1 - e^{-x/\theta}$. Solving this function for x, we obtain the inverse CDF $F_X^{-1}(x)$ as follows:

$$x = F_X^{-1}(x) = -\ln[1 - F_X(x)]$$

In this formula, if $F_X(x)$ is uniformly distributed as **U**(0,1), then x will be exponentially distributed as **E**(θ). Because the unit uniform distribution is symmetrical and ranges between 0 and 1, it makes no difference whether $1 - F_X(x)$ or $F_X(x)$ appears in brackets in this formula. Thus, the following equation generates a single-parameter exponential distribution **E**(θ) when $F_X(x)$ is allowed to have a uniform distribution **U**(0,1):

$$x = -\theta \ln[F_X(x)]$$

The computer simulation program in Chapter 8 (Figure 8.4, page 198) can be modified to generate the exponential distribution **E**(2) by inserting the computer statement U = –2∗ln(U) as line #335 between line #330 and line #340 to make the necessary exponential conversion. When the program is run, we set A = 0, B = 1, and M = 1 so that U will have the unit uniform distribution **U**(0,1), and then line #335 will modify U to be exponentially distributed.

Ordinarily, when dealing with real data, we are not as fortunate as in this illustrative example. Usually, we begin with a set of observations in which the distributional form is unknown. By examining the nature of the physical processes generating the data, we may be able to gain insights about the stochastic process responsible for the data. Graphical techniques also may help us select an appropriate probability model. Once a probability model has been selected, there still is the problem of determining the values of its parameters. Such a procedure often is called "fitting" a probability model to the observations. A third step, equally important, is to determine how well the model fits the observations, or its "goodness-of-fit." In summary, the model-fitting procedure consists of three steps:

1. Select a candidate probability model.
2. Estimate the values of its parameters.
3. Determine the model's goodness-of-fit.

One formal procedure for determining how well a particular probability model fits a data set is the chi-square goodness-of-fit test. This test is useful for grouped data, allowing one to determine the probability with which the candidate probability model can be rejected. To apply the chi-square test, one compares the number of observations O_j observed in interval j with the expected number of observations E_j predicted by the assumed model. If there are m intervals in all, the following statistic is computed:

$$\chi^2 = \sum_{j=1}^{m} \frac{(O_j - E_j)^2}{E_j}$$

The test is nonparametric; that is, this statistic will have a probability distribution that is approximately chi-square (Table 8.2), regardless of the underlying distribution of the observations. To apply the test, one consults published tables of the chi-square distribution,[12] using $m - r - 1$ degrees of freedom, where r is the number of parameters estimated in Step 2 above. As with the construction of a histogram, judgment must be used in choosing the number of intervals into which the data are to be sorted. A general rule is that there should be at least five observations in any cell. If fewer than five observations occur in any interval, then adjacent intervals should be combined to enlarge the count to five or more.

We illustrate this approach by applying it to the example of 10,000 observations shown in the histogram (Figure 3.10). Suppose these data came from a real physical process. In Step 1, after examining the shape of the histogram, we might conclude that the exponential probability model is a reasonable candidate for this process. In Step 2, we then must estimate the exponential probability model's single parameter θ. One approach is to use the arithmetic mean of the observations as an estimate of the arithmetic mean of the model. The arithmetic mean calculated from the 10,000 observations was 2.0053 $\mu g/m^3$, and the arithmetic standard deviation was s = 2.0154 $\mu g/m^3$. Setting the arithmetic mean of the model equal to the arithmetic mean of the observations is called the "method of moments." When parameters are estimated by the method of moments, one begins with the first moment and continues to the higher moments until all the parameters are determined.

Because these observations were generated by computer simulation, we know the true mean to be $\theta = 2$, and the observed mean for the 10,000 observations ($\bar{x} = 2.0053$) happens to be extremely close to the true mean. We shall illustrate the approach by using the true mean and by setting $\bar{x} = \theta = 2$. To apply Step 3 using the chi-square test, we must calculate the expected number of observations in each interval. Using the same interval width as in the histogram (Figure 3.10),

$\Delta x = 0.5$ μg/m³, the probability that an observation will occur in the first interval is calculated as follows:

$$F_X(0.5) = 1 - e^{-0.5/2} = 1 - 0.7788 = 0.2212$$

Then the expected number of observations in the first interval will be E_1 = (10,000)(0.2212) = 2,212 observations. A total of O_1 = 2,234 observations actually appeared in this interval, and the chi-square component for the first interval is computed as follows:

$$\frac{(O_1 - E_1)^2}{E_1} = \frac{(2234 - 2212)^2}{2212} = \frac{(22)^2}{2212} = 0.219$$

In general, the expected number of observations in any interval j is computed as follows, in which n is the total number of observations, and m is the number of intervals:

$$E_j = n[F_X(j\Delta x) - F_X(\{j-1\}\Delta x)] \quad \text{for } j = 1, 2, \ldots, m$$

If E_{j-1} denotes the expected number of observations in the interval just before the jth interval, then the above expression also can be written more simply as follows, in which $E_0 = 0$:

$$E_j = nF_X(j\Delta x) - E_{j-1} \quad \text{for } j = 1, 2, \ldots, m$$

Applying the above expression to calculate the expected number of observations in the second interval $(j = 2)$, we obtain:

$$E_2 = 1000[F_X(1.0)] - 2212 = 3935 - 2212 = 1723$$

For this interval, the chi-square component will be $(1700 - 1723)^2/1723 = 0.307$. The chi-square components for all the intervals in the histogram have been computed (Table 3.7). To avoid any cells with fewer than 5 observations, all observations equal to or greater than 12.5 μg/m³ were grouped into one large interval with 18 observations. Most of the intervals in the table have chi-square components of less than 1.0, and only two intervals, j = 16 and j = 23, have chi-square components of 4.0 or more. The sum of the chi-square components (bottom of table) gives $\chi^2 = 21.170$. In all, there are m = 26 intervals, and we are estimating r = 1 parameters, so there are 26 – 1 – 1 = 24 degrees of freedom. Referring to published tables of the chi-square distribution,[12] we compare our chi-square sum with χ^2 computed from the chi-square CDF (Table 3.8). Because our result of 21.17 does not exceed any of the values in the table, there is no basis for rejecting the exponential probability model as the correct one for this data set.

Not rejecting a model does not imply that we automatically accept the model. Some other model with characteristics similar to the exponential probability model actually may be the correct one. However, in this example, we know that

Table 3.7. Example Showing Application of Chi-Square Goodness-of-Fit Test to Exponential Probability Model with 10,000 Observations

j	x_j	Observed No. O_j	Expected No. E_j	Chi-Square χ^2
1	0.5	2234	2212	0.219
2	1.0	1700	1723	0.307
3	1.5	1315	1341	0.504
4	2.0	1071	1045	0.647
5	2.5	795	814	0.443
6	3.0	643	634	0.128
7	3.5	489	493	0.032
8	4.0	398	385	0.439
9	4.5	300	299	0.003
10	5.0	240	233	0.210
11	5.5	185	182	0.049
12	6.0	143	141	0.028
13	6.5	105	110	0.227
14	7.0	81	86	0.291
15	7.5	76	67	1.208
16	8.0	37	52	4.327
17	8.5	30	40	2.500
18	9.0	31	32	0.031
19	9.5	31	24	2.042
20	10.0	16	20	0.800
21	10.5	17	15	0.267
22	11.0	16	11	2.272
23	11.5	15	9	4.000
24	12.0	6	7	0.143
25	12.5	6	6	0.000
	≥12.5	20	19	0.053
	Totals:	10000	10000	21.170

the observations actually were generated on the computer to be single-parameter exponential **E**(2).

Suppose the set of observations in this example gave $\chi^2 = 37.7$ rather than 21.17. Then we would be justified in rejecting the assumed model with a probability of $P = 0.95$. This result implies that chi-square values larger than 37.7 are rare for the correct model, so the model is not likely to be correct. Stated another way, a true exponential model will give values of χ^2 greater than 37.7 in only 5 out of 100 cases, so it does not seem likely that the exponential probability model could be responsible for these observations. Of course, our particular data set could be among the 5 rare cases. Nevertheless, it is customary to state that,

Table 3.8. χ^2 Values for the Chi-Square CDF with 24 Degrees of Freedom[12]

Probability (CDF):	0.750	0.900	0.950	0.975	0.990	0.995
χ^2:	29.3	34.4	37.7	40.6	44.3	46.9

for $\chi^2 = 37.7$ with 24 degrees of freedom, we are 95% certain that the assumed model is not the correct one.

Like other probabilistic goodness-of-fit tests, the chi-square test allows one to determine the probability that a hypothesized probability model did not give rise to the observed data set. Rejection implies that the assumed model is incorrect, but acceptance does not necessarily mean that the assumed model actually *is correct.* For example, if five candidate models are tried, and three are rejected, which of the remaining two is correct? Some other model, not yet tested, actually may be the appropriate model for the observations. Although called a goodness-of-fit test, the chi-square statistic becomes larger as the fit becomes poorer and provides a measure of how poorly the model fits the observations. Thus the chi-square test actually may be viewed as a "badness-of-fit" test.

With real environmental data sets, the typical finding is that all candidate probability models are rejected by goodness-of-fit tests. For example, Ott, Mage, and Randecker[13] applied the lognormal probability model to a U.S. nationwide cross section of ambient carbon monoxide data sets and found that the model was rejected by the chi-square test with a probability greater than $P = 0.995$ in all years and cities considered. Similarly, Kalpasanov and Kurchatova[14] applied the same model to Bulgarian air quality data and found that it was rejected by another useful goodness-of-fit test, the Kolmogorov-Smirnov test.

One reason that has been suggested for the rejection of probability models when applied to environmental data is the large size of these data sets, often more than 8,000 observations. Presumably, large data sets contain anomalies that cause higher than usual rejection rates with goodness-of-fit tests. The example given above would challenge this reasoning: the exponential probability model is not rejected by the chi-square test, even though the 10,000-observation data set is quite large. Another reason for such models to be rejected is that environmental observations are obtained from physical and chemical measurement systems that themselves introduce error. If an environmental variable has a particular distribution and it is measured with a system that introduces error with another distributional form, then the resulting observations will be a mixture of the two distributions. Goodness-of-fit tests will give higher rejection rates than if the true (undistorted) observations were available.

Probabilistic goodness-of-fit tests are important if we are interested in making predictions about future events, or forecasts, since we must be able to identify a stable stochastic process to make valid future predictions. Alternatively, one may wish empirically to fit a model to a given data set, with no interest in making forecasts. An analyst may have an even more restricted goal: to fit a model to some limited range of the observations. For such empirical fits, the model is treated simply as an equation for compactly and concisely representing the frequency characteristics of the observations, instead of a probabilistic technique.

How can one determine the best values of the parameters of a model for empirically fitting it to a set of observations? Once the values have been determined, how does one determine how well the model fits the observations? Because the human eye readily can detect a straight line, one approach is a graphical technique in which the observations appear as a straight line whenever the model fits them well. For the exponential model, for example, we seek a graphical technique in which truly exponentially distributed observations appear as a straight line when plotted.

Consider the single-parameter exponential CDF,

$$F_X(x) = 1 - e^{-x/\theta}$$

Solving this equation for x and taking logarithms of both sides, we obtain the following expression:

$$\ln[1 - F_X(x)] = -\frac{1}{\theta}x$$

This equation is of the standard linear form $y = mx + b$, in which $y = \ln[1 - F_X(x)]$, $m = -1/\theta$, and $b = 0$. Thus, if the *logarithm* of the quantity $1 - F_X(x)$ is plotted on the (vertical) y-axis with x plotted on the (horizontal) x-axis, then the result will be a straight line. Use of semi-logarithmic plotting paper accomplishes this task. Because the paper already is scaled as logarithmic on the vertical axis, one merely plots the quantity $1 - F_X(x)$ on the vertical axis.

To illustrate the technique with the 10,000 observations of the above example, we plot [1 – *cumulative frequency*] on the vertical (logarithmic) axis (Figure 3.11). The grouped observations in each interval are listed on the histogram (Figure 3.10) and in tabular form (Table 3.7). In the first interval, $0 \le x < 0.5$, the number of observations appearing was $O_1 = 2234$. The proportion of the observations appearing in the first interval was 2234/10000 = 0.2234, and we plot 1 – 0.2234 = 0.7766 at the intercept $x = 0.5$. In the second interval, $O_2 = 1700$ observations, and the cumulative frequency becomes 0.1700 + 0.2234 = 0.3934, so we plot the quantity 1 – 0.3934 = 0.6066 at the intercept $x = 1.0$. We continue plotting these frequencies for intervals up to and including $x = 14$, which includes all but 10 observations. This last point corresponds to a cumulative frequency of 0.9990 and is plotted at the vertical intercept of 1 – 0.9990 = 0.0010. (The counts for intervals above 12.5 do not appear in Table 3.7 but are listed on the histogram in Figure 3.10.)

We see that the resulting plot of these observations, represented by dots, appears relatively linear on this graph (Figure 3.11). It would be possible, by trial and error, to fit a straight line to these points visually that passes through the point $1 - F_X(x) = 1.0$ at $x = 0$. The single-parameter exponential probability model gives $F_X(x) = 0$ for $x = 0$, so all straight lines that pass through this point have single-parameter exponential distributions. To obtain the value of the parameter θ, which will define the slope of this line, we need to find the intercepts of only one point on this line. Choosing the intercepts $1 - F_X(x) = 0.030$ at $x = 7$, we can solve the above equation for θ to obtain the following value:

$$\theta = \frac{x}{\ln\left(\dfrac{1}{1 - F_X(x)}\right)} = \frac{7}{\ln\left(\dfrac{1}{0.03}\right)} = \frac{7}{3.5066} = 1.996$$

In this example, we know that the observations were generated on the computer from a single-parameter exponential distribution **E**(2). Thus the graphically estimated value of $\theta = 1.996$ is very close to the true value of $\theta = 2$. If this graph

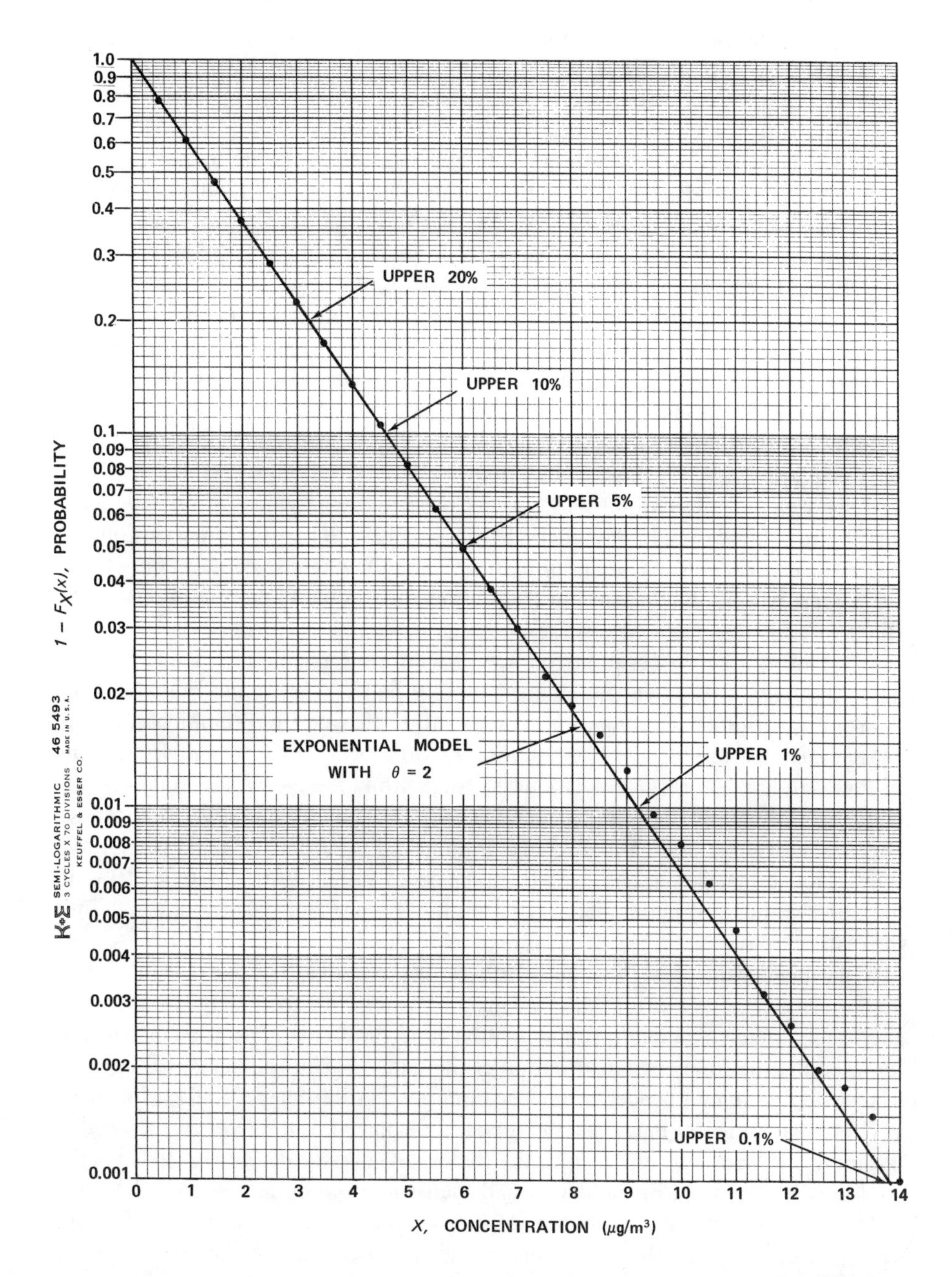

Figure 3.11. Plot of 10,000 simulated observations from a single-parameter exponential distribution, along with the model itself, illustrating the construction of exponential probability paper.

were interpreted probabilistically, then the vertical axis would give the probability that the random variable X exceeds the value of x appearing on the horizontal axis; that is $P\{X > x\}$. If, on the other hand, this graph is interpreted as an empirical plot, then the line is just a smooth curve designed to represent the cumulative frequencies of the observations for any value of x. That is, it serves an interpolative or "smoothing" function.

In Figure 3.11, notice that the upper 20% [$F_X(x) = 0.80$] and upper 10% [$F_X(x) = 0.90$] of the observations appear on the left half of the graph, and the middle and right-hand portions of the graph represent values in the tail of the distribution. For example, all points to the right of the middle of the graph, above $x = 7$, comprise the upper 3% of the observations but reflect about 50% of this diagram. By comparison, the vertical bars on the histogram in Figure 3.10 approach zero rapidly, and observations above $x = 7$ are almost invisible. This example illustrates that semi-logarithmic plots give considerable emphasis to the tail of the distribution. This emphasis on the tail occurs because probability is plotted on a logarithmic scale, and several logarithmic cycles, each representing an order of magnitude, can be used.

In the lower-right portions of Figure 3.11, we see some minor deviations from the line representing the exponential model. These deviations are understandable, because very few observations are present in the extreme values at the tail of the distribution. For example, the interval from $x = 12.5$ to $x = 13.0$ contains only 2 observations. The interval adjacent to this one on the left side contains only 6 observations, while the adjacent interval on the right contains only 3 observations (Figure 3.10).

By plotting the observations on semi-logarithmic paper in this manner, we have, in effect, created exponential probability paper. Curran and Frank[15] apply the semi-logarithmic plotting technique to sulfur dioxide (SO_2) ambient air concentrations measured at the Continuous Air Monitoring Project (CAMP) station in Philadelphia, PA (Figure 3.12). The observations are 24-hour average values collected throughout the year in 1968. On this semi-logarithmic plot, these points appear relatively straight at SO_2 concentrations below about 150 parts-per-billion (ppb). Notice that deviations from linearity occur primarily in the upper 10% of the observations (all points below 10^{-1} on this graph) and thus lie in the tail of the distribution.

If the histogram of a given set of observations looks like the one in Figure 3.10—a mode at the origin and a long tail extending to the right—then the single-parameter exponential model often is a good candidate for empirically representing the frequency distribution of the data set. Unfortunately, histograms of observed pollutant concentrations seldom possess this shape. Rather, for most environmental quality data sets, the mode is displaced to the right of the origin. Beginning at zero near the origin, the histogram bars typically rise to a single mode just to the right of the origin and then show a long tail extending to the right. An example is the histogram of hourly average CO concentrations measured for a year in San Jose, CA (Figure 3.13), which exhibits the characteristic *unimodal* (single-mode), right-skewed shape.

Figure 3.13 illustrates another important point about plotting measurements on a histogram. The measuring instrument and data system operating at this station report the CO readings only in whole integer values. Thus, the data sheets report 1 zero, 911 readings of 1 ppm, 2297 readings of 2 ppm, and so on. Because of the rounding, it is reasonable to assume that all the values reported as 1 ppm, for example, actually lie in the range of 0.5 ppm to just less than 1.5 ppm. Thus, the measurement system automatically groups the data into intervals, regardless of whether we want them grouped, and the histogram should be plotted with inter-

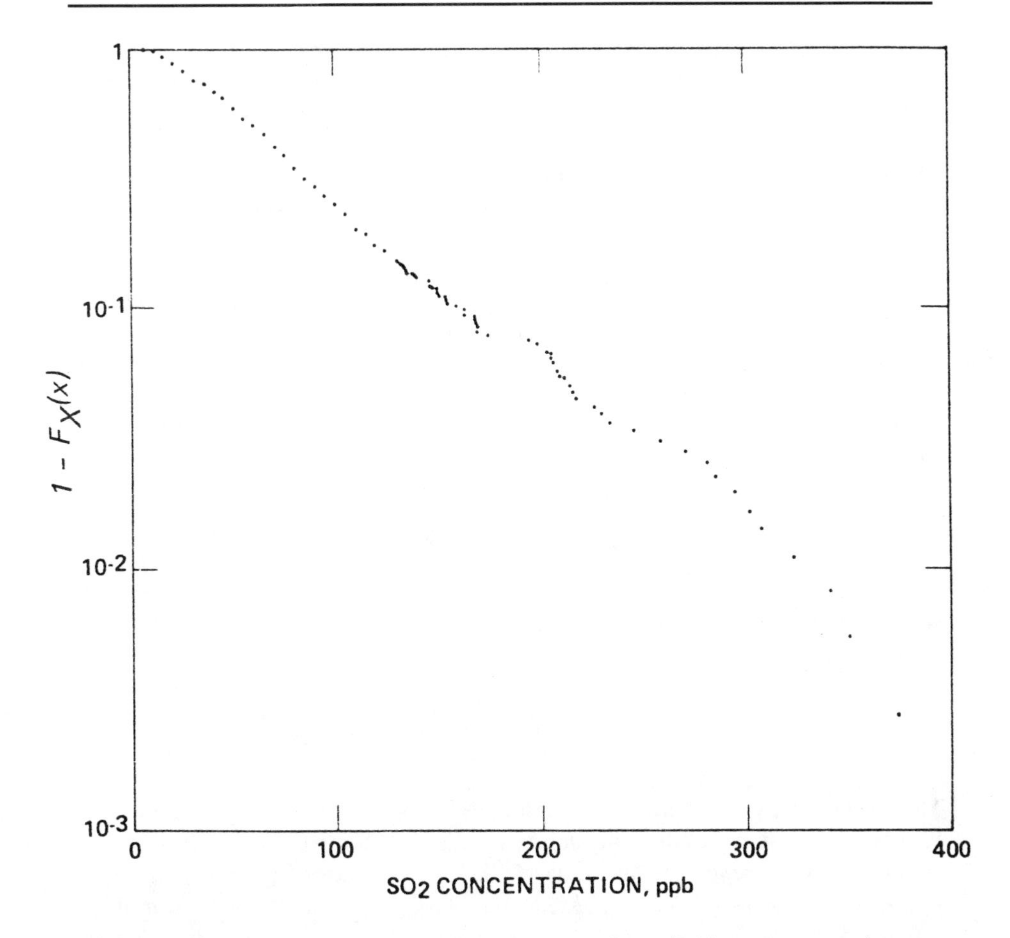

Figure 3.12. Ambient sulfur dioxide concentrations (24-hour averages) measured at the CAMP station in Philadelphia, PA, in 1968, plotted on semi-logarithmic paper (Source: Curran and Frank[15]).

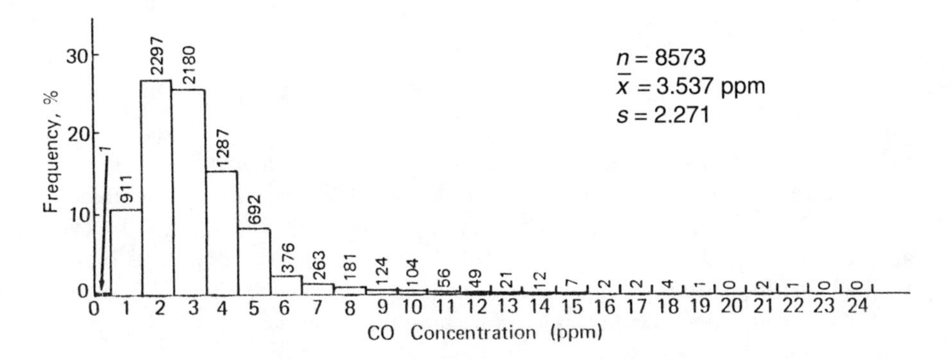

Figure 3.13. Histogram of hourly average carbon monoxide measurements in San Jose, CA, for the 12-month period from June 1970 to May 1971.

vals offset by 0.5 ppm. Thus, the 1 ppm interval is plotted such that it represents the concentration range $0.5 \le x < 1.5$. Ordinarily, a year consists of 24 hours/day × 365 days = 8,760 hours, but usually some of the hours are missing due to instrument calibration and other factors. In this CO data set, 187 observations were missing, giving a total of 8,573 readings. Usually, these missing values are distributed evenly over the year, and the histogram still can be treated as though it is representative of the entire year.

Suppose we seek to develop a distribution with greater flexibility than the exponential, one that is right-skewed and has a single mode that is offset to the right of the origin. One approach is to introduce a substitution into the exponential model. The CDF of the single-parameter exponential probability model was of the following form: $F_X(x) = 1 - e^{-(x/\theta)}$. Suppose we replace the quantity (x/θ) by the more complicated expression $(x/\theta)^\eta$. The resulting CDF will have the following form:

$$F_X(x) = 1 - e^{-\left(\frac{x}{\theta}\right)^\eta}$$

If we differentiate this CDF with respect to x, we obtain the following expression for the PDF:

$$f_X(x) = \frac{\eta}{\theta}\left(\frac{x}{\theta}\right)^{\eta-1} e^{-\left(\frac{x}{\theta}\right)^\eta}$$

This is known as the Weibull distribution **W**(η,θ) (Table 3.9). It is named for Waloddi Weibull, a Swedish physicist, who used it to represent the distribution of the breaking strength of materials. As noted by Johnson and Kotz,[16] several physical reasons have been suggested to explain why certain random variables should have a Weibull distribution. However, as used in practice, the Weibull distribution usually is a form that merely provides more flexibility and variety of shapes than some other models. Its two parameters consist of a scale parameter θ and a shape parameter η. For values of η of 1 or less, the mode of the Weibull distribution lies at the origin. For $\eta = 1$, the Weibull distribution (Figure 3.14) is the same as the single-parameter exponential distribution, **E**(θ). For values of η greater than 1, the mode x_m lies to the right of the origin and is given by the following expression:

$$x_m = \theta(1 - 1/\eta)^{1/\eta}$$

For $\eta = 2$, the Weibull distribution's mode lies at $x_m = \theta\sqrt{2}$ and the curve is very right-skewed. As η increases, the distribution appears more and more symmetrical. At values of η above 3, its right-skewness is difficult to detect by the eye, and it resembles the normal distribution (Chapter 7). An advantage of the Weibull distribution over some other distributions is that the CDF is relatively easy to evaluate with a hand calculator, because the equation of $F_X(x)$ is a simple expression involving only a power function.

Because the adverse effects of environmental pollution usually are associated with very high concentrations, interest customarily focuses on the tails of these

Table 3.9. The Weibull Distribution $\mathbf{W}(\eta,\theta)$

Probability Density Function:

$$f_X(x) = \begin{cases} \dfrac{n}{\theta}\left(\dfrac{x}{\theta}\right)^{\eta-1} e^{-\left(\frac{x}{\theta}\right)^{\eta}}, & x \geq 0,\ \eta > 0,\ \theta > 0 \\ 0, & \text{elsewhere} \end{cases}$$

Expected Value:

$$E[X] = \theta a$$

Variance:

$$Var(X) = \theta^2(b - a^2)$$

Coefficient of Skewness:

$$\sqrt{\beta_1} = \frac{c - 3ab + 2a}{(b - a^2)^{3/2}}$$

Coefficient of Kurtosis:

$$\beta_2 = \frac{d - 4ac + 6ba^2 - 3a^4}{(b - a^2)^2}$$

$$a = \Gamma\left(1 + \frac{1}{\eta}\right),\ b = \Gamma\left(1 + \frac{2}{\eta}\right),\ c = \Gamma\left(1 + \frac{3}{\eta}\right),\ d = \Gamma\left(1 + \frac{4}{\eta}\right)$$

$$\text{where } \Gamma(a) = \int_0^{\infty} u^{a-1} e^{-u} du \text{ or, for integers, } \Gamma(a) = (a-1)!$$

distributions. Bryson[17] has compared the shapes of the tails of various distributions with that of the exponential, which he selects as his reference shape. He calls a distribution that goes to zero more rapidly than the exponential "light tailed." In contrast, he calls a distribution that goes to zero less rapidly than the exponential "heavy tailed." As the random variable X increases to arbitrarily large values, the PDF of a heavy-tailed distribution has more area under the remaining portion of the curve than the PDF of the exponential distribution. Thus, the probability of extremely high concentrations is greater for heavy-tailed distributions than for light-tailed distributions. Curran and Frank[15] compare the tails of three distributions—the Weibull, exponential, and lognormal—by plotting them on semi-logarithmic paper (Figure 3.15). As one would anticipate, the exponential distribution plots as a straight line on this paper. For values of η greater than 1, the Weibull distribution is light-tailed and, as concentration increases, shows downward curvature. In contrast, the lognormal distribution is

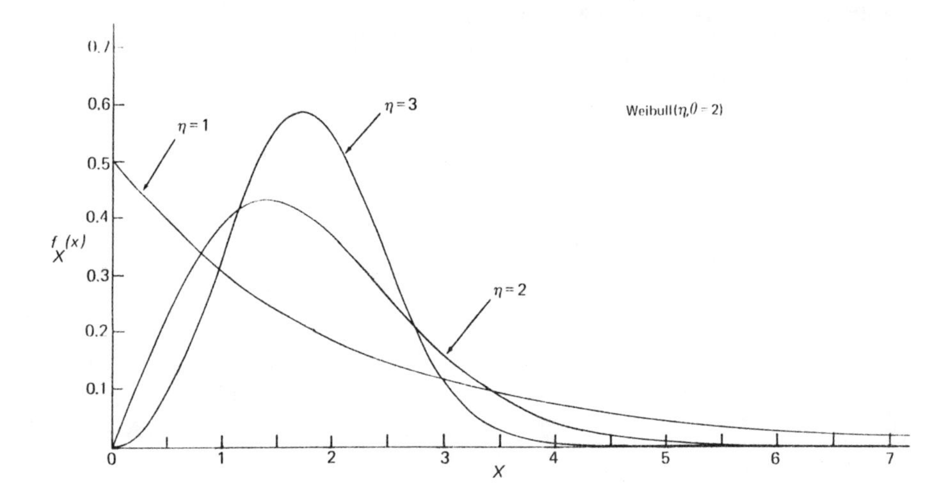

Figure 3.14. Probability density function of the Weibull probability model $\mathbf{W}(\eta,\theta)$ for $\theta = 2$ and different values of η, showing several shapes that are possible and the tendency of the distribution to approach a symmetrical, bell-shaped curve for large η.

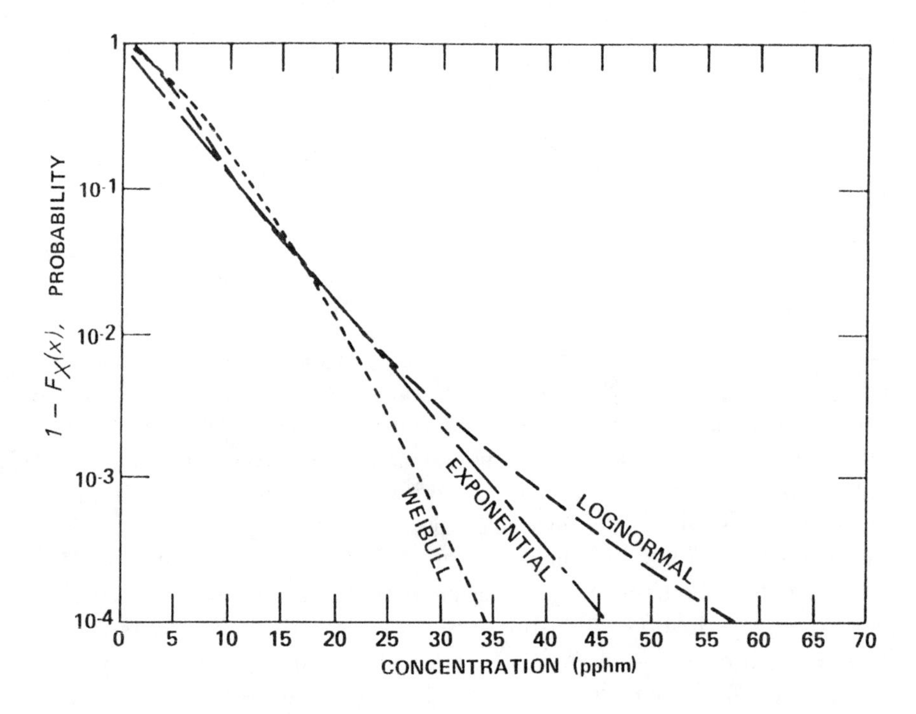

Figure 3.15. Plots of the Weibull, exponential, and lognormal distributions on semi-logarithmic paper, illustrating distributions that are "light-tailed" and "heavy-tailed" relative to the exponential (Source: Curran and Frank[15]).

heavy-tailed and exhibits gradual upward curvature with increasing concentration, relative to the straight line of the exponential distribution.

It is instructive to compare the earlier logarithmic-probability plot (Figure 3.9) of the observations from a purely exponential process with the semi-logarithmic plot (Figure 3.11) of observations from the same process. The earlier plot is based on 100,000 observations and the later plot is based on 10,000 observations, but the number of observations does not affect the characteristic shapes of the plots. Notice that the horizontal axis of logarithmic-probability paper is cumulative frequency rather than concentration, and the vertical axis is (the logarithm of) concentration. On logarithmic-probability paper, distributions that are lognormal plot as straight lines, while distributions that are lighter tailed than the lognormal (such as the exponential or Weibull) show concave downward curvature. Thus, the 100,000 observations from an exponential distribution in Figure 3.9 show gradual concave downward curvature that is quite evident to the eye. For the Weibull distribution, the downward curvature generally would be more striking.

Tail Exponential Method

Although the exponential model may fit the tail of a distribution fairly well, it does not necessarily also fit the rest of the distribution. For example, most observed distributions of environmental concentrations have a single mode (that is, a characteristic "hump") offset to the right of the origin, but the mode of the single-parameter exponential model lies at the origin. Because decision-makers usually are interested in the very high concentrations—those at the tail of the distribution—some investigators have sought to improve the tail fit by fitting the exponential model only to the upper range of the distribution of observations. For such "tail fits," the two-parameter exponential model introduced earlier (Figure 3.5), which has the following CDF, can be used:

$$F_X(x) = 1 - e^{-\frac{x-\alpha}{\theta}} \text{ for } x \geq \alpha$$

To fit this model empirically to a set of observations, one must find values for the two parameters α and θ. One approach is to plot the frequency distribution of the observations on semi-logarithmic paper and to fit a straight line to the distribution, using two points on the line to determine the parameters. This approach sometimes is called the "method of quantiles" or "method of fractiles," since it relies on the values of the variate associated with various cumulative frequencies.

Let x_1 denote the concentration observed at some particular cumulative frequency f_1; similarly, let x_2 denote the concentration observed at some higher cumulative frequency f_2. The first point is obtained by setting the model's CDF equal to the first frequency, or $F_X(x_1) = f_1$. The second point is evaluated in a similar manner, or $f_X(x_2) = f_2$. For the two-parameter exponential model, this approach yields the following two equations:

$$f_1 = 1 - e^{-\frac{x_1-\alpha}{\theta}} \qquad f_2 > f_1,$$

$$f_2 = 1 - e^{-\frac{x_2-\alpha}{\theta}} \qquad x_2 > x_1$$

If we take natural logarithms of these two equations, subtract one from the other, and then solve for θ, we obtain the following result:

$$\theta = \frac{x_2 - x_1}{\ln(1 - f_1) - \ln(1 - f_2)}$$

To compute α, we can use either of the following expressions:

$$\alpha = \theta \ln(1 - f_1) + x_1 \quad \text{or} \quad \alpha = \theta \ln(1 - f_2) + x_2$$

These expressions allow us to calculate the two parameters α and θ for the two-parameter exponential model **E**(α,θ) by the method of quantiles. We first calculate θ using the coordinates of two points on the straight line; then we substitute θ into the above expression to obtain α.

Which two points on the straight line should be used? Curran and Frank[15] choose the quantiles corresponding to the 0.90 and 0.99 cumulative frequencies of the observations. If we substitute $f_1 = 0.90$ and $f_2 = 0.99$ into the above expressions, we obtain the following equations:

$$\theta = 0.43429(x_2 - x_1)$$

$$\alpha = -2.30260 + x_1 \quad \text{or} \quad \alpha = -4.60520 + x_2$$

These serve as useful expressions for calculating parameters for the two-parameter exponential distribution by the method of fractiles using Curran and Frank's approach.

An example using real observations helps to illustrate this approach. If we plot the cumulative frequencies from the histogram of 8,573 hourly average CO concentrations measured in San Jose, CA, shown in Figure 3.13 on semi-logarithmic paper, the resulting dots show reasonable straightness in the middle range (4 ppm to 12 ppm; Figure 3.16). Each dot in Figure 3.16 corresponds to the cumulative frequency (that is, the sum) of the histogram bars in Figure 3.13. Unfortunately, no dots lie precisely on the points corresponding to $f_1 = 0.90$ and $f_2 = 0.99$ (that is, 0.1 and 0.01 on the vertical axis). How, then, do we obtain the proper values for x_1 and x_2? One approach is to plot a smooth curve by eye through the dots (raw data line in Figure 3.16) and then to fit the model to the points on this line. Using this approach, we obtain $x_1 = 6.35$ ppm and $x_2 = 11.8$ ppm (small circles shown on Figure 3.16). Substituting these values into the above expressions gives θ = 2.37 and α = 0.9 ppm, after rounding. With these parameters, the two-parameter exponential model has the following CDF:

$$F_X(x) = 1 - e^{-\frac{x-0.9}{2.37}} \quad \text{for } x \geq 0.9$$

This equation plots as a straight line in Figure 3.16, intercepting 1.0 on the vertical axis at the point $x = \alpha = 0.9$ ppm. The dotted line through the observations does not deviate too greatly from the straight line model. However, the actual difference between the model and the observations may be relatively large but may appear small to the eye on the logarithmic scale. For example, at $x = 2.5$ ppm, the model

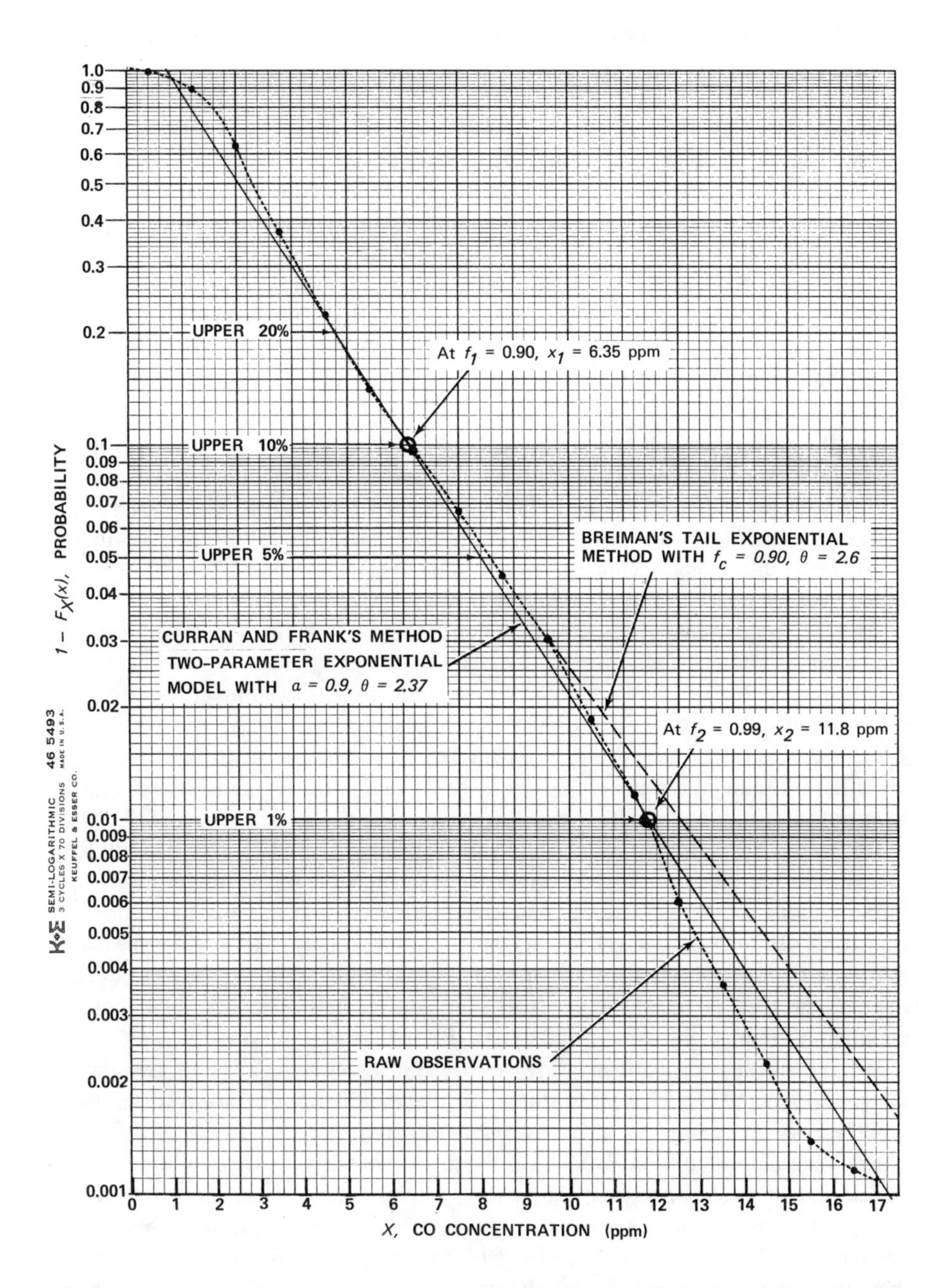

Figure 3.16. Fitting the two-parameter exponential model to the tail of the distribution of hourly carbon monoxide observations measured in San Jose, CA, from June 1970 to May 1971, using the empirical method of Curran and Frank.[15]

predicts a cumulative frequency of $F_X(2.5) = 1 - 0.51 = 0.49$, but the observed cumulative frequency is $(1 + 911 + 2297)/8573 = 0.37$ from Figure 3.13 on page 67. At some points, the fit to the tail is fairly good. At $x = 8.5$, the model predicts a cumulative frequency of $F_X(8.5) = 1 - 0.04 = 0.96$, and the observed cumulative frequency is $0.955 \simeq 0.96$. At $x = 10.5$, the model predicts a cumulative frequency of $F_X(10.5) = 1 - 0.017 = 0.983$, and the observed cumulative frequency is 0.982.

Comparing the original histogram (Figure 3.13) with the semi-logarithmic plot (Figure 3.16) causes one to wonder, "Where has the mode in the interval $1.5 \leq x < 2.5$ gone?" In the semi-logarithmic plot, the slight hump and curvature of the dotted line at the upper left of Figure 3.16 now reflects the mode. This lack of prominence of the mode again illustrates the considerable emphasis that semi-logarithmic plotting gives to the tail of a distribution at the expense of the bulk of the distribution.

Berger, Melice and Demuth[18] sought to find a statistical distribution "... which fits properly the high concentration levels, for example, the concentrations higher than the 95th percentile." They examined 24-hour average ambient SO_2 concentrations from January 1977 to March 1979 at 12 monitoring stations in Gent, Belgium. When the cumulative frequencies of observed SO_2 concentrations were fit to a lognormal distribution, they found that the extreme observations were overestimated. This result is consistent with the heavy-tailed property of the lognormal distribution noted by Bryson[17] and Curran and Frank[15] (Figure 3.15). Berger, Melice, and Demuth[18] found that the two-parameter exponential distribution, on the other hand, fit the SO_2 observations very well. Simpson, Butt, and Jakeman[19] found the two-parameter exponential distribution gave a good fit to the higher values of SO_2 measured near a power plant in the Upper Hunter Valley of New South Wales, Australia.

Breiman, Gins, and Stone[20–22] sought to develop a method for estimating the higher concentrations of an observed data set when some of the observations are missing. For example, how does one estimate the expected value of the 99th percentile or of the second highest daily air pollutant concentration when the data for some days in the year are missing? They call their approach the "tail exponential" method. To apply the approach, the analyst first selects an upper cutoff cumulative frequency—usually the upper 10% or 20% of the observations—and then the concentration corresponding to this cutoff frequency. Then a two-parameter exponential model is fit to the portion of the distribution above the cutoff point, ignoring all observations below the cutoff frequency.

Breiman, Gins, and Stone[20] consider the tail exponential method to be "universal" in that it does not require the analyst to make the difficult decision about whether a particular data set can be represented by a lognormal distribution, a Weibull distribution, or any other particular family of distributions. They tested their approach using computer simulation to generate observations from five different distributions: an exponential, lognormal, and three different Weibull forms. They compared the tail exponential fit with a Weibull model fit by examining estimated and observed values at the 99.45th percentile for both models and all five data sets. The tail exponential estimate using 10% of the upper tail was better for all distributions than the Weibull estimators, even when the underlying distribution was actually Weibull. It was significantly better for the non-Weibull distributed data. They conclude that many distributions that are not exponential behave as though they are tail exponential beyond a certain point:

> Unfortunately, the sample distribution function is not a good estimate of the upper tail of the distribution. The alternative is to fit some family of distributions to the valid data and draw the missing data from the fitted distribution. This may be a Pandora's Box, as there are numerous vying families of distributions: lognormal, exponential, gamma, Weibull, 3-parameter lognormal, etc. The standard way of deciding which family to use is to plot the data on the various corresponding probability papers and see which one looks most like a straight line. Actually, since only the upper tail of the distribution is relevant, the question is on which type of probability paper does the upper tail, say the upper 10% to 20%, look most like a straight line. Examination of a large number of such plots revealed that the "straightness" of the upper tail was, to a large extent, insensitive to the type of family plotted against. If the data was going to be "straight" for any, it was straight for all. If not straight for one, it was usually not straight for any.[20]

In effect, they are defining "tail generation" as a stochastic process unto itself, regardless of the original distribution producing the tail, for the class of distributions likely to generate environmental pollutant concentrations.

Because Breiman's tail exponential method treats tail generation as a stochastic process, it is not merely an empirical technique like Curran and Frank's curve-fitting approach; rather, it has a probabilistic interpretation. Thus, Breiman, Gins, and Stone[20] use a probabilistic approach—*maximum likelihood estimation* (MLE)—to estimate the main parameter of their distribution. It is useful to discuss the principles behind the MLE approach, which is important for most probability models.

Suppose we are given n observations which are known to come from a particular distribution, but we don't know the parameters of that distribution. Using these n observations, what is the most likely value of the distribution's parameters? If we treat each observation as an independent outcome, then the probability of the joint occurrence of this sample of size n will be the product of n PDFs of this distribution, with each PDF evaluated at the value of an observation. More formally, we have obtained n observations $x_1, x_2, \ldots, x_n$ in the sample. We then form the "likelihood function" L by multiplying the PDFs for the distribution evaluated at each observation:

$$L = f_X(x_1) f_X(x_2) \ldots f_X(x_n)$$

We then must find the values of the parameters that maximize L. Often, the MLE estimators for a distribution can be found by differentiating L and setting the result equal to zero.

To apply the MLE approach to the single-parameter exponential probability model, we simplify the exponential PDF by substituting $\theta = 1/\lambda$, obtaining the following expression: $f_X(x) = \lambda e^{-\lambda x}$. Then the likelihood function for n observations is written as follows:

$$L = \lambda e^{-\lambda x_1} \lambda e^{-\lambda x_2} \ldots \lambda e^{-\lambda x_n} = \lambda^n e^{-\lambda \sum_{i=1}^{n} x_i}$$

Taking the first derivative of this expression with respect to λ, we obtain the following:

$$\frac{\partial L}{\partial \lambda} = n\lambda^{n-1} e^{-\lambda \sum_{i=1}^{n} x_i} - \lambda^n \left(\sum_{i=1}^{n} x_i \right) e^{-\lambda \sum_{i=1}^{n} x_i}$$

Setting this expression equal to zero to find the value of λ that maximizes L, solving for λ, and substituting $\theta = 1/\lambda$, we obtain the following result:

$$\theta = 1/\lambda = \frac{1}{n} \sum_{i=1}^{n} x_i$$

Thus, the MLE estimator of the parameter θ of the single-parameter exponential probability model is the arithmetic mean of the observations.

To apply the two-parameter exponential model using Breiman's tail exponential method, we first choose a cutoff frequency f_c and estimate the corresponding cutoff concentration x_c. Then we compute the arithmetic mean $\bar{x}$ of the tail observations—those larger than x_c—and the MLE estimator of θ will be $\bar{x} - x_c$. These steps of the tail exponential method are summarized as follows:

1. Select a cutoff cumulative frequency f_c (Usually $f_c = 0.80$ or $f_c = 0.90$.)
2. Estimate the value x_c corresponding to f_c.
3. Compute the arithmetic mean $\bar{x}$ of the observations above f_c; then $\theta = \bar{x} - x_c$.

The CDF of the tail for the tail exponential method then will be written as follows:

$$F_X(x) = 1 - (1 - f_c)e^{-(x - x_c)/\theta} \quad \text{for } x > x_c$$

To apply the tail exponential method to the San Jose CO concentration data (histogram in Figure 3.13), we choose the cutoff point at $f_c = 0.90$. The same graphical approach used to illustrate Curran and Frank's approach above can be used to estimate the 90-percentile concentration: $x_c = x_1 = 6.35$ ppm. Recalling that all values on this histogram actually are integers that lie in the middle of the intervals, the tail will consist of all integer values above 6.35 ppm; that is, all values of 7 ppm or more. From this histogram, we see that there are 263 values of 7 ppm, 181 values of 8 ppm, 124 values of 9 ppm, etc., giving a total of 829 values in the tail. Using these data to compute the mean for the tail, we obtain $\bar{x} = 8.95$ ppm. Then the MLE estimate of the parameter θ will be $\bar{x} - x_c = 8.95 - 6.35 = 2.6$ ppm, and Breiman's tail exponential method yields the following equation:

$$F_X(x) = 1 - 0.1e^{-(x - 6.35)/2.6} \quad \text{for } x > 6.35$$

If we plot this equation as $1 - F_X(x)$ in Figure 3.16 (straight dashed line), the resulting line has a different slope than Curran and Frank's approach. An important reason for this difference is that the MLE approach gives heavy weight to those concentrations between 7 and 9 ppm, where many observations are present. Curran and Frank's approach, on the other hand, fits the extreme tail frequency of $f_2 = 0.99$, where few observations happen to be present.

The findings reported by Breiman, Gins, and Stone[20–22] when sampling from various distributions imply that the tail exponential method provides a good estimate of the expected values of the various quantiles in the tail. That is, if this same stochastic process were repeated over and over, we can imagine that the tails would lie above and below Breiman's tail exponential line, but their average would tend to approach this line. Thus, the tail exponential fit is more representative of the underlying process generating the observations than a fit by Curran and Frank's method of quantiles.

Comparing these two approaches illustrates an important difference in model-fitting concepts. The San Jose CO data may be viewed as just one realization from an underlying stochastic process. Curran and Frank's method fits this particular realization quite well, but it does little more. Like the methods of Simpson, Butt, and Jakeman[19] and Larsen[11] (Chapter 9), their empirical model should not be used to make probability statements about pollutant concentrations. For example, it is incorrect to claim that a CO concentration of 11.8 ppm will be exceeded with probability $P = 1 - 0.99 = 0.01$ simply because the Curran and Frank line passes through 11.8 ppm at the frequency 0.99 (Figure 3.16). On the other hand, Breiman's MLE fit seeks to identify an underlying probabilistic model, so probability statements are appropriate. For example, substituting $F_X(x) = 0.99$ into Breiman's tail exponential fit above and solving for x yields 12.3 ppm. From this, we can say, "There is 99% chance that the CO concentration will be less than 12.3 ppm," or, "The CO concentration should exceed 12.3 ppm with a probability $P = 0.01$."

As discussed in Chapter 4, the Environmental Protection Agency (EPA) has specified the U.S. national standard for photochemical oxidant (ozone) in a probabilistic form. The standard is attained when the expected number of days per year in which the ozone concentration exceeds 0.12 ppm is "less than or equal to 1".[23] Prior to controlling the sources of air pollution in a particular location, there will be some ozone concentration x_d for which the expected number of exceedances is equal to 1. The concentration x_d is known as the "design value," since it allows regulatory officials to determine the percentage reduction in sources that will be needed to meet the desired standard x_s. If the standard is to be met in the future from year to year, as it is supposed to be, then it is especially important that the design value be based not on the chance occurrences within a particular data set but on the underlying process itself. Breiman's tail exponential method allows one to estimate x_d as the expected value of the appropriate quantile, thus providing a more stable and accurate estimate than the empirical tail fitting approaches of Larsen,[11] Curran and Frank,[15] and Simpson, Butt, and Jakeman.[19]

Crager[24] applied Breiman's tail exponential method to atmospheric ozone data from 26 air monitoring stations covering a 17-year period in the San Francisco Bay area. He found that the tail exponential method gave better estimates of the design values than Larsen's empirical method. In another study, Crager[25,26] developed asymptotic confidence intervals for estimates from the tail exponential method. He points out two important advantages of the tail exponential method:

> There are two main advantages to the tail exponential method. First, it is more accurate; the method was found to reduce the root mean square error of the estimated design value by a factor of more than 2 over Larsen's method. Second, and more important, the tail exponential design value estimator comes with easily computed asymptotic confidence intervals for the design value, which may be used to judge the accuracy of the estimate; no associated confidence intervals have been found using Larsen's design value estimator, and the problem of finding such intervals appears difficult at best.[24]

For most data sets that were tested, the tail exponential model fit the observations above the 25 percentile quite well; for some, taking a logarithmic transformation improved the fit.

Crager[24] compared the MLE approach for estimating θ with another approach: minimization of the sum of the squares of the distances between the observations and the point on the tail exponential line, or "least squares." The ozone data ordinarily are rounded to a few significant figures and therefore are very discrete. He found that the least squares approach was more robust to this rounding problem than the MLE approach. With the rounded data, the least squares estimators retained their theoretical confidence intervals better than the MLE approach; that is, the 95% confidence interval actually contained the true design value very close to 95% of the time. Crager[25,26] showed that asymptotic confidence intervals could be calculated using the normal distribution (Chapter 7). Both the least squares and MLE approaches performed better than the empirical approach.

The EPA recommends the use of three years of data to compute design values. In the San Francisco Bay area, high ozone levels do not occur in the "winter" season (November through March), but are restricted to the 212-day "summer" period from April 10 to November 7. To apply an empirical approach, such as the one patterned after Larsen,[11] one can combine three years of data for the summer periods and interpolate graphically to determine the concentration corresponding to the cumulative frequency $f = 1 - 1/212 = 0.0995$. To apply the tail exponential method, on the other hand, one estimates θ from the combined data either by the method of MLE or by least squares. Then one uses the exponential probability model's CDF to find $x_d = F_X^{-1}(0.0995)$. Thus, the tail exponential method takes advantage of probability theory and uses probability models to predict expected values and confidence intervals, while the empirical approach has none of these advantages.

Suppose we obtain a sample of n independent observations from a particular distribution and select the maximum value. Then we select another n independent observations from the same distribution and obtain another maximum. If this process is repeated indefinitely, the resulting maxima will approach a particular distributional form, the "extreme value" distribution. Extreme value processes have been investigated by Gumbel,[27,28] and the extreme value probability model sometimes is called a Gumbel distribution. Typically it involves the double exponential form:

$$F_X(x) = 1 - e^{-e^{-(x-\alpha)/\theta}}$$

Roberts[29,30] reviews extreme value theory for its applicability to air pollution problems, and Kinnison[31] applies it to a variety of problems, including hydrology (forecasting floods), environmental pollution, and the breaking strength of materials. Monthly maximum hourly ozone concentrations may be modeled by an extreme value process, although serial dependency in the data could violate the required independence assumption. Leadbetter[32] indicates that the classical extreme value theory still is valid, provided the dependence is not "too strong".

An important concept in extreme value theory is the "return period," which is the expected time interval between exceedances of a particular value. In environmental problems, the return period might be the average number of days between exceedances of a standard. When the ozone standard is attained exactly, for example, the expected number of exceedances is 1 per year, so the return period is 365 days.

If $F_X(x)$ is the CDF of pollutant concentrations, and x_s denotes a particular concentration of interest, then the return period T_s of exceedance of x_s will be defined as follows:

$$T_s = \frac{1}{1 - F_X(x_s)}$$

Since the tail exponential method can provide the CDF $F_X(x_s)$ needed in the above return period equation, Breiman's tail exponential approach may be viewed as an alternative to classical extreme value theory. Crager[26] derives confidence intervals for return periods, and he notes that the tail exponential method has several advantages to extreme value approaches: it is easy to apply and one readily can verify that the distribution of each year's observations is tail exponential using probability plots. In contrast, it is doubtful that the maxima of different stations and different years can be viewed as a sample from a single extreme value distribution, so extreme value theory becomes difficult to apply and to verify. The tail exponential method, on the other hand, offers a simple, robust, and fairly general method for modeling environmental quality data probabilistically.

SUMMARY

This chapter presents several important concepts and their related terminology. There are two types of probability models: those for discrete random variables and those for continuous random variables. Discrete probability models use a probability mass function (PMF) and continuous probability models use a probability density function (PDF). The cumulative distribution function (CDF)—either the sum of the PMF or the integral of the PDF—allows the probabilities to be computed for specific ranges of a random variable. Examples of discrete probability models are the discrete uniform distribution **U**(p) and the geometric distribution **GM**(p). Examples of continuous probability models are the continuous uniform distribution **U**(a,b), the single parameter exponential distribution **E**(θ), the two-parameter exponential distribution **E**(α,θ), and the Weibull distribution **W**(η,θ).

The expected value, also called the average or mean, provides a measure of central tendency of a probability model. Other measures of central tendency include the median and the mode. The moments of a probability model give information about its properties. The expected value is the first moment about the origin. The moment generating function is useful for computing the moments about the origin of some probability models. The second moment about the mean is known as the variance of the model, providing a measure of the spread of the distribution. Other parameters, such as the coefficients of skewness and kurtosis, provide measures of the shape of the probability model.

This chapter also discusses how to calculate various statistics for a set of observations. A statistic allows the properties of the observations to be expressed parsimoniously. Statistics for central tendency include the mean, median, and mode. The standard deviation and its square, the variance, reflect the dispersion, or spread, of the observations. The concept of degrees of freedom is important in its computation, which can be done efficiently on computers using one-pass techniques. The coefficient of variation, or ratio of the standard deviation to the mean, is a useful measure of the relative dispersion. The coefficients of skewness and kurtosis of the data reflect the shape of the distribution of observations.

Graphical techniques—such as histograms and frequency plots—convey rich descriptive information about the observations. Both grouped and ungrouped data can be plotted, although the techniques differ. If cumulative frequencies of the observations are plotted on special probability papers, then the straightness of the plot indicates whether the distribution resembles a particular probability model. Formal techniques for fitting probability models to a set of observations include graphical fitting by eye, the method of quantiles, the method of moments, and the method of maximum likelihood estimation (MLE). Formal goodness-of-fit tests, such as the chi-square test, allow one to determine probabilistically how poorly a given model fits the observations.

Breiman's tail exponential method, with parameters estimated by MLE, seeks to reflect the underlying processes generating the tails of environmental data sets (that is, the upper 10- or 20-percentiles). Studies of air quality data show that the tail exponential method produces more accurate estimates of regulatory design values than the alternative empirical methods (for example, the methods of Larsen; Curran and Frank; and Simpson, Butt, and Jakeman). The tail exponential method is robust with respect to uncertainties about various distributions, and it allows confidence intervals to be calculated for predicted design values.

PROBLEMS

1. In the vehicular emission control example given in this chapter, what is the probability that an emission control device fails after 20 years? [Answer: 0.12158]
2. Suppose that the average life expectancy of a type of emission control device is 15 years. If failures are assumed to be independent with a fixed failure probability, what is the probability distribution of the year in which the device fails? [Answer: **GM**(0.93333)] Plot the PMF and CDF for this distri-

bution. What proportion of the new devices will fail within the first five years? [Answer: 29.2%]

3. If the PDF of the uniform distribution is $f_X(x) = 1/(b - a)$, show that the CDF is given by $F_X(x) = (x - a)/(b - a)$.
4. For the case in which $a = 1/3$ and $b = 3/4$ in Problem 3, what is the probability $P_X\{1/2\}$? [Answer: The probability is undefined.] What is the probability $P\{X \leq 1/2\}$? [Answer: $F_X(1/2) = 2/5$]
5. Suppose that the exponential distribution is used in place of the geometric distribution to represent the time to failure of an emission control device. If the exponential CDF at each integer $X = 1, 2, \ldots$ is to be approximately the same as the geometric CDF at the corresponding integer $K = 1, 2, \ldots$, what is the parameter θ of this exponential distribution? [Answer: 9.4913] Plot the resulting PDF and PMF on the same graph. Plot the two CDFs on the same graph. Notice that it is possible, using the exponential CDF, to calculate probabilities for any portion of a year.
6. Discuss the fundamental differences between one-pass and two-pass techniques for computing statistics.
7. Derive the expressions given in this chapter for the single-pass formulas for s_3 and s_4.
8. Calculate s_3 and s_4 from the observations given in Table 3.5 using the two-pass formulas and compare your answers with the results given in this chapter for the single-pass formulas. Compute $\sqrt{b_1}$ and b_2.
9. Calculate the geometric mean for the values given in Table 3.5 by the standard formula. Check your answer by taking logarithms of the 24 values, adding them, dividing by 24, and taking the anti-logarithm of the result.
10. Discuss the difference between two approaches for plotting observations on probability paper: individual frequency plotting and grouped data plotting. When should each approach be used?
11. Explain how random variables following a particular distribution are generated on the computer using the inverse transformation method. Illustrate your explanation with the two-parameter exponential distribution. Write a computer program to generate the exponentially distributed random variable **E**(2,3). Use this program to generate 1,000 values on the computer and sort the values into intervals to form a histogram.
12. Derive the CDF of the Weibull distribution from its PDF. Derive the formula for the Weibull mode x_m. Plot the Weibull PDF and CDF for the cases of $\theta = 2$ and $\eta = 4, 5, 6$, and 7.
13. Discuss how a random variable with a Weibull distribution with $\eta > 1$ should appear when it is plotted on logarithmic-probability paper.
14. Derive the equations giving the parameters for the two-parameter exponential distribution when two quantiles are known.
15. Using Breiman's tail exponential method with the San Jose CO data given in this chapter, find the CO concentration corresponding to an exceedance probability of 0.001. [Answer: $F_X^{-1}(0.999) = 18.323$]
16. If the straight line depicting Breiman's tail exponential method in Figure 3.16 were extended to the top of the graph, what value of x would correspond to 1 on the vertical axis? [Answer: $F_X^{-1}(0) = 0.363$]

17. Using the rules for expected values and the mathematical definition of the variance, derive the rules for combining variances of mutually independent random variables (Rules 3 and 4).

REFERENCES

1. Pielou, E.C., *An Introduction to Mathematical Ecology* (New York: John Wiley and Sons, 1969).
2. Keyfitz, N., *Introduction to the Mathematics of Population* (Reading, MA: Addison-Wesley, 1968).
3. Keyfitz, Nathan, "Finding Probabilities from Observed Rates or How to Make a Life Table," *The American Statistician*, 27–33 (February 1970).
4. Keyfitz, Nathan, "On Future Population," *J. American Statistical Assoc.* 67(338):347–363 (June 1972).
5. Keyfitz, N., and W. Flieger, *Population, Facts and Methods of Demography* (San Francisco, CA: W.H. Freeman Co., 1971).
6. Barlow, Richard E., and Joao L.M. Saboia, "Bounds and Inequalities on the Rate of Population Growth," Operations Research Center, College of Engineering, University of California, Berkeley, CA, ORC 73–14 (August 1973).
7. Shedler, G.S., "Generation Methods for Discrete Event Simulation: Uniform and Nonuniform Random Numbers," IBM Research Laboratory, San Jose, CA, Report No. RJ2789(35455) (1980).
8. Kendall, Sir Maurice, and Alan Stuart, *The Advanced Theory of Statistics* (New York: Macmillan Publishing Co., 1977).
9. "Catalog and Price List," Technical and Engineering Aids for Management (TEAM) Co., Box 25, Tamworth, NH 03886.
10. Mage, David T., personal communication, Washington, DC, 1984.
11. Larsen, Ralph I., "A Mathematical Model for Relating Air Quality Measurements to Air Quality Standards," U.S. Environmental Protection Agency, Research Triangle Park, NC, Publication No. AP–89 (November 1971).
12. Beyer, William H., ed., *Handbook of Tables for Probability and Statistics*, 2nd ed., (Boca Raton, FL: CRC Press Inc., 1968).
13. Ott, Wayne R., David T. Mage, and Victor W. Randecker, "Testing the Validity of the Lognormal Probability Model: Computer Analysis of Carbon Monoxide Data from U.S. Cities," U.S. Environmental Protection Agency, Washington, DC, EPA–600/4–79–040, NTIS PB88–246053 (June 1979).
14. Kalpasanov, Y., and G. Kurchatova, "A Study of the Statistical Distributions of Chemical Pollutants in Air," *J. Air Poll. Control Assoc.*, 26:981–985 (October 1976).
15. Curran, Thomas C., and Neil H. Frank, "Assessing the Validity of the Lognormal Model When Predicting Maximum Air Pollution Concentrations," Paper No. 75–51.3 presented at the 68th Annual Meeting of the Air Pollution Control Association, Boston, MA (June 1975).

16. Johnson, Norman L., and Samuel Kotz, *Continuous Univariate Distributions-1* (Boston, MA: Houghton Mifflin Co., 1970).
17. Bryson, Maurice C., "Heavy-Tailed Distributions: Properties and Tests," *Technometrics* 16(1):61–68 (February 1974).
18. Berger, A., J.L. Melice, and C.L. Demuth, "Statistical Distributions of Daily and High Atmospheric SO_2 Concentrations," *Atmos. Environ.*, 16(12): 2863–2677 (1982).
19. Simpson, R.W., J. Butt, and A.J. Jakeman, "An Averaging Time Model of SO_2 Frequency Distributions from a Single Point Source," *Atmos. Environ.*, 18(6):1115–1123 (1984).
20. Breiman, Leo, John Gins, and Charles Stone, "Statistical Analysis and Interpretation of Peak Air Pollution Measurements," Technology Service Corp., Santa Monica, CA, TSC-PD-A190-10 (December 1979).
21. Breiman, L., C.J. Stone, and J.D. Gins, "New Methods for Estimating Tail Probabilities and Extreme Value Distributions," Technology Service Corp., Santa Monica, CA, TSC-PD-A226-1 (December 1979).
22. Breiman, L., C.J. Stone, and J.D. Gins, "Further Development of New Methods for Estimating Tail Probabilities and Extreme Value Distributions," Technology Service Corp., Santa Monica, CA, TSC-PD-A243-1 (January 1981).
23. "National Primary and Secondary Ambient Air Quality Standards," *Federal Register*, 44:8202 (February 8, 1979).
24. Crager, Michael R., "Exponential Tail Estimates of Bay Area Design Values and Expected Return Periods," SIMS Technical Report No. 6, Department of Statistics, Stanford University, Stanford, CA (October 1982).
25. Crager, Michael R., "Exponential Tail Quantile Estimators for Air Quality Data, Part I: Asymptotic Confidence Intervals for Extreme Quantiles," SIMS Technical Report No. 4, Department of Statistics, Stanford University, Stanford, CA (September 1982).
26. Crager, Michael R., "Exponential Tail Quantile Estimators for Air Quality Data, Part II: Estimator Robustness; Confidence Intervals for Combined Distributions," SIMS Technical Report No. 5, Department of Statistics, Stanford University, Stanford, CA (September 1982).
27. Gumbel, E.J., "Statistical Theory of Extreme Values and Some Practical Applications," National Bureau of Standards, Applied Math Series No. 33 (1954).
28. Gumbel, E.J., *Statistics of Extremes* (New York: Columbia University Press, 1958).
29. Roberts, E.M., "Review of Statistics of Extreme Values with Applications to Air Quality Data, Part I. Review," *J. Air Poll. Control Assoc.*, 29(6):632–637 (June 1979).
30. Roberts, E.M., "Review of Statistics of Extreme Values with Applications to Air Quality Data, Part II. Applications," *J. Air Poll. Control Assoc.*, 29(7):733–740 (July 1979).
31. Kinnison, Robert R., *Applied Extreme Value Statistics* (New York: MacMillan Publishing Co., 1985).

32. Leadbetter, M.R., “On Extreme Values of Sampled and Continuous Stochastic Data,” SIMS Technical Report No. 10, Department of Statistics, Stanford University, Stanford, CA (August 1977).

4 Bernoulli Processes

Consider a process consisting of several repeated trials. Each trial has only one of two possible outcomes, either event A or the null event $\bar{A}$. On each trial, the probability of occurrence of the event will be the same, and the events are assumed to be independent. If our interest focuses on the *total number of times* that event A occurs on a specified number of repeated trials, then the result is a *Bernoulli process.*

Examples of real-world phenomena that usually are modeled as Bernoulli processes involve discrete events and counts of the total number of occurrences of the event of interest:

1. A fair coin is flipped 10 times. What is the probability that 4 "heads" will appear?
2. Four persons occupy an office at the same time; the probability that a person "is smoking" is the same for each person, and smoking activity is assumed to be independent. At a given instant of time, find the probabilities associated with each of the following: (a) one person is smoking, (b) two persons are smoking, and (c) three persons are smoking.
3. When a metropolitan area is in compliance with an air quality standard, the expected number of exceedances per year is 1.0 or less. If the air quality of an area is such that the expected number of exceedances is exactly 1.0, what is the probability that (a) more than one exceedance will occur in a year, (b) more than 3 exceedances will occur in 3 years, (c) 6 or more exceedances will occur in 3 years?
4. A field survey must be designed to investigate whether pesticide contamination is present in the soil of a given area. When contamination is present, past experience indicates that samples collected in the area will give positive readings (concentrations above a threshold limit) with a known probability. How many soil samples should be collected in the area to give 95% assurance that a positive reading is not accidentally overlooked?
5. The probability that a particle arrives at a particular location in space is known and is the same for all particles, and arrivals are assumed to be independent. Suppose that 10 particles are released; what is the probability that (a) 2 particles will arrive, (b) 3 particles will arrive, and (c) 5 particles will arrive?

In the coin flipping example above, the concept of an independent trial is very straightforward. Each time the coin is flipped, a trial occurs, and the random variable of interest is the total number of heads that appear on a run of 10 successive flips. The concept of an independent trial may be less obvious in the smoking example. Here, the smoking status of each person (whether they are smoking or not smoking) at some arbitrary instant of time is treated as a trial. It is assumed that the smoking activity of one person does not influence the smoking activity of another person, and the random variable of interest is the total number of persons smoking at the particular time of interest. In the air quality example, each successive day is analogous to each successive coin flip in the coin flipping example. However, the probability that a fair coin comes up "heads" usually is assumed to be $p = 1/2$, while the probability that an air quality standard is violated on a particular date could be very different from $p = 1/2$. An expected value of 1.0 exceedance per year would correspond to the daily probability of an exceedance of $p = 1/365$. In the example describing the arrival of particles at a particular location in space, the probability of arrival of any single particle also could be very different from $p = 1/2$, depending on the distance between the source and the arrival location and the intervening factors.

CONDITIONS FOR BERNOULLI PROCESS

The above examples all share the common properties that (a) the probability of occurrence of the event of interest is the same for all trials, (b) all trials are independent, and (c) the quantity of interest is the total number of successful outcomes. That is . . .

> A Bernoulli process is one concerned with the count of the total number of independent events, each with the same probability, occurring in a specified number of trials.

The quantity of interest in a Bernoulli process can be represented as the discrete random variable K, which is a count of the total number of events (i.e., successful outcomes) occurring in n trials. More formally, in a given realization of n trials, K will denote the total number of times that the event A occurs, where $K = \{0, 1, 2, \ldots, n\}$ denotes the sample space. The probability that event A occurs on each trial will be denoted by $P\{A\} = p$, and the corresponding probability of the null event $\bar{A}$ will be denoted by $P\{\bar{A}\} = 1 - p = q$.

DEVELOPMENT OF MODEL

Prior to discussing a general model for Bernoulli processes, it is instructive to present an example of a familiar situation that might be modeled as a Bernoulli process, partly because it gives insight into the model and partly because the basic model can be developed from it by mathematical induction.

Example: Number of Persons Engaged in Cigarette Smoking

Assume that, within a given population, we do not know who is a smoker and who is not. Assume further, that, at a given instant of time, the probability that a person chosen at random from this population is actively smoking is $P\{A\} = 1/9$, where A denotes the act of smoking.* Now suppose that an office room contains two such persons. Here the act of smoking by one person is assumed to have no influence (i.e., it neither encourages nor discourages) the act of smoking by the other person. Such independence may be a justifiable assumption if, for example, human smoking activity is governed more by basic physiological needs than by social influences.

Let A_1 denote the event that Person 1 is smoking and A_2 denote the event that Person 2 is smoking. Then, $P\{A_1\} = P\{A_2\} = 1/9$. At a given instant of time, any one of three possible outcomes can occur: (a) zero persons are smoking, (b) one person is smoking, or (c) two persons are smoking. That is, the sample space will be $K = \{0, 1, 2\}$, where K is the total number of persons smoking. In the first case, neither person is smoking, and the probability that $K = 0$ will be computed as the intersection of two independent null events:

$$P\{K=0\} = P\{\bar{A}_1\bar{A}_2\} = qq = (8/9)(8/9) = 0.790$$

The probability that exactly one person is smoking must be computed by taking into account the combinations of the two different ways in which this situation can occur. First, Person 1 may be smoking while Person 2 is not smoking:

$$P\{A_1\bar{A}_2\} = pq = (1/9)(8/9) = 0.099$$

Second, Person 1 may be not smoking while Person 2 is smoking:

$$P\{\bar{A}_1 A_2\} = qp = (8/9)(1/9) = 0.099$$

The event that "one person is smoking" is the combination of these two possibilities, and so the probability that one person is smoking is, by the rules governing the union of two independent events, the sum of these two probabilities:

$$P\{K=1\} = pq+qp = 2pq = 0.099 + 0.099 = 0.198$$

Finally, the probability that two persons are smoking at the same time is computed as the intersection of the two independent smoking events:

*It is assumed that the persons are totally anonymous; nothing is known about their previous smoking history or habits, and everyone has a finite probability of engaging in smoking.

$$P\{A_1A_2\} = pp = (1/9)(1/9) = 0.012$$

By the above reasoning, we have computed the probabilities associated with all elements comprising the sample space:

$$P\{K=0\} = q^2 \quad = 0.790$$

$$P\{K=1\} = 2pq = 0.198$$

$$P\{K=2\} = p^2 \quad = 0.012$$

$$\text{Total}: \ 1.000$$

If these values are plotted, it is seen that the resulting probability distribution is right-skewed and has its mode at the origin (Figure 4.1).

Next, to carry the example further, suppose that three persons are present in the office; what will be the resulting probability distribution for the total number of persons smoking at some arbitrary instant of time? Here, the sample space consists of $K = \{0, 1, 2, 3\}$, and we again obtain the probabilities associated with each of these elements by considering the various combinations in which the events can occur. Specifically, let A_1 denote the event that Person 1 is smoking; let A_2 denote the event that Person 2 is smoking; and let A_3 denote the event that Person 3 is smoking. If, as before, the smoking probabilities for the three persons are equal and are given by $P\{A_1\} = P\{A_2\} = P\{A_3\} = 1/9$, and if the smoking activities are assumed to be independent, then the probability that no one is smoking is computed as $P\{K = 0\} = P\{\bar{A}_1\bar{A}_2\bar{A}_3\} = (8/9)(8/9)(8/9) = (8/9)^3 = 0.7023$. Similarly, for the maximum value of K, the probability that everyone is smoking at the same time is computed as $P\{K = 3\} = P\{A_1A_2A_3\} = (1/9)(1/9)(1/9) = (1/9)^3 = 0.0014$. Computation of the probabilities for the intermediate cases—$K = 1$ and $K = 2$—is more complicated, be-

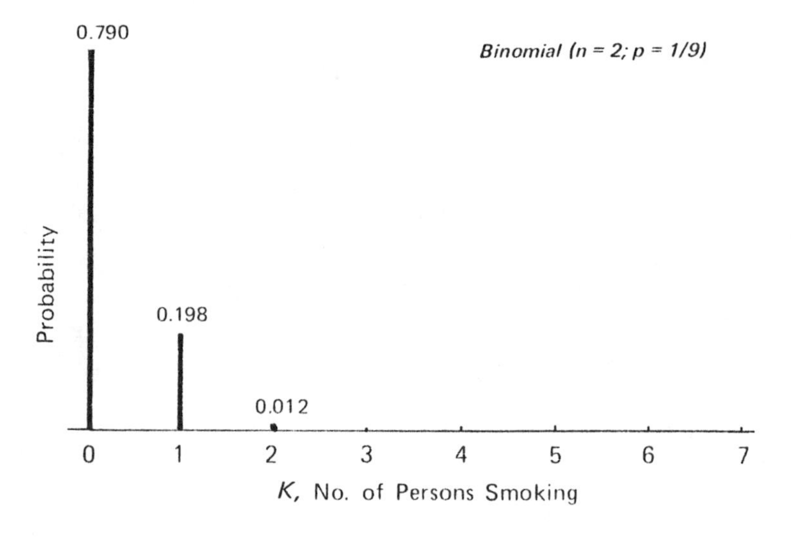

Figure 4.1. Probability distribution for the number of persons smoking in a two-person office.

cause we must take into account all the combinations in which these values of K can occur. For example, there are three possible cases in which *one person* is actively smoking: the events $\{A_1\bar{A}_2\bar{A}_3\}$, $\{\bar{A}_1A_2\bar{A}_3\}$, and $\{\bar{A}_1\bar{A}_2A_3\}$. Substituting the appropriate probabilities for each case, and properly considering the union of three independent events, the probability associated with one person smoking is computed as $P\{K = 1\} = pqq + qpq + qqp = (1/9)(8/9)(8/9) + (8/9)(1/9)(8/9) + (8/9)(8/9)(1/9) = 0.0878 + 0.0878 + 0.0878 = 0.2634$. Similarly, for the case in which $K = 2$, three possible cases can occur in which two persons are smoking at the same time: $\{A_1A_2\bar{A}_3\}$, $\{A_1\bar{A}_2A_3\}$, and $\{\bar{A}_1A_2A_3\}$. Thus, $P\{K = 2\} = ppq + pqp + qpp = (1/9)(1/9)(8/9) + (1/9)(8/9)(1/9) + (8/9)(1/9)(1/9) = 0.01097 + 0.01097 + 0.01097 = 0.0329$. In summary, the above reasoning has made it possible to compute the following probabilities for all elements in the sample space $K = \{0, 1, 2, 3\}$:

$$P\{K=0\} = q^3 = 0.7023$$

$$P\{K=1\} = 3pq^2 = 0.2634$$

$$P\{K=2\} = 3p^2q = 0.0329$$

$$P\{K=3\} = p^3 = 0.0014$$

$$\text{Total}: \ 1.0000$$

When these values are plotted as a probability distribution (Figure 4.2), it is seen that the shape is right-skewed, and the mode is at the origin.

If the same reasoning is applied to a four-person office, the following probabilities are obtained (see next page):

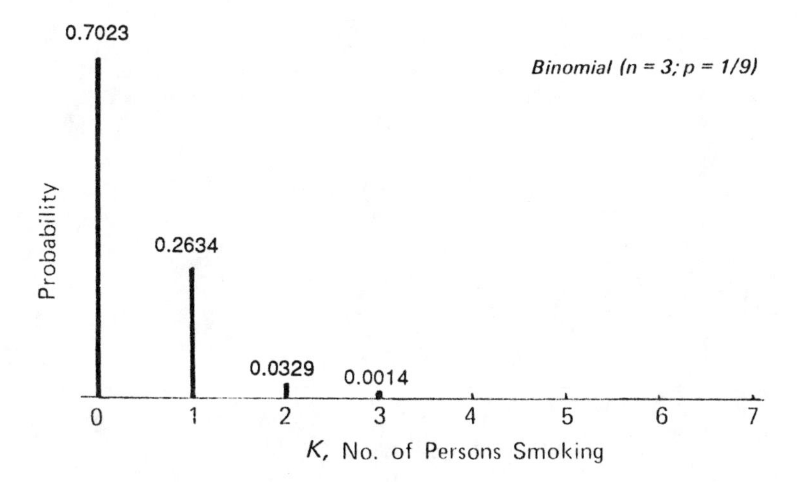

Figure 4.2. Probability distribution for the number of persons smoking in a three-person office.

$$P\{K = 0\} = \quad q^4 \quad = 0.62430$$

$$P\{K = 1\} = 4q^3p \quad = 0.31215$$

$$P\{K = 2\} = 6q^2p^2 = 0.05853$$

$$P\{K = 3\} = 4qp^3 \quad = 0.00488$$

$$P\{K = 4\} = \quad p^4 \quad = 0.00015$$

$$\text{Total}: \quad 1.00000$$

Like the two- and three-person office, the four-person office also yields a right-skewed distribution for this value of p (Figure 4.3), and the mode of the distribution (outcome with the greatest probability) occurs for the case in which no one is smoking, or $K = 0$. However, as n increases, the distribution will exhibit greater mass to the right of the origin, and the mode eventually will shift to the right of the origin. For example, if n is set to 9, for a 9-person room, the mode shifts from the origin to $K = 1$, and the distribution exhibits probabilities that decline rapidly as K increases (Figure 4.4). Most of the mass of the distribution appears in the values of K less than 4. For example, the probability that three or fewer persons are smoking, or the sum of the first four probabilities listed on the figure, gives the following result: $P\{K \leq 3\} = P\{K = 0\} + P\{K = 1\} + P\{K = 2\} + P\{K = 3\} = 0.3464 + 0.3897 + 0.1949 + 0.0568 = 0.9878$. Due to the considerable right-skewness of this distribution, the probability that K is greater than 3 is only $P\{K > 3\} = 1 - 0.9878 = 0.0122$. Large values of K have extremely low probabilities. For example, the probability that all 9 persons chosen at random are smoking is given by $P\{K = 9\} = 0.000000003$.

In the above examples, we have assumed that the probability is $p = 1/9$ that a person chosen at random from the population at a particular instant of time is engaged in the act of smoking. It is of interest to consider a simple practical interpretation of these examples. Consider a large office building consisting of

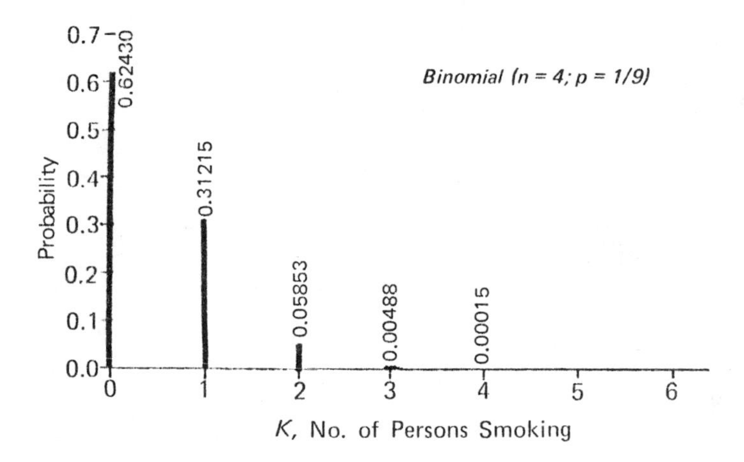

Figure 4.3. Probability distribution for the number of persons smoking in a four-person office.

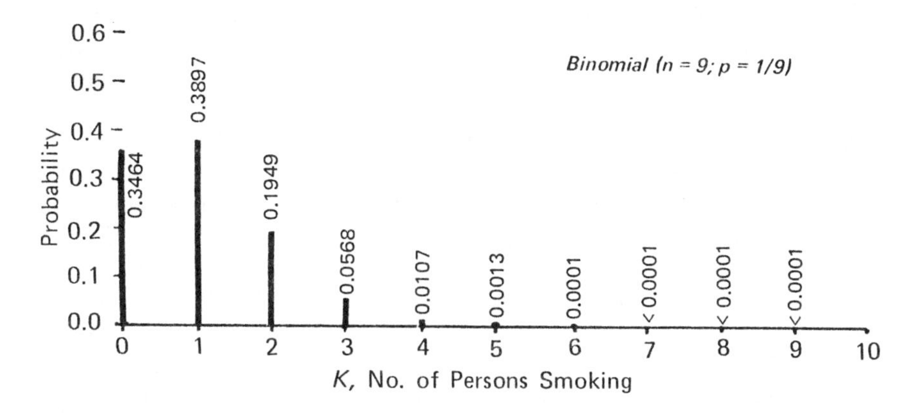

Figure 4.4. Probability distribution for the number of persons smoking in a 9-person office.

thousands of rooms, some of which are occupied by just two people. If, at some arbitrary instant, it were possible to take a snapshot of the two-person offices, the expected percentages derived from Figure 4.1 would be as follows: no one would be smoking in 79% of the offices; one person would be smoking in 19.8% of these offices, and two persons would be smoking in 1.2% of these two-person offices. Similarly, if there are offices in this building occupied by three persons, the probabilities given in Figure 4.2 would apply. For the three-person offices, the expected percentages are as follows: no one will be smoking in 70.23% of the rooms; one person will be smoking in 26.34% of the rooms; two persons will be smoking in 3.29% of the rooms; and three persons will be smoking in 0.14% of the rooms. Finally, in the four-person offices in this building, Figure 4.3 can be used to compute expected percentages, which, when rounded off, gives the following values: no one will be smoking in 62.4% of the four-person offices; one person will be smoking in 31.2% of these offices; two persons will be smoking in 5.9% of these offices; and three or more persons will be smoking in only 0.488% + 0.015% = 0.6% of these offices. Of course, real office buildings may segregate smokers from nonsmokers, or may prohibit smoking in certain locations. Nevertheless, such a probabilistic analysis of smoking activities in real office buildings may provide useful information for designing ventilation systems and can be useful for understanding and modeling indoor air pollution problems.

Development of Model by Inductive Reasoning

As n becomes larger than 4, computation of the individual probabilities by the above approach of considering all combinations becomes tedious. We therefore seek a general model, or a single equation, for computing the probability distribution for any value of n and p. Suppose that a table is developed listing all the individual terms involving p and q that were computed above for the cases $n = 1, 2, 3$, and 4, along with two additional cases, $n = 5$ and $n = 6$ (see Table 4.1). Is any pattern evident?

Notice that the first term, for the case $K = 0$, in every row consists of p raised to the nth power, while the last term, for the event $K = n$, consists of q raised to the nth power. Notice, also, that if we move our eyes horizontally along each row of terms, the sum of the exponents of p and q in a given term always is equal to n, regardless of the value of K. Finally, notice that p is raised to the Kth power in every term. Thus, it becomes possible, simply by inspection, to predict the exponents of p and q in each term in the table, as well as for additional terms if the table were expanded. If k denotes the values taken on by the random variable K, then all terms involve p and q raised to the following exponents: $p^k q^{n-k}$. This rule even applies to the first and last terms, because, for $k = 0$, it gives q^n, and, for $n = 0$, it gives p^k.

It is necessary, if a general model is to be obtained, also to find a rule for generating the integer coefficients that appear in front of the p's and q's. If the table is examined carefully, and if more terms are generated, it becomes possible, by inspection, to guess the next coefficient of each term. A rule that will allow one to compute all the integer coefficients in the table, arrived at by fitting the results to the observed pattern of coefficients, is as follows: For the nth row and the kth column, the coefficient is given by the expression $n!/k!(n-k)!$, where $n!$ denotes the factorial product $n(n-1)(n-2)\ldots 1$. It is customary to use the following notation for this expression, which sometimes is called the *combinatorial law*:

$$\binom{n}{k} = \frac{n!}{k!(n-k)!}$$

Thus, for $n = 5$ and $k = 2$, this relationship gives the following result, which is the same value listed in the table:

Table 4.1. Expansion of Bernoulli Process for n from 1 to 6.

		K						
		0	1	2	3	4	5	6
n	1	q	p					
	2	q^2	$2qp$	p^2				
	3	q^3	$3q^2p$	$3qp^2$	p^3			
	4	q^4	$4q^3p$	$6q^2p^2$	$4qp^3$	p^4		
	5	q^5	$5q^4p$	$10q^3p^2$	$10q^2p^3$	$5qp^4$	p^5	
	6	q^6	$6q^5p$	$15q^4p^2$	$20q^3p^3$	$15q^2p^4$	$6qp^5$	p^6

$$\binom{5}{2} = \frac{5 \cdot 4 \cdot \not{3} \cdot \not{2} \cdot \not{1}}{1 \cdot 2 \cdot (\not{3} \cdot \not{2} \cdot \not{1})} = 10$$

Because of the many factors in the numerator that tend also to appear in the denominator and to cancel, it is possible to formulate a relatively simple computational procedure for this expression: In the numerator, write $n(n-1)(n-2)$ . . . such that there are exactly k factors, and write just $k!$ in the denominator. That is, for $k = 4$, write just $n(n-1)(n-2)(n-3)$ in the numerator, giving a total of 4 factors in the product, and write just 4! in the denominator. When 4! is expanded as the product 4·3·2·1, the denominator also will contain a total of 4 factors. In general, the number of factors in the numerator always will be equal to the number of factors in the denominator, and this fact can be used to check the computation.

If the above relationship for the integer coefficient is multiplied by the above relationship for the exponents of p and q, we obtain a general expression for the probability distribution of the random variable K, where n and p are parameters of the distribution:

$$P\{K = k\} = \binom{n}{k} p^k q^{n-k} = \binom{n}{k} p^k (1-p)^{n-k}$$

This expression is the probability mass function (PMF) of the binomial distribution.

BINOMIAL DISTRIBUTION

In the above discussion, we employed inductive reasoning to develop rules for the coefficients and exponents of a successive number of terms, thus arriving at an equation for the binomial distribution. This model also can be developed by considering the rules of combinations and permutations.

A specific sequence of k successes and $n - k$ failures in a single trial of independent events will be given as $p^k(1-p)^{n-k} = p^k q^{n-k}$. By the rules of combinations and permutations, as described in most textbooks on college algebra, there will be $n!/[k!(n-k)!]$ equally likely sequences in which k successes and $(n-k)$ failures can occur on n trials. Thus, by the addictive law of mutually exclusive events, the probability of exactly k successes in n independent trials, given a probability of success p in a single trial, is given by the binomial distribution (Table 4.2).

Consider the expansion of $(p + q)^n$ as follows:

$$(p+q)^n = p^n + \binom{n}{1} p^{n-1} q + \binom{n}{2} p^{n-2} q^2 + \ldots + q^n$$

Notice that the individual terms obtained from this expression are the same terms as those obtained from the binomial distribution, except that they are expressed here as a sum. In other words, the right-hand side of this expression is the same result that would be obtained if the terms in any horizontal row in

Table 4.1 were added together for $K = 0, 1, 2, \ldots, n$. In general, the result of adding the probabilities associated with all elements of the sample space $K = \{0, 1, 2, \ldots, n\}$. where K has a binomial distribution, is equivalent to $(p + q)^n$. We can use this result to show that the sum of terms of a binomial distribution must always total 1. Substituting $q = 1 - p$ into the above expression, we obtain the following:

$$(p+1-p)^n = (1)^n = 1$$

This result shows that the sum of the terms of a binomial distribution, for any value of n, is 1. As discussed earlier in this book (Chapter 3), this result is a necessary property of any probability distribution.

The binomial distribution will be right-skewed if the coefficient of skewness, $\sqrt{\beta_1}$, is positive. This condition will be met if the numerator in the expression for $\sqrt{\beta_1}$, listed in Table 4.2 is positive, or if $(1 - 2p) > 0$. If we set $(1 - 2p) = 0$ and then solve the p, we find that $p = 1/2$. For this case, $\sqrt{\beta_1} = 0$, and the binomial probability model will be symmetrical. If $p > 1/2$, then $\sqrt{\beta_1}$ will be negative, and the distribution will be left-skewed. If $p < 1/2$, as it was in all the smoking examples presented above (Figures 4.1–4.4), then $\sqrt{\beta_1}$ is positive, and the distribution is right-skewed. Thus, the binomial distribution can be either left-skewed, symmetrical, or right-skewed, depending on the value of just one parameter, p. If p is held constant, however, and the other parameter of the model, n, becomes increasingly large, the magnitude of $\sqrt{\beta_1}$ becomes smaller because $\sqrt{n}$ appears in the denominator. Thus, if n increases without limit for a given value of p, the mean, np, moves to the right of the origin, and the distribution ultimately appears symmetrical, because $\sqrt{\beta_1}$ approaches 0 asymptotically.

In this book, the binomial probability model usually will be represented by the notation $\mathbf{B}(n,p)$, where n and p are the values of the two parameters. In Fig-

Table 4.2. The Binomial Distribution $\mathbf{B}(n, p)$

Probability Mass Function (PMF):

$$P_K\{k\} = \binom{n}{k} p^k q^{n-k}, \quad k = 1, 2, \ldots, n; \; q = 1 - p$$

Expected Value:

$$E[K] = np$$

Variance:

$$Var(K) = npq$$

Coefficient of Skewness:

$$\sqrt{\beta_1} = \frac{1-2p}{\sqrt{npq}}$$

Coefficient of Kurtosis:

$$\beta_2 = 3 + \frac{1-6pq}{npq}$$

ure 4.1, for example, the distribution of smokers in a two-person office would be represented as **B**(2,1/9). Similarly, the distribution of the number of persons engaged in the act of smoking in Figures 4.2, 4.3, and 4.4 are plots of the binomial probability models **B**(3,1/9), **B**(4,1/9), and **B**(9,1/9), respectively.

In all of these figures, the vertical (probability) axis has the same scale, so it is possible to compare the binomial distributions to see how the shape changes for different values of the parameters. In Figures 4.1 to 4.4, the probability $p = 1/9$ is held constant, but n increases from 2 to 9. Because $p < 1/2$, the distribution is right-skewed in all cases, but the mode of this distribution moves to the right of the origin as n changes from $n = 4$ to $n = 9$. Eventually, if n increased further, the distribution would shift further to the right, and, although the right-skewness would continue to be evident, the distribution increasingly would exhibit greater symmetry. Even for small values of n, however, the long right "tail" associated with the right-skewness is quite pronounced. This long right tail results because the probabilities become extremely small as 1/9 is raised to greater and greater powers of K. Because of these small probabilities, it has been necessary to list them to several decimal places.

If the value of n is held constant while p is varied, a number of different distributional shapes are generated, as shown by the plots for the **B**(9,1/9), **B**(9, 1/2), and **B**(9, 8/9) probability models (Figures 4.4, 4.5, and 4.6). If $p = 1/2$, the binomial model generates a symmetrical probability distribution (Figure 4.5). If $p = 8/9$ (Figure 4.6), the same probability distribution as for the case in which $p = 1/9$ results (Figure 4.4), except that the horizontal axis is reversed. Thus, the two cases yield the same numerical probabilities, except in reverse order, and the plotted distributions complement each other by appearing as mirror images. These examples show the flexibility of the binomial probability model and the variety of distributional shapes it can take on.

The above presentation of probability distributions for the number of persons smoking in rooms of fixed size was intended to illustrate how the binomial distribution could be developed by applying basic principles of probability theory to a very simple case. The example was intended to be instructive rather than a typical practical application. However, the binomial distribution has broad applicabil-

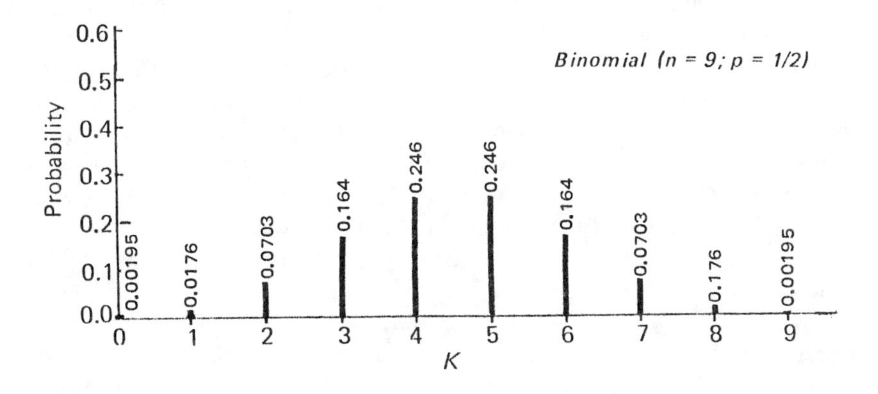

Figure 4.5. Example of symmetrical "bell-shaped" distribution which results from the binomial probability model when $p = 1/2$.

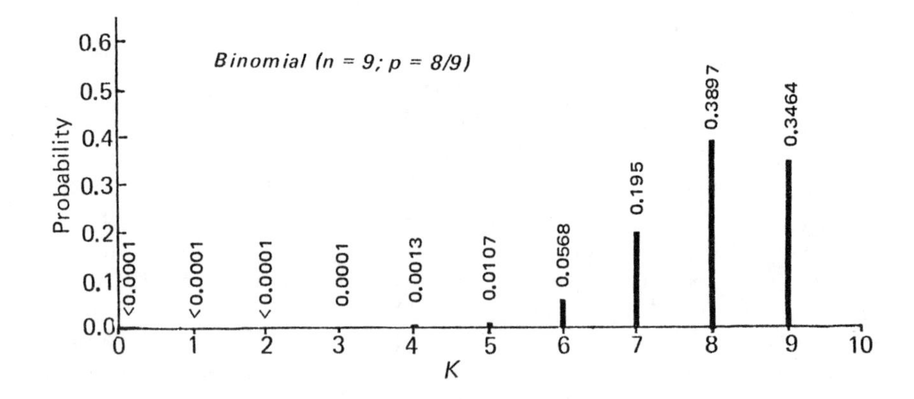

Figure 4.6. Example of left-skewed distribution generated by the binomial probability model when $p = 8/9$.

ity to a variety of problems in probability theory, and many useful applications can be found in the statistical and engineering literature. The binomial distribution also is well suited for a variety of environmental control problems, and has found increasing use in interpreting environmental standards and in regulatory decision-making.

APPLICATIONS TO ENVIRONMENTAL PROBLEMS

In the field of air pollution control, the original National Ambient Air Quality Standards (NAAQS) promulgated in the U.S. in 1971 specified an upper limit of ambient pollutant concentration that was "not to be exceeded more than once per year" at a given monitoring location. In this deterministic form of the standard, air pollution control officials ordinarily focused their concern on the second highest value observed during the year at a given location. If the second highest value exceeded the NAAQS, then the location was judged not to be in compliance with the standard.

One disadvantage of this deterministic form of the air quality standard is that no allowance is made for missing observations. For example, if just two or three measurements were available from a given location, it would make little sense to compare the second highest reading from this site with the second highest reading from a site in which a full year of observations were available. A second disadvantage of the deterministic form is that it does not accommodate very rare or unusual events. For example, very unusual meteorological conditions may cause excessively high air pollutant concentrations in an occasional year, even though this location may be well controlled in other years. Because the deterministic form is quite sensitive to these fluctuations, regulatory agencies increasingly have moved toward the practice of specifying their standards in a probabilistic form.

For example, the original NAAQS for photochemical oxidant (ozone), adopted by the EPA in 1971, specified a deterministic format in which not more than one hourly average value per year was allowed to exceed 0.08 ppm.[1] In 1979 however, the EPA revised the numerical value of this standard to 0.12 ppm ozone, but it also changed the manner in which the standard is specified from a deterministic form to a probabilistic form based on the "expected number of days" that the numerical value is exceeded.[2,3] With the newer probabilistic form, the ambient air quality standard is attained when the expected number of days per calendar year that maximum hourly ozone concentrations exceed 0.12 ppm is less than or equal to 1.

We can express the revised probabilistic form of the standard formally as follows. Let X denote the maximum hourly ozone concentration observed on a particular 24-hour period, and let x_s denote the numerical value of the ozone standard (in this case, $x_s = 0.12$ ppm). If the event $\{X > x_s\}$ occurs on that date, we count this as *one exceedance of the standard.* Our interest focuses on the total number of days during the calendar year on which an exceedance occurs. Thus, let K be a random variable denoting the total number of exceedances occurring over a given time period of interest, such as a year. Then, for a given location to be in compliance with the probabilistic form of the standard, the expected number of exceedances per year $E[K]$ must be less than or equal to 1.0:

$$E[K] \leq 1.0$$

Although K is an integer random variable, the expected value is a decimal, and "1.0" is used here to mean "1.0000...". If the expected number of exceedances is greater than 1.0; that is, if $E[K] > 1.0$, then the location is not in compliance, and the standard is violated.

Use of the expected value of K as the basic form of the standard is desirable from a statistical standpoint. However, expected values usually are difficult to estimate from real world conditions and from actual data. Indeed, because of the asymptotic property of expected values, determining with absolute certainty whether a standard is met at a given site would require consideration of an infinite number of days. As discussed by Curran and Cox,[3] the EPA therefore has recommended that the standard be considered as averaged over three years. In this approach, which is described in an EPA guidelines document on this topic,[4] the number of exceedances in each of three successive years are averaged together, and the standard is exceeded at this site if the resulting average is greater than 1.0:

> Consequently, the expected number of exceedances per year at a site should be computed by averaging the estimated number of exceedances for each year of data during the past three calendar years. In other words, if the estimated number of exceedances has been computed for 1974, 1975, and 1976, then the expected number of exceedances is estimated by averaging those three numbers. If this estimate is greater than 1, then the standard has been exceeded at this site.[4]

Some individual years will have missing observations, and the EPA recommends that the number of exceedances for these years be estimated by counting the number of exceedances which actually were observed and dividing by the number of observations available, giving, in effect, an "exceedance rate." If this exceedance rate is multiplied by the number of observations possible, the result is an estimate of the number of exceedances that would have occurred if all observations were present. This approach implicitly assumes that the exceedance rate for the missing days is the same as the exceedance rate observed on the valid days. If all observations are present, then the exceedance rate must be less than or equal to 1 in 365 (excluding leap year) for the site to be in compliance with the ozone standard. Let r denote the exceedance rate (number of exceedances per unit time). Then a given site will be in compliance with the standard if $r \leq 1/365$ (i.e., $r \leq 0.00274$) exceedances per year. Notice that r is similar to a probability, but it is conceptually different in that it is a rate. The exceedance rate is considered in greater detail in our discussion of Poisson processes (Chapter 5).

Consider a hypothetical monitoring site which is on the borderline in that it just meets (but does not exceed) the ambient air quality standard. That is, assume that the expected number of exceedances at this site is exactly 1.0, or $E[K] = 1.0$. Some years will experience 0 exceedances, and some years will experience more than 1 exceedance, but the expected number of exceedances per year will be 1.0, and the expected number of exceedances during any 100-year period will be 100. If the number of exceedances experienced on many 100-year periods were averaged, we would expect the average ultimately to approach 100.

We can imagine the existence of a probability distribution that describes the number of exceedances occurring in any particular year. The expected value of this distribution will be $E[K] = 1.0$, but what will be its variance? What probability model is most naturally appropriate for this situation? Once this probability model is selected, what are the probabilities associated with 2, 3, or 4 exceedances in a particular year? Because the standard is exceeded if the average number of exceedances during three years is greater than 1.0, it is important to determine how often 4 or more exceedances occur in any 3-year period, since 4 or more exceedances will cause the 3-year average to exceed 1.0. What is the most desirable probability model for the number of exceedances in a 3-year period? Using this probability model, what are the probabilities associated with 4, 5, or 6 or more exceedances during any 3-year period?

Probability Distribution for the Number of Exceedances

Javits[5] has suggested that the number of daily exceedances of air quality standards per year should be treated as a Bernoulli process, and he has applied the binomial distribution to interpretation of the ozone standard. In this approach, daily exceedances are assumed to be independent events, and the 365 days of each calendar year are assumed to be a series (i.e., a run) of 365 independent Bernoulli trials. Javits discusses the basis for these assumptions in some detail, and the following development is based on the concepts presented in his paper.[5]

Let K be a random variable describing the total number of exceedances encountered in a 1-year period, and let M be a random variable representing the total number of exceedances counted in a 3-year period. Consider the situation described above in which air quality is at the critical point in which it just exactly meets (but does not exceed) the statistical form of the air quality standard. This critical point is selected because it facilitates analysis and because it is useful for calculating "design values" in order to estimate how much control is needed to attain the air quality standard. At this critical point, the expected number of exceedances for the 1-year and 3-year time periods will be as follows:

$$E[K] = 1.0 \quad \text{and} \quad E[M] = 3E[K] = 3.0$$

Assume that the number of exceedances occurring during these time periods can be represented as a binomially distributed random variable. We first determine the parameters n and p of this distribution for the two time periods. In the first case, $n = 365$ days (excluding leap year), and, in the second case, $n = (3)(365) = 1095$ days. Using the equation for the expected value of the binomial probability model (Table 4.2), $E[K] = np$. In the first case, we have assumed that $E[K] = 1.0$, so we can solve the equation $np = 1.0$ for p, which gives $365p = 1.0$, or $p = 1/365$. In the second case, $E[M] = 3.0$, so we can solve the equation $1095p = 3.0$ to obtain the probability $p = 3/1095 = 1/365$. Thus, in both cases, the probability p is the same, although the values of n are different. It can be seen why the probability is the same, regardless of the time period chosen, if we consider the exceedance rate. By assuming that the expected number of exceedances per year is 1.0, the exceedance rate will be 1 exceedance per 365 days, or $r = 1/365$, regardless of the time period chosen.

Thus, using the above values of n and p, the probability distribution for the number of exceedances in a 365-day year will be given by the binomial probability model with parameters $n = 365$ and $p = 1/365$, or **B**(365,1/365), where $q = 1 - p = 1 - 1/365 = 364/365$:

$$P\{K = k\} = \binom{365}{k}\left(\frac{1}{365}\right)^k \left(\frac{364}{365}\right)^{365-k}$$

$$\text{for } k = 0, 1, 2, \ldots, 365$$

The initial terms of this distribution can be computed easily with a hand calculator that is capable of raising numbers to powers as high as 365. For example, the probability at $K = 0$ can be computed as follows:

$$P\{K = 0\} = \binom{365}{0}\left(\frac{1}{365}\right)^0 \left(\frac{364}{365}\right)^{365-0} = (1)(1)\left(\frac{364}{365}\right)^{365} = 0.3674$$

Because $k = 0$, the first factor, or combinatorial, becomes 1, since 0! is defined as 1. Similarly, the second factor becomes 1, because any number raised to the power 0 gives 1. This result indicates that, if the ozone air quality standard is met exactly so that $E[K] = 1.0$, then years in which there are no exceedances will occur with probability $p = 0.367$ (rounded off). In other words, in 100 years of observations, we would expect to find $(0.367)(100) = 36.7$ years, on the average, in which zero exceedances occurred. Conversely, the probability that one or more exceedances occur in a given year is computed as $P\{K \geq 1\} = 1 - P\{K = 0\} = 1 - 0.367 = 0.633$.

Using the above probability distribution for K, the probability that just one exceedance occurs in a given year is computed in the same manner:

$$P\{K=1\} = \binom{365}{1}\left(\tfrac{1}{365}\right)^1\left(\tfrac{364}{365}\right)^{365-1} = (365)\left(\tfrac{1}{365}\right)\left(\tfrac{364}{365}\right)^{364} = 0.3684$$

Thus, the probability that exactly one exceedance occurs in any year is given by (rounded off) $p = 0.368$, which is almost the same as the probability that 0 exceedances occur. Combining these results, it is seen that the probability that the number of exceedances per year is either 0 or 1 is given by the sum of these two probabilities, $P\{K = 0\} + P\{K = 1\} = 0.3674 + 0.3684 = 0.7358$. This result gives the probability that "one or fewer exceedances" occurs. Consequently, the probability that "more than one exceedance" occurs per year is obtained by subtracting this result from 1: $P\{K > 1\} = 1 - 0.7358 = 0.2642$. For an integer probability distribution such as this, the event "more than one exceedance" is logically equivalent to the event "two or more exceedances" per year. That is, $P\{K > 1\} = P\{K \geq 2\}$. Thus, this probability implies that, on the average, (100%)(0.2642) = 26.4% of the years will experience two or more exceedances, even though the ozone standard has been attained.

Calculating the probabilities for $K = 2$ and $K = 3$ in the manner described above, we obtain the following:

$$P\{K=2\} = \frac{365 \cdot 364}{1 \cdot 2}\left(\tfrac{1}{365}\right)^2\left(\tfrac{364}{365}\right)^{363} = 0.18419$$

$$P\{K=3\} = \frac{365 \cdot 364 \cdot 363}{1 \cdot 2 \cdot 3}\left(\tfrac{1}{365}\right)^3\left(\tfrac{364}{365}\right)^{362} = 0.06123$$

Notice that the above expressions illustrate the convenient rule mentioned earlier for computing the combinatorial coefficient. When $K = 2$, the numerator should contain two factors, and the denominator should contain two factors. The numerator begins with $n = 365$, and each factor thereafter is decreased by 1, giving 365·364. Conversely, the denominator begins with $k = 1$, and each factor thereafter is increased by 1, giving 1·2. When $K = 3$, the same rule is followed, and there are three factors in the numerator, 365·364·363, and three factors in the denominator 1·2·3. In actual calculations, it is useful to reduce the numerator and denominator by as much as possible by dividing each of them by common factors; that is, by making cancellations.

The probabilities for larger values of K, ranging from 0 to 12, have been computed for convenience (Table 4.3). The column of this table labeled $P_K\{k\}$ lists the individual probabilities associated with integer values of K, and the column labeled $F_K(k)$ is the sum of the probabilities in the previous column, or the cumulative probabilities. The cumulative probability, $F_K(k)$ also is the cumulative distribution function (CDF) of the model. In general, the CDF provides a convenient way to compute the probabilities associated with ranges of the random variable K, since $F_K(k) = P\{K \leq k\}$. Thus, to compute the probability that "one or fewer exceedances" occurs, we write $P\{K \leq 1\} = F_K(1) =$

Table 4.3. Probability Distribution for the Number of Exceedances, Using the Binomial Model with Expected Number of Exceedances of 1.0

	1-Year Period		3-Year Period	
No. of Exceedances k	Probability $P_K\{k\}$	Cumulative Probability $F_K(k)$	Probability $P_M\{k\}$	Cumulative Probability $F_M(k)$
0	0.36737	0.36737	0.04958	0.04958
1	0.36838	0.73576	0.14916	0.19874
2	0.18419	0.91995	0.22414	0.42288
3	0.06123	0.98118	0.22435	0.64723
4	0.01522	0.99640	0.16826	0.81549
5	0.00302	0.99942	0.10086	0.91635
6	0.000498	0.999920	0.05034	0.96670
7	0.000070	0.9999904	0.02152	0.98821
8	0.0000086	0.9999989	0.00804	0.99625
9	0.0000009	0.9999999	0.00267	0.99892
10	0.00000009	0.9999999	0.00080	0.99972
11	0.000000008	0.9999999	0.00022	0.99993
12	0.000000001	0.9999999	0.00005	0.99998

0.73576 from the table. To compute the probability that "more than one exceedance" will occur in a given year, we note that $P\{K > 1\} = 1 - F_K(1)$, so $P\{K > 1\} = 1 - 0.73576 = 0.26424$. This result is the same probability that was computed above for the case of "two or more exceedances," or $P = 0.264$ (rounded off). By using the CDF listed in the table, the computation process becomes more formal and less subject to error.

Examining the probabilities listed for K in this table, and recalling that the number of daily exceedances per year is a random variable that ranges from 0 to 365, it is evident that the resulting probability distribution is extremely right-skewed, with most of the probability mass occurring in the narrow range between $K = 0$ and $K = 4$. By the time that $K = 4$, the CDF gives a value of $F_K(4) = 0.99640$, indicating that 99.6% of the probability mass is accounted for. Thus, when the ozone air quality standard is attained, there will be 99% assurance that 4 or fewer exceedances per year occur, and more than 5 exceedances will occur, on the average, in less than 1% of the years.

If the probability distribution for the random variable K, the number of exceedances per year, is plotted, it becomes necessary to compress the horizontal axis in order to fit the entire 365-day year on the graph (Figure 4.7). Here, compression is achieved by plotting the first 9 values ($K = 0$ to $K = 8$) and the last 7 values ($K = 359$ to $K = 365$), with the intermediate values excluded. Most of the intermediate values of K occur with extremely low probability. For example, $P\{K = 12\} = 0.000000001$, and $P\{K = 20\} = 9 \times 10^{-20}$. Similarly, $P\{K = 35\} = 7 \times 10^{-42}$, and probabilities for higher values of K from 359 to 365 are much smaller than the "0.000000001" upper limits listed on the graph. If the graph had not been compressed and all 365 days were plotted, then the distribution would look as though there were a "blip" of probabilities close to the origin, with an enormously long tail extending to the right.

In the above discussion, it has been assumed that pollutant sources have been controlled so that air quality at the measurement location is at the critical point

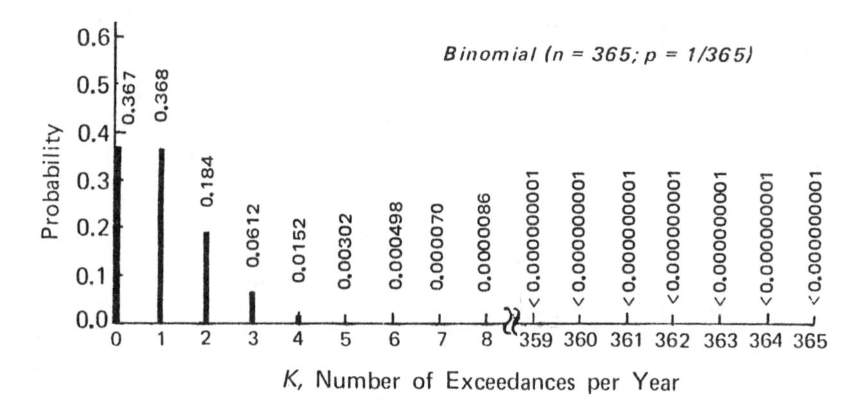

Figure 4.7. Probability distribution for the number of days per year exceeding an air quality standard when the expected number of exceedances per year is one.

in which the expected number of exceedances per year is exactly one, or $E[K] = 1.0$. The standard is attained if compliance efforts either equal or surpass this degree of control, or $E[K] \leq 1.0$. If the control level is surpassed, and $E[K] < 1.0$, then the probabilities associated with the random variable K will be lower than those computed here (Table 4.3 and Figure 4.7). For $E[K] = 1.0$, for example, the probability of more than 4 exceedances per year is computed as $P\{K > 4\} = 1 - 0.9664 = 0.0036$. This result implies that, for an area that just meets the standard, there will be only 36 years in 10,000 in which 5 or more exceedances (more than 4 exceedances) occur. In areas that surpass the standard, and $E[K] < 1.0$, there will be even fewer than 36 years in 10,000 with 5 or more exceedances. Because 5 or more exceedances are so rare, the occurrence of 5 or more exceedances at a real monitoring station during a particular year strongly suggests the location is not in compliance with the statistical part of the standard. That is, it is likely that the expected number of exceedances at such a location is greater than one, or $E[K] > 1.0$.

It is evident that the ozone air quality standard can be viewed as consisting of two parts: (a) a statistical part requiring that the expected number of exceedances be one or less, or $E[K] \leq 1.0$, and (b) a deterministic part specifying that the average number of exceedances observed in any 3-year period be 3 or less. If K_1, K_2, and K_3 denote the number of exceedances observed in the first, second, and third years of a 3-year period, then the deterministic part of this standard requires that the average of these three numbers, $(K_1 + K_2 + K_3,)/3$, be 1.0 or less. This condition can be met only if the total number of exceedances in the 3-year period, $M = K_1 + K_2 + K_3$, is 3 or less. If M has a value of 4 or more, the location is not in compliance with the standard. What is the probability that 4 or more exceedances will occur in any 3-year period? Fortunately, the same probability model can be used to develop the distribution for the number of exceedances in a 3-year period.

Using the Bernoulli process model, the distribution of the total number of exceedances over 3 years, or (3)(365) = 1095 days, will be binomial **B**(1095,1/365) and the random variable M will have the following PMF:

$$P\{M = k\} = \binom{1095}{k}\left(\tfrac{1}{365}\right)^k\left(\tfrac{364}{365}\right)^{1095-k}$$

Use of this model to compute probabilities for large values of M can become tedious, but the first four terms can be computed with relative ease:

$$P\{M = 0\} = \left(\tfrac{364}{365}\right)^{1095} = 0.04958$$

$$P\{M = 1\} = \frac{1095}{1}\left(\tfrac{1}{365}\right)^1\left(\tfrac{364}{365}\right)^{1094} = 0.14916$$

$$P\{M = 2\} = \frac{1095 \cdot 1094}{1 \cdot 2}\left(\tfrac{1}{365}\right)^2\left(\tfrac{364}{365}\right)^{1093} = 0.22414$$

$$P\{M = 3\} = \frac{1095 \cdot 1094 \cdot 1093}{1 \cdot 2 \cdot 3}\left(\tfrac{1}{365}\right)^3\left(\tfrac{364}{365}\right)^{1092} = 0.22435$$

Thus, the probability of the event "4 or more exceedances" is computed from these results as follows:

$$P\{M \le 3\} = P\{M = 0\} + P\{M = 1\} + P\{M = 2\} + P\{M = 3\}$$

$$= 0.04958 + 0.14916 + 0.22414 + 0.22435 = 0.64723$$

$$P\{M \ge 4\} = P\{M > 3\} = 1 - P\{M \le 3\} = 1 - 0.64723 = 0.35277$$

This result also can be obtained directly from Table 4.3, which lists the PMF and the CDF for the first 12 terms of this distribution. The above cumulative probability for $P\{M \le 3\}$ is listed in the table giving the CDF as $F_M(3) =$ 0.64723. The probability that M is 4 or greater is computed in the same manner as indicated above, giving $P\{M \ge 4\} = 0.35277$. The PMF and CDF are plotted in Figure 4.8.

The above result implies that, if the standard is just attained so that the expected number of exceedances is 1.0, then 4 or more exceedances will occur in any 3-year period with probability of approximately $P = 0.353$. Thus, although the statistical basis for the air quality standard has been attained exactly, then actual levels will not be in compliance with the standard in 35.3% of the 3-year periods.

For the case of more than 6 exceedances, Table 4.3 gives $F_M(6) = 0.96670$, which results in a 3-year exceedance probability of $P\{M > 6\} = 1 - F_M(6) = 1 -$ 0.96670 = 0.0333. Thus, if the statistical part of the standard (1.0 expected exceedances per year) has been met, then more than 6 exceedances will occur in 3.3% of the 3-year periods, on the average. As we shall see in the next chapter, the Poisson distribution is a convenient approximation to the binomial distribu-

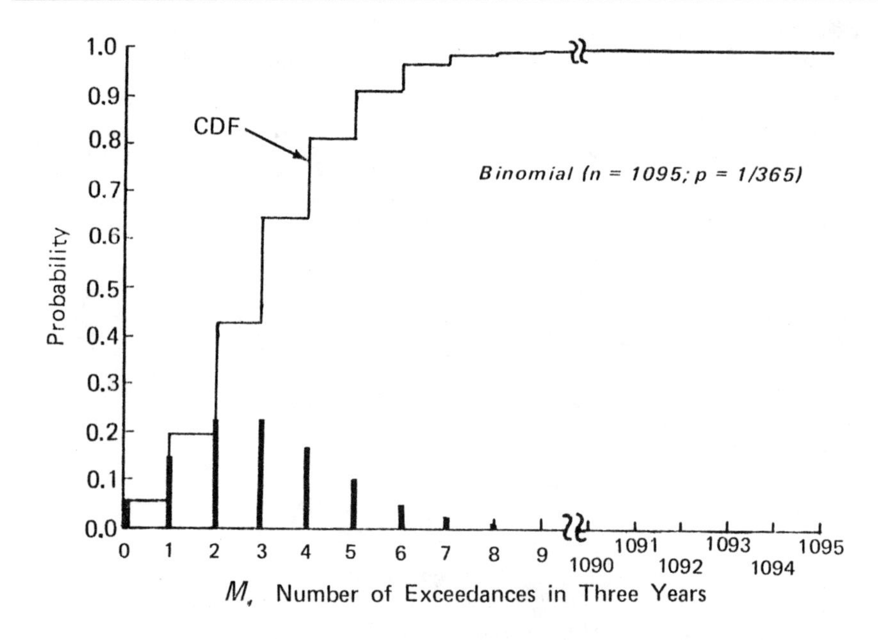

Figure 4.8. Probability distribution for the number of exceedances in a 3-year period when the expected number of exceedances per year is one.

tion for this situation, and Javits[5] obtains the nearly identical result of 3.4% using the Poisson probability model.

From these results, Javits[5] concludes that the frequency with which a given site will be out of compliance with the standard, even though it exactly meets the statistical basis for the standard, is too great. Consequently, he believes that the statistical and compliance components of the EPA standard are not consistent:

> ... Thus in any single 3-yr period in which the expected number of exceedances per year is 1.0, there is better than one chance in three of failing the 3-yr requirement. Obviously, it will not take many consecutive 3-yr periods before violation of the 3-yr requirement is practically assured. However, there is only a 3.4% probability of incurring more than 6 exceedances in a 3-yr period.
>
> ... It has been demonstrated that the two requirements of the ozone NAAQS are inconsistent. If the current design value were attained there remains a sizeable probability of violating the 3-yr requirement in any subsequent 3-yr time period.[5]

This problem also is considered in the next chapter, except that it is modeled using the Poisson distribution.

Robustness of Statistical Assumptions

Because the number of days per year exceeding the standard has been treated as a Bernoulli process, the following assumptions were made implicitly: (a) *stationarity*: the probability of an exceedance does not vary from day to day, and (b) *independence*: daily maximum ozone concentrations are independent. Daily air quality levels are a function of meteorological conditions, which do vary from day to day and from season to season, causing the daily exceedance probabilities to vary also. With regard to the independence assumption, examination of observed data usually reveals that concentrations on successive dates are correlated. These two facts raise doubts about the justification for treating the number of daily exceedances as a Bernoulli process, and it is necessary to discover how sensitive the predicted results are to violations of the assumptions that have been made.

Stationarity

A way to examine the impact of day-to-day variation in the probability of an exceedance would be to apply a model in which the probabilities are allowed to vary but the expected number of exceedances is held constant. For example, the daily exceedance probabilities can be made to vary in some reasonable manner over the year (or from month to month) to account for seasonal change, with the expected number of exceedances held at $E\{K\} = 1.0$. Then the probability distribution of K can be recalculated to see how much the resulting probabilities have changed. If the change in the resulting values is negligible, then we can conclude that neglecting the variation in daily exceedance probabilities is justifiable.

Javits[5] examines how the variation in daily exceedance probabilities affects the predicted results by considering an extreme case. Suppose that the daily probabilities are especially unequal: on two days each year, the probability of an exceedance is $p = 0.5$, and on the remaining days, the probability is $p = 0.0$. Under these assumptions, the expected number of exceedances per year is $E[K] = 2(0.5) = 1.0$. The maximum possible number of exceedances in any 3-year period will be $n = 6$, and the number of exceedances will follow a binomial distribution $\mathbf{B}(6,0.5)$:

$$P\{M = k\} = \binom{6}{k}(0.5)^{6-k}(0.5)^k$$

From this distribution, the probability of 3 or fewer exceedances in any 3-year period will be given by:

$$P\{M \leq 3\} = P\{M = 0\} + P\{M = 1\} + P\{M = 2\} + P\{M = 3\} =$$

$$0.015625 + 0.093750 + 0.234375 + 0.312500 = 0.656875$$

Thus, the probability of 4 or more exceedances will be given by $P\{M \geq 4\} = P\{M > 3\} = 1.00000 - 0.656875 = 0.343125$. This result, obtained from this very simple binomial distribution, gives an expected percentage of 34.3%, which differs by only one percentage point from the original expected percentage of 35.3% obtained from the original binomial distribution **B**(1095,1/365). If the daily probabilities of an exceedance do not vary as much as in the extreme case assumed here, then, as noted by Javits,[5] the result will be even closer to the 35.3% figure. Thus, if the expected number of exceedances per year is 1.0, the probability of 4 or more exceedances in any 3-year period, which causes noncompliance with the ozone standard, will be very close to 35.3%, regardless of whether the daily probabilities remain constant or are allowed to vary.

The probability of 4 or more exceedances in any 3-year period appears to be relatively insensitive to variation in the daily probabilities, provided that the expected number of exceedances is 1.0, but this probability has been derived by summing terms comprising the cumulative distribution function. It is likely that individual PMF terms that were added to form the CDF are not as robust to changes in p as are the CDF values, particularly for the lower values of M. For example, $P\{M = 3\} = 0.224$ in the original distribution (see Table 4.3), but $P\{M = 3\} = 0.313$ in the extreme case model **B**(6,0.5) discussed above. However, the CDF of the original distribution gives $F_M(3) = 0.647$, while the CDF for the above model gives $F_M(3) = 0.657$, which are much closer values. Because the daily probabilities in real situations are unlikely to vary as much as was assumed in the **B**(6,0.5) model, Javits' contention that the binomial distribution, when applied to the number of exceedances of the ozone standard, is relatively insensitive to the nonstationarity of its parameters seems both convincing and remarkable.

Independence

The second assumption, that daily maximum concentrations are independent, is inconsistent with observed environmental phenomena. Meteorological conditions greatly affect air pollutant concentrations, and meteorological conditions tend to be correlated in time. One reason is that the weather changes very slowly. For example, if a storm with high winds occurs during a particular hour of the day, the hour immediately afterward is unlikely to exhibit dry stable air. Rather, the storm usually will persist for several hours or more. Thus, if it is known that high winds are present on a particular hour, the probability of high winds on the following hour will be greater than would otherwise be the case. Because air pollutant concentrations are greatly affected by these weather conditions, concentrations observed on adjacent hours usually will exhibit strong dependencies. If some time interval (i.e., a "lag time") is allowed to elapse between the hours, this dependency decreases.

The ozone air quality standard is based on the highest hourly reading observed during each 24-hour period, so readings may be separated in time by several hours or more. However, it is possible for a particular weather condition to persist long enough to cause maximum concentrations on two or three successive days, and, even though the maxima are several hours apart, they may exhibit some degree of correlation. How will violation of the independence assumption in the Bernoulli process model affect the computed probabilities for the ozone standard?

Javits[5] investigates the impact of violating the independence assumption by considering two extreme cases of dependencies of daily maxima. In the first case, he assumes that each year is made up of consecutive, nonoverlapping pairs of daily observations. The two days in each pair are linked to each other so that, if one day exceeds the specified concentration, the other one does also. Thus, both days either exceed the 0.12 ppm ozone level, or both days fail to exceed the 0.12 ozone level, and all pairs are independent of each other. Therefore, each year consists of half as many independent Bernoulli trials as before, or 365/2 = 182.5 or 182 possible "doublets." Each doublet counts as 2 exceedances; therefore, if the expected number of exceedances per year is 1.0, the corresponding expected number of doublets per year will be 0.5. Let J be a random variable denoting the number of doublets that occur. Then $E[J] = 0.5$ doublets per year; solving for p as before, $np = (365/2)p = 0.5$, which gives $p = 1/365$. Consequently, the probability distribution for the number of doublets occurring per year will be binomial **B**(182,1/365). Since any 3-year period consists of 3(365/2) = 547 maximum possible doublets, the probability distribution for the number of doublets occurring in 3 years will be binomial **B**(547,1/365):

$$P\{J=j\} = \binom{547}{j}\left(\frac{1}{365}\right)^j\left(\frac{364}{365}\right)^{547-j}$$

Here, the probability of 4 or more exceedances in any 3-year period will be the same as the probability of 2 or more doublets:

$$P\{0 \text{ doublets}\} = P\{J=0\} = \left(\frac{1}{365}\right)^0\left(\frac{364}{365}\right)^{547} = 0.22298$$

$$P\{1 \text{ doublet}\} = P\{J=1\} = 547\left(\frac{1}{365}\right)^1\left(\frac{364}{365}\right)^{547} = 0.33416$$

$$P\{2 \text{ or more doublets}\} = P\{J \geq 2\} = 1.00000 - 0.22298 - 0.33416 = 0.44286$$

Thus, for this extreme case of lack of independence, the chance of 4 or more exceedances (2 or more doublets) in a 3-year period will be 44.3%.

In Javits' second extreme case of lack of independence, the exceedances always appear in "triplets," consecutive nonoverlapping groups of three days, in which all three days either exceed the 0.12 ppm ozone level or all three days fail to exceed the 0.12 ppm ozone level. Now each year will consist of one-third as many Bernoulli trials as before. Each triplet counts as 3 exceedances; therefore, if the expected number of exceedances is maintained at 1.0 per year, the corresponding expected number of triplets per year will be 1.0/3 = 1/3. To find p for this distribution, we set $1/3 = np = (365/3)p$, which gives $p = 1/365$, the same result as before. The maximum number of triplets possible in one year will be (365/3) = 121.67 = 122 approximately, and the probability distribution for the number of triplets per year will be binomial **B**(122, 1/365). Similarly, the maximum number of triplets possible in any 3-year period will be 3(365/3) = 365, and the probability distribution of the number of triplets will be **B**(365/1/365). If I is a random variable denoting the number of triplets in any 3-year period, then the probability of 4 or more exceedances (more than one triplet) is computed as follows:

$$P\{0 \text{ triplets}\} = P\{I=0\} = \left(\tfrac{1}{365}\right)^0 \left(\tfrac{364}{365}\right)^{365} = 0.36737$$

$$P\{1 \text{ triplet}\} = P\{I=1\} = 365\left(\tfrac{1}{365}\right)^1 \left(\tfrac{364}{365}\right)^{364} = 0.36838$$

$$P\{\text{more than 1 triplet}\} = 1.00000 - 0.36737 - 0.36838 = 0.26425$$

Thus, for this second extreme case of nonindependence, the chance of 4 or more exceedances (more than 1 triplet) is 26.4%.

For these two assumed cases of extreme dependence of concentrations on adjacent dates, these results indicate that the chance of 4 or more exceedances ranges from 26.4% (for triplets) to 44.3% (for doublets), compared with the value of 35.3% obtained from the original distribution with no dependence. Although this range is significant in absolute terms, it is not very great when one considers how drastic the assumed condition of dependence are. In real situations, the degree of dependence is likely to be much less than for perfectly linked pairs of days or perfectly linked triplets of days, causing the result to be much closer to the 35.3% figure, as pointed out by Javits:[5]

> These probabilities differ from that computed under the independence assumption (35.3%) by only about 9 percentage points. In practice we would expect to find much less serial dependence than these two extreme cases, and correspondingly find better agreement with the probability computed under the independence assumption.[5]

These results have practical implications for the setting of ambient air quality standards. Air pollution control officials ordinarily will judge if a given location is in compliance with the ozone standard by counting the number of exceedances in a 3-year period. If this total is 4 or more, which will make the average number of exceedances for the 3-year period greater than 1.0, then the location will be judged to be out of compliance with the deterministic part of the ozone standard. However, the statistical part of the ozone standard states that the expected number of exceedances per year is to be 1.0. The above results show that, if the statistical part of the ozone standard is just attained, then the deterministic part will be out of compliance approximately one-third (35.3%) of the 3-year periods. It is reasonable to assume that regulatory officials will not be satisfied with a noncompliance rate as high as 35.3%.

Although the EPA has not indicated what it considers to be an acceptable noncompliance rate, most air pollution control agencies probably will seek a noncompliance frequency that is much less than 35.3%, perhaps less than 10%. If they do so, then the expected number of exceedances per year will have to be less than 1.0. As indicated by Javits,[5] an expected number of exceedances per year of 0.5 will produce a noncompliance rate of less than 10%, or assurance that the deterministic part of the ozone standard will be met in at least 90% of the 3-year periods (see Problem 5). Thus, the deterministic part of the EPA standard used for compliance seems inconsistent with its statistical part, and Javits recommends that the two forms be modified to become consistent:

> It is recommended that the EPA modify the ozone NAAQS to eliminate this inconsistency. If the EPA wishes to retain the current definition of the design value, then it should allow 6 exceedances in a 3-yr time period. If the EPA desires instead to retain the 3-yr requirement, then it should redefine the design value to 0.5 exceedances per year.[5]

As indicated earlier in this chapter, for a standard based on 1.0 expected exceedances per year, the probability of 6 or more exceedances will be $P = 0.0333$. For such a standard, there will be at least 95% assurance (actually, $P = 1.0000 - 0.0333 = 0.9667$, giving 96.67% assurance) that fewer than 6 exceedances will occur in any 3-year period.

Regulatory officials usually will seek to impose sufficient control on the sources of air pollution that noncompliance with the deterministic part of the standard (4 or more exceedances in 3 years) is very rare. If they achieve a noncompliance rate of 5% or less, the above analysis indicates that the expected number of exceedances will be much less than 1.0 per year. Consequently, the deterministic part of the ozone standard will produce a much lower number of expected exceedances per year than the EPA assumes. It is important to be aware of this inconsistency between the deterministic and statistical parts of the ozone standard. There is a need, in the development of future air quality standards, for the EPA to specify what the noncompliance rate is to be.

The analysis by Javits[5] applying the binomial distribution is quite impressive because it deals with two serious problems that normally interfere with the application of such models to environmental quality data: (a) nonstationarity of underlying statistical parameters, and (b) nonindependence of concentrations observed close to each other in time. For this particular case, it appears that the Bernoulli process model is relatively robust, or insensitive, to violations of either of these two fundamental assumptions. This robustness is partly a result of the fact that n is very large relative to the expected number of exceedances, causing the parameter p to be very small. As a consequence, much of the probability mass of this distribution lies very close to the origin, and the probability of even a moderate number of exceedances (say, 4 or more) includes so much of the probability mass that changes in the basic assumptions do not appreciably alter the result. Thus, these findings are partly a result of the extreme right-skewness of this distribution and its parameters.

Air pollution control programs in the U.S. rely on uniform ambient air quality standards, the NAAQS, which are set on a national basis. In contrast, water quality standards vary from state to state, and examples of statistical formats that have wide acceptance are difficult to find in the water pollution field. The concepts presented in this chapter will apply to any environmental standard that is based on counts of the number of violations or exceedances during a specified time interval, regardless of whether the concentrations are observed in an air, water, or soil carrier medium. The ozone NAAQS provides an especially good example, however, because its legal definition includes statistical language, the "expected number of exceedances per year," which allows the probability of an exceedance p to be specified exactly.

Bernoulli processes have environmental applications besides the implementation of air and water quality standards. The binomial distribution can be validly applied whenever the conditions for a Bernoulli process are met, and the total number of discrete independent events are being counted. Suppose, for example, that a pesticide is known to be present in the soil of a particular area, and that n soil samples are collected in the area. Assume, further, that the technique available for measuring the pesticide concentration in each sample is able to detect levels accurately only above some minimum detectable limit (MDL), and that all concentrations below this limit are reported as zero levels. When a given sample is above the MDL value, it is recorded as a "positive" reading. Suppose that the probability of a positive reading is known to be $p = 1/3$ for all samples. If we are required to design a sampling program in which one group of samples of size n is to be collected, how large should n be to achieve 95% assurance that a positive reading will not be incorrectly overlooked?

Let K be a random variable representing the number of positive readings found in a sample of size n. Assuming that the observations are independent and can be treated as a Bernoulli process, the probability distribution for the number of positives will be $\mathbf{B}(n,1/3)$. If only two samples are collected, the resulting distribution $\mathbf{B}(2,1/3)$ will yield the following probabilities for the number of positive readings: $P\{K = 0\} = 0.444$, $P\{K = 1\} = 0.444$, and $P\{K = 2\} = 0.111$. If three samples are collected, the probability distribution will be $\mathbf{B}(3,1/3)$, and the probability of zero positive readings will be $P\{K = 0\} = 0.276$. Thus, with just two samples, there will be a 44% chance that no positives will be found, and, for three samples, there will be a 28% chance that no positives will be found. To achieve 95% assurance that a positive will not be overlooked, n must be large enough to make $P\{K = 0\} \leq 0.05$. The probability $P\{K = 0\}$ is the first term in any binomial distribution:

$$P\{K = 0\} = \binom{n}{0} p^0 (1-p)^{n-0} = (1-p)^n$$

The critical value of n can be found by setting $P\{K = 0\} = (1 - p)^n = 0.05$, and solving for n. Taking natural logarithms, the following result is obtained:

$$n = \frac{\ln(0.05)}{\ln(1-p)} = \frac{\ln(0.05)}{\ln(2/3)} = 7.39$$

Thus, n must be larger than 7.38 to obtain the desired level of assurance, so we choose $n = 8$. The resulting probability distribution $\mathbf{B}(8,1/3)$ is right-skewed (Figure 4.9), and the probability that no positive reading appears in a group of 8 samples is $P\{K = 0\} = 0.039$, which meets the desired criterion (0.05 or less).

Of course, in actual measurement situations, the investigator usually does not have the value of p readily available. However, it may be possible to estimate p from historical records or from past experience. Possibly, p can be estimated from a knowledge of the physical setting itself; for example, information about the amount of pesticide applied to the area and the mixing volume of the soil. The value of p also can be estimated directly from measurements by collecting groups of samples of size n and by finding the value of p in the probability model $\mathbf{B}(n,p)$ which generates a distribution that best fits the histograms of the observations. Detailed discussion of such estimation procedures is beyond the scope of this book.

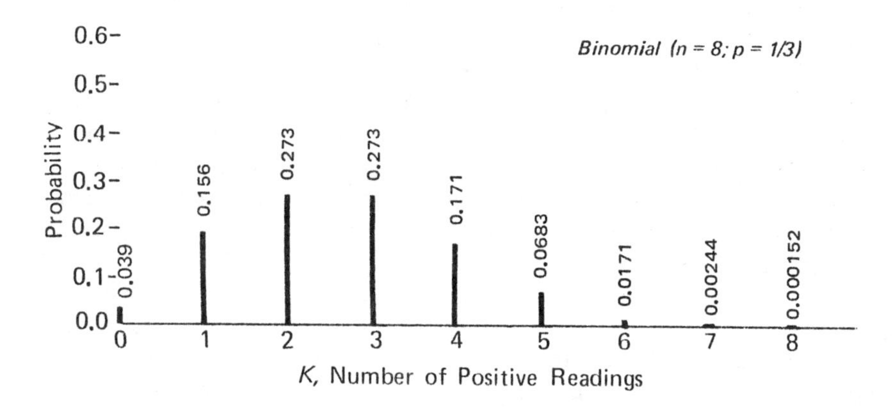

Figure 4.9. Probability distribution for the number of positive readings of a pesticide (values above the MDL) if 8 soil samples are collected and the probability of a positive reading in each sample is $p = 1/3$.

Another application of Bernoulli processes to environmental problems is in counting the number of particles that arrive at a given sampling point when a "puff" of particles has been released at some source. In such applications, the particles are assumed not to interact in any fashion, and arrivals of particles are assumed to be independent and to have the same probability. Once a particle arrives, it is counted once and absorbed, so it cannot be counted again. If $n = 1000$ particles have been released, and if the arrival probability of each particle at an observation location is known to be $p = 1/100$, then the probability distribution for the number of arrivals, after all particles have reached their destinations and stopped moving, will be **B**(1000,1/100). Arrival phenomena usually are treated as Poisson processes (Chapter 5), but they also can be formulated as a Bernoulli process for this particular set of conditions. Particle arrival phenomena are of interest in studies of the underlying mechanisms that govern the diffusion and transport of pollutants in the environment.[6]

COMPUTATION OF B(n,p)

Because of the simplicity of the basic equation of the binomial distribution, the probabilities for **B**(n,p) can be computed with relative ease for most applications. Hand calculators, such as the various models produced by Hewlett-Packard, contain built-in functions that can raise the parameter p to any power 1,2, ... ,n. Many have a built-in factorial function that can be used to compute $n!$, $k!$, and $(n-k)!$ in the combinatorial coefficient, but each of these values can become very large, and care must be taken to see that they do not overflow the capacity of the calculator. One way to avoid numerical overflow is to avoid the built-in factorial function and to divide the numerator and denominator by as many common factors as possible prior to computation. Although the PMF can be computed with relative ease using such a calculator, evaluation of the CDF requires summation of the individual PMF terms, and it may be advisable to compute the CDF by writing a program that adds up successive PMF terms on a programmable calculator. One convenient way to compute the combinatorial coefficient in each successive PMF term is by the following algorithm: (a) increment k by 1; (b) decrement n by 1; (c) compute the new ratio n/k and multiply the combinatorial coefficient by this ratio. The modified combinatorial coefficient is stored each time a PMF term is computed so the process can be repeated when the next term is computed. Each PMF term is computed first by multiplying the combinatorial coefficient by p^n and then by $(1-p)^{n-k}$. As each PMF term is computed, it is added to the cumulative sum of the previous PMF terms in order to form the CDF. If a printer is available, a useful table can be generated by listing k, the PMF, and the resulting CDF side-by-side. If all n terms comprising the distribution are generated, the CDF values ultimately will approach 1.0000... exactly, allowing for roundoff error in the right-most digits.

A practical BASIC computer program for computing terms of the binomial probability model on a person computer such as the IBM-PC* is given (Figure 4.10). It can generate tables similar to Table 4.3, but all computations are in dou-

*IBM is a registered trademark of International Business Machines, Inc.

```
100 CLS : PRINT : PRINT : PRINT
110 PRINT "PROGRAM TO COMPUTE TERMS OF BINOMIAL PROBABILITY MASS FUNCTION"
120 PRINT "          (ALL VALUES ARE COMPUTED IN DOUBLE PRECISION)"
130 PRINT
140 DEFDBL A-Z
150 INPUT "enter p,n"; P, N
160 PRINT "probability p =", P
170 PRINT "and n =", N
180 L = N + 1
190 PRINT "the total number of terms possible = "; L
200 PRINT "how many terms should be computed?"
210 INPUT "j ="; J
220 FACTOR = 1
230 M = N
240 Q = 1 - P
250 REM k counts main computation loop; stops when k=j
260 K = 0
270 IF K = J THEN GOTO 420
280 REM if k=0, p=q^n to avoid division by zero
290 IF K = 0 THEN GOTO 370
300 R = M / K
310 M = M - 1
320 PROB = P ^ K
330 QROB = Q ^ M
340 FACTOR = FACTOR * R
350 VALUE = FACTOR * PROB * QROB
360 GOTO 380
370 VALUE = Q ^ N
380 SUM = SUM + VALUE
390 PRINT K, VALUE, SUM
400 K = K + 1
410 GOTO 270
420 END
```

Figure 4.10. BASIC computer program for computing the PMF and the CDF for the binomial probability model.

ble-precision arithmetic to assure greater accuracy as the CDF approaches 1. The program asks the user to specify p and n, and, because not all the terms of the distribution may be required, the program also asks the user to specify how many terms are to be generated. The program lists the values for K alongside the PMF and CDF on the computer's monitor screen, beginning with values of $K = 0,1,\ldots$. If these values are to be printed out on paper, the user should replace the statement "PRINT" with "LPRINT" in lines #160, #170, and #390 of the program.

PROBLEMS

1. Consider a Bernoulli process consisting of three independent trials. If the probability of a successful outcome on any trial is $p = 1/4$, what is the probability associated with either 0, 1, 2, or 3 successful outcomes? [Answers: 0.421875, 0.421875, 0.140625, 0.015625] Identify this distribution and plot its PMF and CDF. [Answer: **B**(3, 1/4)]
2. Consider the probability distribution **B**(9,1/9) for the number of persons engaged in smoking in a 9-person office (Figure 4.4). Calculate the mean, variance, and coefficients of skewness and kurtosis for this distribution. [Answers: 1, 8/9, $7/\sqrt{72}$, 249/72]
3. Consider the probability distribution **B**(9,1/2) that is similar to the one above but is symmetrical (Figure 4.5). Calculate the mean, variance, and coefficients to skewness and kurtosis for this distribution. [Answers: 9/2, 9/4, 0, 249/72] Compare these results with the ones calculated above and discuss qualitatively the reasons for the similarities and differences.
4. Consider the probability distribution **B**(9,8/9) that is similar to the one in Problem 2 but has its axis reversed, thus forming a mirror image (Figure 4.6). Calculate the mean, variance, and coefficients of skewness and kurtosis for this distribution. [Answers: 8, 8/9, $-7/\sqrt{72}$, 249/72]. Compare these results with those calculated for Problems 2 and 3 and discuss qualitatively the reasons for the similarities and differences.
5. Suppose that the statistical part of the ozone air quality standard is modified so that the expected number of exceedances per year is 0.5 or less. If M denotes the number of exceedances in any 3-year period, what is the probability distribution of M when the standard is just attained and $E[M] = 0.5$. [Answer: **B**(1095,1/730)] Compute the probabilities associated with $M = 0$, 1, 2, or 3. [Answers: 0.2229, 0.3348, 0.2512, 0.1256]. What is the probability that 3 or fewer exceedances will occur in any 3-year period. [Answer: 0.9345] Discuss why this result implies there will be at least a 90% compliance rate with the deterministic part of the ozone standard when the expected number of exceedances per year is 0.5 or less.
6. Suppose it is desired to achieve a compliance rate with the deterministic part of the ozone air quality standard (average number of exceedances in any 3-year period of 1.0 or less) of at least 95%. If the statistical part of the standard is set to $E[M] = 0.4$ expected exceedances per year, compute the compliance rate and show that it meets the desired goal. [Answer: 96.7%]

7. Javits[5] examines the robustness of the Bernoulli process model to serial dependencies by examining two cases of extreme dependence: linked groups of two days (doublets) and linked groups of three days (triplets). Consider another case of extreme dependence: consecutive, nonoverlapping groups of four days (quadruplets), all of which either exceed or fail to exceed the specified concentration level. If the expected number of exceedances is 1.0 per year, what is the probability distribution for the number of quadruplets in (a) a one-year period, and (b) a three-year period [Answers: **B**(91,1/365), **B**(273,1/365)]. Show that the chance of 4 or more exceedances will be 53.7% for this very extreme case of dependence, compared to 35.3% for the case of independent observations.
8. Suppose an ambient air quality standard is promulgated that states that the expected number of exceedances per year is to be 5.0 or less. If the expected number of exceedances is exactly 5.0 and a Bernoulli process is assumed, what will be the probability distribution of the number of exceedances in (a) any one-year period, and (b) any three-year period? [Answers: **B**(365,5/365), **B**(1095, 5/365)]
9. In a given 10-day period, the probability of violating a water quality standard at a particular location on a stream is assumed to be the same each day, and violations are assumed to be independent. If one sample is collected each day over a 10-day period and $p = 1/3$, what is the probability that (a) the water quality standard will be violated at least once, (b) that it will be violated 2 or more times, and (c) that it will be violated 5 or more times? [Answers: 0.9827, 0.8960, 0.2131] Plot the PMF and CDF for this distribution.
10. A survey must be designed to measure the concentration of a toxic organic compound in the breath of human subjects. If past experience indicates that a person's breath concentration of the compound will be above a threshold concentration with probability p (or greater) if the area is contaminated, show that the following equation gives the minimum number of samples that must be collected to have 99% assurance that a reading above the threshold is not accidentally overlooked:

$$n_{min} = \frac{\ln(0.01)}{\ln(1-p)}$$

If $p = 1/3$, what is n_{min}, and what will be the minimum number of people to be sampled and the probability distribution of the number of people whose breath will be above the threshold level? [Answers: 11.36, 12, **B**(12,1/3)]

REFERENCES

1. "National Primary and Secondary Ambient Air Quality Standards," *Federal Register*, 36:8186 (April 30, 1971).
2. "National Primary and Secondary Ambient Air Quality Standards," *Federal Register*, 44:8202 (February 8, 1979).

3. Curran, Thomas C., and William M. Cox, "Data Analysis Procedures for the Ozone NAAQS Statistical Format," *J. Air Poll. Control Assoc.* 29(5):532–534 (May 1979).
4. U.S. Environmental Protection Agency, "Guidelines for the Interpretation of Ozone Air Quality Standards," Report No. EPA-450/4-79-003, OAQPS No. 1.2-108, Office of Air Quality Planning and Standards, Research Triangle Park, NC (January 1979).
5. Javits, Jarold S. "Statistical Interdependencies in the Ozone National Ambient Air Quality Standard," *J. Air Poll. Control Assoc.* 30(1):58–59 (January 1980).
6. Ott, Wayne, "A Brownian Motion Model of Pollutant Concentration Distributions," SIMS Technical Report No. 46, Department of Statistics, Stanford University, Stanford, CA (April 1981).

5 Poisson Processes

In the previous chapter on Bernoulli processes, the random variable of interest was the total number of successful outcomes observed when a specified number of trials took place. Because the two parameters of the binomial distribution, n and p, were assumed constant, the process was stationary. No explicit assumptions were made about the rate at which the trials occurred, and time entered only indirectly as the period over which the trials obviously must have taken place. Thus, if n Bernoulli trials took place over some elapsed time period, it made no difference whether they all occurred at the beginning of the period, the end of the period, or were equally spaced over the period.

Now consider a process which operates continuously in time but does not involve independent trials. Here, the random variable of interest is the number of events occurring over some observation time period of interest. In general, the longer the observation period, the more events will be expected. Because of the importance of time in this process, each occurrence of an event usually is described as an "arrival." Thus, interest focuses on the total number of arrivals occurring during the observation time period. If the observation period becomes extremely long, the number of arrivals divided by the observation time period will approach a single parameter, called the "arrival rate." Indeed, the arrival rate is the only parameter describing this process, and it fully characterizes the process. The arrival rate is assumed to remain constant with time, and thus the process is stationary. Because the process is stationary, it makes no difference when the beginning of the observation period is chosen; only the duration of the observation period affects the number of arrivals observed. Such a process is called a *Poisson process.*

Examples of real-world phenomena that can be modeled as Poisson processes involve discrete events that are counted over some time period of interest:

1. On a given day, the number of vehicles arriving at a particular intersection is assumed to be governed by a fixed rate of 40 vehicles per hour. What is the probability distribution for the number of vehicles that arrive in any 3-minute period? What is the probability that more than 5 vehicles will arrive in any 3-minute period?
2. An hourglass passes sand at a rate of 1 grain per second. In any 3-second period, what is the probability distribution of the number of grains of sand that will pass?
3. When an effluent is discharged continuously into a stream, past experience indicates that a water quality standard is violated downstream at a rate of 3 violations per day. In any given 24-hour

period, what is the probability of observing (a) no violations, (b) 2 or more violations, (c) 5 or more violations at the downstream location?

4. If the average exceedance rate of an ambient air quality standard is 1.0 exceedance per year, what is the probability of 4 or more exceedances in any 3-year period?
5. At a particular point in space, particles of a tracer pollutant arrive at a constant average rate of 2 particles per minute. In any 5-minute period, what is the probability that (a) no particles arrive, (b) 2 or fewer particles arrive, (c) 5 or more particles arrive?

In all of the above examples, the number of arrivals during the observation period is assumed to be independent of the number of arrivals prior to the observation period. That is, the process has no memory. The average arrival rate is a property of the process itself, and thus it is independent of the point in time at which the observation period begins. Surprisingly, these assumptions are sufficient to fully specify a particular probability distribution for this process, the Poisson distribution.

CONDITIONS FOR POISSON PROCESS

The above problem can be formulated using the following common assumptions: (a) the quantity of interest is the number of events (arrivals) occurring during an observation period; (b) the number of arrivals during the observation period is independent of the *number of arrivals* prior to the observation period; (c) for any observation period, the distribution of the number of arrivals will be independent of the *time* at which the observation period begins. In summary, ...

> A Poisson process describes the total number of independent events occurring during a specified observation period in which the event arrival rate is fixed.

Let N_t denote the count of the number of arrivals occurring during the time period $(0,t)$. If a graph is drawn of N_t, each new arrival increments the total count by a unit "jump" (Figure 5.1). If the observation period is of duration s, we are interested in the probability distribution of the random variable $N = N_{t+s} - N_t$. The conditions for the Poisson process are summarized mathematically as follows.

A counting process $N = \{N_t; t \geq 0\}$ is Poisson if the following conditions are met: (a) each jump in N_t is of unit magnitude; (b) for any $t,s \geq 0$, $N = N_{t+s} - N_t$ is independent of past arrivals* $\{N_u; u \leq t\}$; for any $t,s \geq 0$, the distribution of $N = N_{t+s} - N_t$ is independent of t. The first condition implies that two arrivals cannot occur at exactly the same time, although a pair of arrivals can occur very close together in time. The second condition implies that the process is

*Here N_u represents the count of all past arrivals for the interval $0 < u \leq t$.

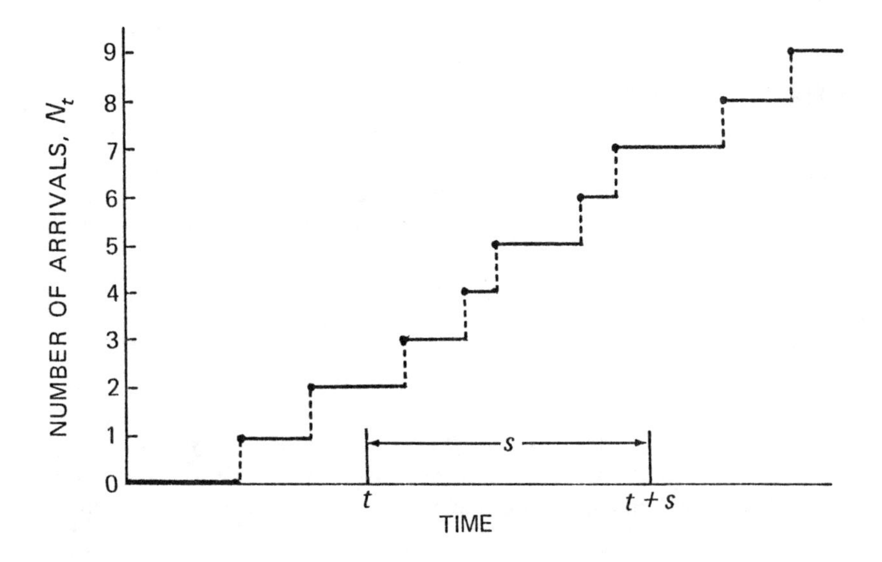

Figure 5.1. Each arrival in a Poisson process causes a "jump" of unit magnitude in the count of arrivals. An observation period of duration *s* is shown beginning at time *t*.

memoryless; the number of arrivals occurring in the past does not affect the number of arrivals occurring during the observation period. The third assumption indicates that the process is stationary and implies that a single probability distribution can be found that meets the first two conditions. If such a probability model exists, what form will it have?

DEVELOPMENT OF MODEL

In Condition (b) above, the number of arrivals in the observation time period of interest $(t, t+s)$ is assumed to be independent of the number of arrivals occurring prior to time t. Thus, if a particular number of arrivals occurred in the interval $(0,t)$; say, zero arrivals, it would not affect the number of arrivals occurring in the observation period $(t, t+s)$. Consider the special case in which zero arrivals occur in both the interval $(0,t)$ and $(t, t+s)$. Because the overall interval $(0, t+s)$ is made up of these two adjacent intervals, then zero arrivals also must occur in the overall interval. By the product law for independent events, the probability of zero arrivals in the overall interval must equal the product of the probabilities for each of the two sub-intervals:

$$P\{N_{t+s} = 0\} = P\{N_t = 0\}P\{N_{t+s} - N_t = 0\}$$

By Condition (c), which requires stationarity, the probability of zero arrivals in the interval $(0,s)$ is the same regardless of where the interval is translated in time $(t,t+s)$:

$$P\{N_s = 0\} = P\{N_{t+s} - N_t = 0\}$$

Substituting this equation into the one above, we have:

$$P\{N_{t+s} = 0\} = P\{N_t = 0\}P\{N_s = 0\}$$

To meet these conditions, we must find a single probability model in which *multiplication* of the respective probabilities for two adjacent time segments causes the time to be incremented *additively*. That is, we seek the function $f(t) = P\{N_t = 0\}$ that satisfies the following equations:

$$f(t)f(s) = f(s+t); \quad 0 \leq f(t) \leq 1$$

The exponential function has this "independence-time additivity" property, because the product of two exponential functions with s and t and the positive constant λ in their exponents is a third exponential with the sum $s + t$ contained in its exponent:

$$e^{-\lambda t}e^{-\lambda s} = e^{-\lambda(t+s)}$$

Further, the exponential is the only function with this property. Thus, the probability that there will be zero arrivals in any observation period of duration s is given by the following probability model:

$$P\{N_{s+t} - N_t = 0\} = e^{-\lambda s}$$

As s becomes large, the probability that there will be zero arrivals gradually approaches zero (Figure 5.2). Conversely, the probability of *at least one* arrival

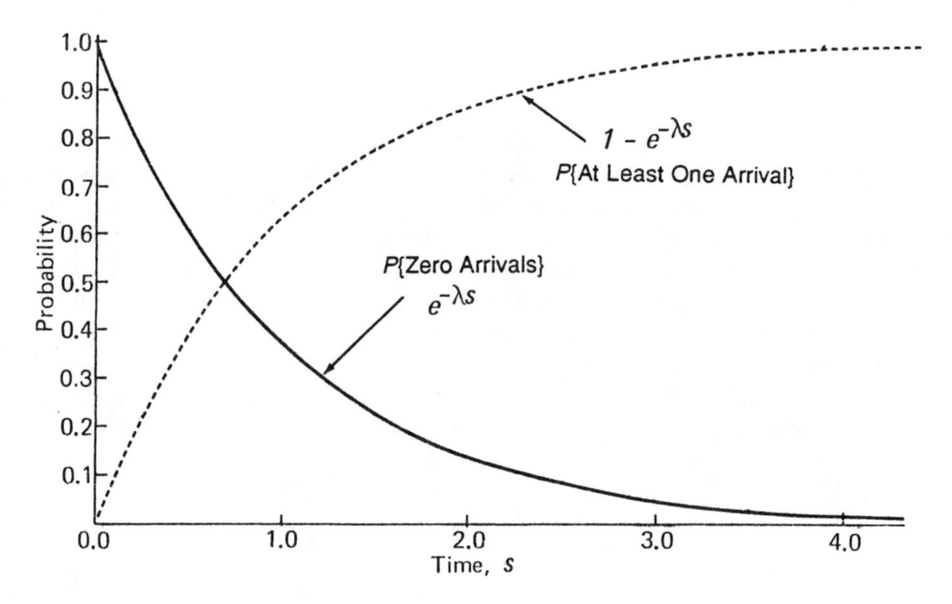

Figure 5.2. Probability of zero arrivals in a Poisson process during an observation period of duration s, where $\lambda = 1$.

gradually approaches 1 as s becomes extremely large, as indicated by the following equation:

$$P\{N_{s+t} - N_t \geq 1\} = 1 - e^{-\lambda s}$$

The exponential probability model arises naturally from the above set of conditions. However, it applies only to "zero arrivals," and not to any arbitrary number of arrivals (say, 2, 3, or 4 arrivals) during the observation period. To calculate the probabilities for these cases, we must find the probability model for the integer random variable $N_{t+s} - N_t$ that naturally arises from the stated conditions.

The correct probability model, if it exists, must meet the requirement that the number of arrivals in two adjacent segments combine as a sum; that is, $N_{t+s} = N_t + (N_{t+s} - N_t)$. It also must meet the independence-time additivity requirement that multiplication of the probabilities for two adjacent time segments equals the probability for the addition of times. By using probability generating functions, which are beyond the scope of this book, Cinlar[1] shows that the following probability model meets these requirements, in which λ is a positive constant and $N_{t+s} - N_t$ is the number of arrivals in the observation period $(t,t+s)$:

$$P\{N_{t+s} - N_t\} = \frac{e^{-\lambda s}(\lambda s)^n}{n!}, \quad n = 0,1, \ldots$$

This result is known as the *Poisson distribution* (Table 5.1.).

Usually, it is difficult to apply the above conditions (a), (b), and (c) to real-

Table 5.1. The Poisson Distribution **P**(λt)

Probability Mass Function (PMF):

$$P_{N_t}\{n\} = \frac{e^{-\lambda t}(\lambda t)^n}{n!}, \quad n = 0,1,2,\ldots; \quad \lambda > 0, \ t \geq 0$$

Expected Value:

$$E[N_t] = \lambda t$$

Variance:

$$Var(N_t) = \lambda t$$

Coefficient of Skewness:

$$\sqrt{\beta_1} = \frac{1}{\sqrt{\lambda t}}$$

Coefficient of Kurtosis:

$$\beta_2 = \frac{1 + 3\lambda t}{\lambda t}$$

world problems to determine whether an actual process can be treated as Poisson or not. Fortunately, these conditions imply a set of auxiliary conditions which are more useful for practical purposes. Although no particular interpretations of the constant λ has been discussed, it is actually the "arrival rate" of the process (number of arrivals per unit time). The stationarity condition (c) implies that the arrival rate λ is constant. If a particular process is observed for some time period s, then the average arrival rate $(N_{t+s} - N_t)/s$ will approach λ as s becomes large. The expected number of arrivals is proportional to the time duration of the observation period:

$$E[N_{t+s} - N_t] = \lambda s$$

The practical conditions for a Poisson process are summarized formally as follows:

(a) Each jump is of unit magnitude.

(b) For any $t,s \geq 0$, $\quad E[N_{t+s} - N_t | N_u; u \leq t] = \lambda s$

The first condition requires, as before, that two or more arrivals cannot occur at exactly the same time. The second condition requires that the expected number of arrivals be equal to the product of a constant arrival rate and the duration of the observation period. The second condition holds regardless of the point in time at which the observation period begins; that is, regardless of the number of arrivals N_u at any prior point in time, $u \leq t$.

Because this process is independent of the time chosen to begin counting arrivals, it makes no difference whether the observation period begins at the origin $t = 0$ or at any other time t. The important consideration is the duration s of the observation period. Thus, the observation period can begin at the origin, as interval $(0,s)$, or it can begin at any arbitrary time t, as interval $(t,t+s)$.

To simplify the notation used in the remainder of this chapter, we substitute t for s. Thus, we consider a Poisson process beginning at the origin and extending over an observation period of duration t, as interval $(0,t)$.

POISSON DISTRIBUTION

The above discussion presents theoretical conditions for a Poisson distribution to arise naturally in a real physical process. The Poisson probability model satisfies these conditions for any time $t \geq 0$ in which $N_t = \{0, 1, 2, \ldots\}$ is the number of arrivals in the time segment $(0,t)$. The Poisson probability mass function has just one parameter, the arrival rate λ (Table 5.1).

If the terms of the Poisson probability mass function are added together, the following series results:

$$\sum_{n=0}^{\infty} \frac{e^{-\lambda t}(\lambda t)^n}{n!} = e^{-\lambda t}\left(1 + \frac{\lambda t}{1!} + \frac{(\lambda t)^2}{2!} + \frac{(\lambda t)^3}{3!} + \ldots\right)$$

The expansion in parentheses on the right-hand side of the equation is the Taylor series for $e^{\lambda t}$. Thus, the result of summing all the terms in the Poisson distribution is $e^{-\lambda t} e^{\lambda t} = e^0 = 1$, a required condition for any probability distribution.

The expected value of the distribution is computed by multiplying every term in the distribution by n:

$$E[N_t] = \sum_{n=0}^{\infty} n \frac{e^{-\lambda t}(\lambda t)^n}{n!} = e^{-\lambda t}\left(0 + \frac{\lambda t}{1!} + \frac{2(\lambda t)^2}{2!} + \frac{3(\lambda t)^3}{3!} + \frac{4(\lambda t)^4}{4!} + \ldots\right)$$

$$= e^{-\lambda t}(\lambda t)\left(1 + \frac{\lambda t}{1!} + \frac{(\lambda t)^2}{2!} + \frac{(\lambda t)^3}{3!} + \ldots\right) = \lambda t$$

A computation similar to this one shows that the variance of the Poisson probability model is equal to its mean, $Var(N_t) = \lambda t$.

The coefficient of skewness $\sqrt{\beta_1}$ contains λt in the denominator and therefore, $\sqrt{\beta_1}$ cannot be negative. Thus, the Poisson distribution always is right-skewed. If λt increases without limit, then $\sqrt{\beta_1}$ ultimately will approach zero, and, like most distributions of this kind, the Poisson distribution will become increasingly symmetrical (and normal) in appearance.

Notice that the two parameters λ and t in the Poisson probability model may be written as just one parameter $\rho = \lambda t$, allowing the probability mass function to be expressed in the following compact notation:

$$P_N\{n\} = \frac{e^{-\rho}\rho^n}{n!}, \quad n = 0, 1, 2, \ldots; \; \rho \geq 0$$

This compact form of the Poisson model is useful for computational purposes. In most practical applications, the more general form is used, because λ usually is treated as a fundamental constant of the process, and one is interested in the distributions that arise for different time periods t. In general, the notation $\mathbf{P}(\rho)$ denotes a random variable with a Poisson distribution and parameter $\rho = \lambda t$.

EXAMPLES

The most typical application of the Poisson process is to queues of persons waiting to be served. For example, past experience might indicate that people arrive at a bank teller's window with an arrival rate of λ persons per hour that is approximately constant during the morning hours. For some practical problem of interest to management, it may be important to compute the probability distribution of the number of persons arriving in any 5-minute period, and a Poisson model may be assumed for this purpose. Such computations may be helpful, for example, in determining how many tellers should be available to avoid long lines of customers at any window.

An analogous process is the number of vehicles arriving at a traffic intersection. Suppose that traffic counts on one of the streets indicates that the arrival rate on Friday afternoons between 1:00 pm and 4:00 pm is approximately constant at 40 vehicles per hour. This conclusion might be based on records of hourly traffic counts on this roadway on many previous Fridays. If this process is assumed to be Poisson, what is the probability distribution for the number of vehicles that arrive between 2:00 pm and 2:01 pm?

For this case, $\lambda = 40/60 = 2/3$ vehicles per minute, giving $\rho = (2/3)(1 \text{ minute}) = 2/3$, giving the Poisson probability mass function **P**(2/3):

$$P\{N_{2:01} - N_{2:00} = n\} = \frac{e^{-2/3}(2/3)^n}{n!}$$

Similarly, the probability distribution for the number of vehicles arriving between 2:01 pm and 2:04 pm is obtained for the parameter $\rho = (2/3)(3 \text{ minutes}) = 2$, giving the Poisson distribution **P**(2):

$$P\{N_{2:04} - N_{2:01} = n\} = \frac{e^{-2}(2)^n}{n!}$$

Finally, consider the number of vehicles arriving between 2:00 pm and 2:04 pm, which gives the parameter $\rho = (2/3)(4 \text{ minutes}) = 8/3$ and the probability distribution **P**(8/3):

$$P\{N_{2:04} - N_{2:00} = n\} = \frac{e^{-8/3}(8/3)^n}{n!}$$

Because the above distribution contains the two time segments from 2:00 pm to 2:01 pm and from 2:01 pm to 2:04 pm, its parameter, $\rho = 8/3$, will be the sum of individual parameters for the two segments; that is, $\rho = 2/3 + 2 = 8/3$.

This result illustrates the intrinsic independence-time additivity property of the Poisson process, a characteristic discussed above in development of the model. This important property can be illustrated graphically (Figure 5.3). Consider two adjacent observation time intervals (t_1, t_2) and (t_2, t_3). Each interval is assumed to have the same arrival rate λ. Let the time duration of the first segment be represented by $a = t_2 - t_1$; then the probability distribution of the num-

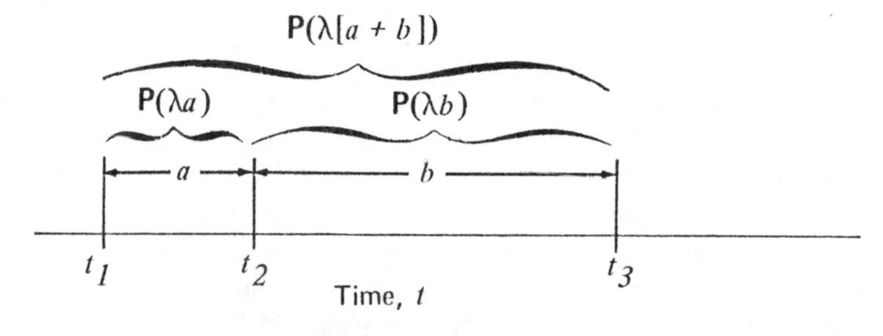

Figure 5.3. Illustration of the independence-time additivity property of the Poisson process. If the probability distribution of arrivals for time segment (t_1, t_2) is **P**(λa) and for time segment (t_2, t_3) is **P**(λb), then the probability distribution of arrivals for the entire segment (t_1, t_3) will be **P**($\lambda[a + b]$).

ber of arrivals in this interval will be Poisson $\mathbf{P}(\lambda a)$. Similarly, let the time duration of the second segment be represented by $b = t_3 - t_2$; then the probability distribution of the number of arrivals during the second segment will be Poisson $\mathbf{P}(\lambda b)$. By the independence-time additivity property, the probability distribution for the number of arrivals during the total time segment (t_1,t_3) will be Poisson with its parameter equal to the sum of the two individual parameters; that is, $(\lambda a + \lambda b) = (\lambda[a + b])$.

The notation N_t was used above to emphasize that the random variable of interest in a Poisson process is the number of arrivals over some observation time period $(0,t)$. In the following examples, we simplify the notation by dropping the subscript t, using N to denote the number of arrivals over the observation period $(0,t)$.

Many common physical processes can be modeled as Poisson. Consider, for example, a carefully constructed hourglass in which sand particles pass from the top compartment to the bottom compartment through an extremely small opening. Suppose one is interested in the count of the number of particles arriving in the bottom compartment of the hourglass. In an actual hourglass, the arrival rate may change slightly as the top compartment becomes empty, but we shall assume that the arrival rate is constant. If the arrival rate is 1 grain per second, what is the probability distribution for the number of grains that arrive in any 1-second period?

The resulting probability distribution will be Poisson with $\rho = \lambda t =$ (1 arrival/second)(1 second) = 1 arrival, or $\mathbf{P}(1)$:

$$P\{N = n\} = \frac{e^{-1}(1)^n}{n!} = \frac{e^{-1}}{n!}$$

Using this PMF to compute probabilities for different cases, the probabilities of zero arrivals and of one arrival are obtained as follows:

$$P\{N = 0\} = \frac{e^{-1}}{0!} = 0.36788$$

$$P\{N = 1\} = \frac{e^{-1}}{1!} = 0.36788$$

Notice that the probabilities for these two cases, $N = 0$ and $N = 1$, are the same and that the sum is 0.73576, indicating that approximately 73.6% of the probability mass is contained in the first two terms of this distribution. As a consequence, the Poisson distribution $\mathbf{P}(1)$ is extremely right-skewed (Figure 5.4).

Suppose a 3-second period is considered; what will be the probability that no grains of sand arrive in the bottom compartment of the hourglass? What will be the probability that just one grain of sand arrives in any 3-second period? For a 3-second period, the resulting distribution will be Poisson with $\rho = \lambda t =$ (1 arrival/second)(3 seconds) = 3 arrivals, or $\mathbf{P}(3)$:

$$P\{N = n\} = \frac{e^{-3}(3)^n}{n!}$$

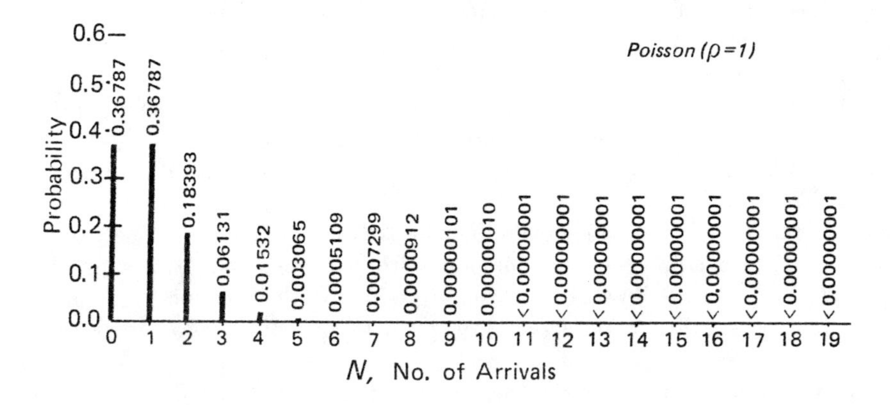

Figure 5.4. Probability distribution showing the first 19 terms of the Poisson probability model **P**(1).

Using this PMF, the probability of zero arrivals and of one arrival are computed as follows:

$$P\{N = 0\} = \frac{e^{-3}(3)^0}{0!} = 0.04979$$

$$P\{N = 1\} = \frac{e^{-3}(3)}{1!} = 0.14936$$

The resulting distribution also is right-skewed (Figure 5.5), but the right-skewness is less pronounced than for the case **P**(1) plotted in Figure 5.4.

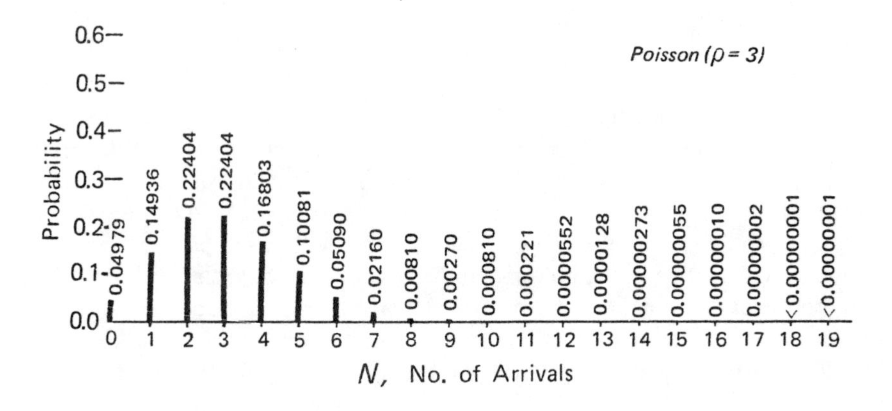

Figure 5.5. Probability distribution showing the first 19 terms of the Poisson probability model **P**(3).

APPLICATIONS TO ENVIRONMENTAL PROBLEMS

The problem of computing probabilities associated with the number of times a concentration limit is exceeded (or an environmental standard is violated) is sufficiently important that it deserves detailed consideration. Under certain conditions, it is possible to express such problems in terms of an average exceedance rate of r exceedances (or violations) per unit time. In the environmental application of Bernoulli processes (Chapter 4), the binomial probability parameter was set equal to the exceedance rate; that is, $p = r$. In similar environmental applications of Poisson processes, the Poisson arrival rate is set equal to the average exceedance rate; that is, $\lambda = r$. Using the same notation as before without the subscript t, we let K be a random variable denoting the number of exceedances, and the Poisson PMF is written as follows:

$$P\{K = k\} = \frac{e^{-rt}(rt)^k}{k!}$$

Here t will be the duration of the observation period over which the exceedances are counted. In the Bernoulli processes, t did not appear explicitly, but it was assumed that the exceedances occurred during some finite time period.

Suppose that a water pollutant is discharged into a stream continuously and that the number of violations per unit time of a water quality standard can be treated as approximately constant (analogous to a constant arrival rate). If, for example, violations occur at a rate of 1 violation per week, then the probability distribution of the number of violations occurring in any 1-week period will be Poisson $\mathbf{P}(1)$. If 3-week periods are considered, then the probability distribution will be Poisson $\mathbf{P}(3)$. These probability distributions were just plotted (Figures 5.4 and 5.5). It is remarkable that such a descriptive distribution, which can embrace so many different situations, can be developed from fairly simple assumptions of a constant exceedance rate and independence of events.

Probability Distribution for the Number of Exceedances

In Chapter 4, the number of daily exceedances of the ozone National Ambient Air Quality Standard (NAAQS) was modeled as a Bernoulli process. It is instructive to consider the same problem, now treating it as a Poisson process. This approach was suggested by Javits.[2]

As before, let K denote the number of exceedances in a 1-year period and M denote the number of exceedances in any 3-year period. Consider the critical point at which air quality just meets (but does not exceed) the air quality standard. When this borderline situation occurs, the following conditions are met:

$$E[K] = 1.0 \quad \text{and} \quad E[M] = 3E[K] = 3.0$$

Because there are 365 days in a year (ignoring leap year), we compute the exceedance rate by setting $E[K] = 1.0 = rt = (r)(365 \text{ days})$. Solving for r, we obtain $r = 1/365$ exceedances per day. Substituting this value into the Poisson environmental model given above, we obtain the following PMF:

$$P\{K=k\} = \frac{e^{-t/365}(t/365)^k}{k!}, \quad k=0,1,2, \ldots$$

If one is interested in the probability distribution for 1-year periods, then substituting t = 365 days into the above relationship gives the Poisson distribution **P**(1). To calculate the probability distribution for 3-year periods, we substitute t = (3 years)(365 days per year) = 1095 days into this equation, obtaining the Poisson distribution **P**(3). Again, these happen to be the same probability distributions that were just discussed and plotted (Figures 5.4 and 5.5).

Finally, to compare probabilities obtained from the binomial and Poisson distributions for this example, probabilities for values of K and M ranging from 0 to 12 have been computed and listed (Table 5.2) in exactly the same tabular format as in Chapter 4 (Table 4.3, page 101). The first two columns give the PMF $P_K\{k\} = P\{K = k\}$ and the CDF $F_K(k) = P\{K \leq k\}$ for 1-year periods, or the distribution **P**(1). The second two columns give the corresponding PMF and CDF values for 3-year periods, or the distribution **P**(3). Comparison of this table with the earlier one reveals that the Poisson and binomial distributions give almost exactly the same results in this problem. For example, the CDF $F_K(1) = P\{K \leq 1\} = 0.73576$ in this table, which is identical to the result listed in the earlier table based on the binomial distribution. Thus, the probability of "more than one exceedance" in a given year will be the same as before by both models: $P\{K > 1\} = 1 - P\{K \leq 1\} = 1 - F_K(1) = 1 - 0.73576 = 0.26424$. If these tables were computed to more than five places, more differences would be evident, but the overall agreement between the two tables is fairly good. The probability of 0 exceedances in a 3-year period is $P_M\{0\} = 0.04979$ by the Poisson distribution and $P_M\{0\} = 0.04958$ by the binomial distribution. If these two values are rounded off to 2 significant figures, both answers will be the same,

Table 5.2. Probability Distribution for the Number of Exceedances, Using the Poisson Model with the Expected Number of Exceedances of 1.0

	1-Year Period		3-Year Period	
No. of Exceedances k	Probability $P_K\{k\}$	Cumulative Probability $F_K(k)$	Probability $P_M\{k\}$	Cumulative Probability $F_M(k)$
0	0.36788	0.36788	0.04979	0.04979
1	0.36788	0.73576	0.14936	0.19915
2	0.18394	0.91970	0.22404	0.42319
3	0.06131	0.98101	0.22404	0.64723
4	0.01533	0.99634	0.16803	0.81526
5	0.00307	0.99941	0.10082	0.91608
6	0.000511	0.999917	0.05041	0.96649
7	0.0000728	0.9999898	0.02160	0.98810
8	0.0000091	0.9999989	0.00810	0.99620
9	0.0000010	0.9999999	0.00270	0.99890
10	0.00000010	0.9999999	0.00081	0.99971
11	0.00000001	1.0000000[a]	0.00022	0.99992
12	0.00000000	1.0000000[a]	0.00006	0.99998

[a]Rounded off.

0.050. If we consider the probability of "more than three exceedances" in a 3-year period, then $P\{M > 3\} = 1 - P\{K \le 3\} = 1 - F_M(3) = 1 - 0.64723 = 0.35277$, the same result—to 5 places—that was obtained from the binomial distribution in Chapter 4. Thus, if the Poisson model were used in Chapter 4 instead of the binomial model and the results were rounded to 3 places, the conclusions about the expected percentages would be the same: 26.4% of the 1-year periods would have more than 1 exceedance, and 35.3% of the 3-year periods would have more than 3 exceedances, even though the statistical goal of the standard—$E[K] = 1.0$—were met exactly.

These calculations illustrate the degree to which the Poisson model provides a good approximation of the binomial model for the case in which p is small and n is large. For values of K from 0 to 5 in this example, the PMF values for the two models will be the same if they are rounded to 2 places. The CDF, because it is a sum of many individual probabilities, shows greater agreement between the two models: for values of K from 1 to 12, the CDF values will be identical if they are rounded to 3 places. For values of M from 0 to 7 in this example, the PMFs will be the same if they are rounded to 2 places, and for values of M from 1 to 5, the PMFs will be the same if they are rounded to 3 places. In general, the results given in these two tables usually will match if they are rounded to 2 places and often will match if they are rounded to 3 places. A few of the entries are identical to 5 places. In this example, the binomial model for K is **B**(1/365,365) and the binomial model for M is **B**(1/365,1095). With such a small probability as $p = 1/365$, it is evident that we are modeling a "rare event," and the Poisson model is especially well-suited for modeling rare events. In addition, because each exceedance must occur on a different day, no two exceedances (i.e., Poisson arrivals) can occur at the same time, another physical property suggesting treatment of this problem as a Poisson process. Interestingly, one theoretical rationale suggests treating this problem as a Bernoulli process and another theoretical rationale suggests treating it as a Poisson process. The choice between the two rationales depends upon the perspective from which the problem is viewed. In the Bernoulli process, the problem is seen as consisting of 365 independent trials per year, with probability $p = 1/365$ of obtaining an exceedance on any trial. In the Poisson process, the problem is seen as consisting of independent events (exceedances) in time, occurring with a constant arrival rate of $\lambda = r = 1/365$ exceedances per day, and usually counted over a 1-year or 3-year observation period. Regardless of which model is selected, the resulting probabilities computed from the two probability models—the binomial and Poisson—will give approximately the same answers for this set of parameters. The Poisson model also may be viewed as a computational approximation to the binomial model in this situation.

Robustness

Application of either of these models assumes that the underlying process is stationary and independent. Our experience tells us that these two assumptions are not realistic for the physical situation being modeled—daily exceedances of standards—but the answers computed from these models still will be correct if they can be shown to be substantially unaffected by the nonstationarity and nonindependence present in this physical situation. In Chapter 4, the binomial model was used to demonstrate that computed probabilities were fairly robust in the face of violations of the stationarity and independence assumptions.

For the stationarity assumption, an extreme case proposed by Javits[2] was considered: the probability of an exceedance was assumed to be $p = 0.5$ on just 2 days of the year and $p = 0.0$ on all the other days of the year, satisfying the requirement that $E[K] = 1.0$. In this extreme case of nonstationarity, the binomial model gave the result that 4 or more exceedances in 3 years would occur with probability 0.343. This is very close to the probability 0.353 obtained from the binomial model for the extreme case of stationarity in which $p = 1/365$ was assumed for every day of the year. Thus, if $E[K] = 1.0$, extreme nonstationarity did not alter the overall conclusion that 4 or more exceedances in 3 years would occur in about 35% of the 3-year periods. This stationarity test is easier to carry out with the binomial model than with the Poisson model.

To test the independence assumption, another extreme case—linked adjacent days—suggested by Javits[2] was considered, and here the Poisson model performs more effectively than the binomial model. Two possible situations with extreme dependencies were considered: (a) linked pairs of days ("doublets") and (b) linked sets of 3 days each ("triplets"). For doublets, both days in the pair either exceed the 0.12 ppm ozone standard or both days fail to exceed the standard. For triplets, the three adjacent days either all exceed the 0.12 ppm ozone standard or they all fail to exceed the standard.

Let J be a random variable denoting the number of doublets occurring in the time period of interest. In any 1-year period, the number of doublets possible will be $n = 365/2 = 182.5$.* If the standard is met exactly, with 1.0 expected exceedance per year, then the expected number of doublets will be $E[J] = 0.5$ doublets per year, since each doublet counts as 2 exceedances. In the Poisson probability model, the parameter $\rho = rn = (r)(182.5) = 0.5$. Solving for the exceedance rate r, we obtain $r = 0.5/182.5 = 1/365$. In any 3-year period, the number of doublets possible will be $(3)(182.5) = 547.5$ doublets. Thus, for a 3-year period, the Poisson parameter will be $\rho = (r)(547.5) = (1/365)(547.5) = 1.5$, and the probability distribution for the number of doublets will be $\mathbf{P}(1.5)$. Using this Poisson model to compute the probability of 2 or more doublets (4 or more exceedances) in a 3-year period, the following result is obtained:

$$P\{0 \text{ doublets}\} = P\{J = 0\} = \frac{e^{-1.5}(1.5)^0}{0!} = 0.22313$$

$$P\{1 \text{ doublet}\} = P\{J = 1\} = \frac{e^{-1.5}(1.5)^1}{1!} = 0.33470$$

$$P\{2 \text{ or more doublets}\} = P\{J \geq 2\} = 1.00000 - 0.22313 - 0.33470 = 0.44217$$

This result of 0.44217 can be compared with the answer of 0.44286 obtained with the binomial model in Chapter 4. Both models give the same answer, when rounded off, that the chance of 4 or more exceedances (2 or more dou-

*A year can contain only 182 doublets, but the decimal is included to facilitate computation.

blets) in a 3-year period is 44%. (Notice that 3 exceedances cannot occur because doublets occur only in pairs.)

In the second extreme example of dependency suggested by Javits,[2] exceedances always appear in triplets, or linked sets of 3 days each. Let I be a random variable denoting the number of triplets. In any 1-year period, the number of triplets possible will be $n = 365/3 = 121.67$. If the standard is met exactly, with 1.0 expected exceedance per year, then the expected number of triplets will be $E[I] = 1/3$, because each triplet counts as 3 exceedances. In the Poisson probability model, we set $E[I] = \rho = (r)(121.67) = 1/3$. Solving for the exceedance rate r, we obtain $r = 1/365$, the same result as above. In any 3-year period, the maximum number of triplets possible will be $(3)(121.67) = 365$ triplets. For 3-year periods, the Poisson parameter will be $\rho = rn = 1$, and the probability distribution for the number of triplets will be $\mathbf{P}(1)$. This distribution was presented earlier in this chapter (Figure 5.4 and Table 5.2). From those results, the probability of 0 triplets will be $P\{I = 0\} = 0.36788$, and the probability of 1 triplet will be $P\{I = 1\} = 0.36788$. Thus, the probability of more than 1 triplet in 3 years is given by $P\{I > 1\} = 1.00000 - 0.36788 - 0.36788 = 0.26424$. This result compares favorably with the value of 0.26425 obtained in Chapter 4 using the binomial distribution.

In summary, this analysis considers the borderline case in which the statistical goal of the ozone standard is met exactly: the expected number of exceedances is 1.0 per year. If the daily maxima are assumed to be independent, which can be called "singlets," then both the Poisson and binomial models give the same result: the chance of 4 or more exceedances in 3 years will be 35.3%. If an extreme case of nonstationarity is considered, the binomial model indicates that this result will change to 34.3%. Because this change is not great, the result appears robust to violations of the stationarity assumption. If the daily maxima are assumed to be dependent and linked as doublets, then the chance of 4 or more exceedances in 3 years will be 44.2% by the Poisson model (44.3% by the binomial model). If the daily maxima are assumed to be linked as triplets, then the chance of 4 or more exceedances in 3 years will be 26.4% by both models. Thus, the chance of 4 or more exceedances in 3 years ranges from about 26% to about 44% for these extreme violations of the stationarity and independence assumptions, compared with about 35% for the stationary, independent case. As noted by Javits,[2] these assumed violations of the stationarity and independence assumptions are sufficiently severe that it is likely that real world conditions will give results closer to the 35% figure.

This analysis is important because it shows how a theoretically derived model, such as the binomial or Poisson, can be applied to real world situations in which some fundamental conditions for the theory may be violated to a degree. For any probability model intended for application to actual problems, a similar robustness analysis usually is necessary and should be carried out. Such a robustness analysis shows whether violations of the basic assumptions invalidate conclusions based on the model, and, if so, by how much.

The above robustness analysis indicates that the Poisson probability model can be applied validly to problems about the number of violations of environmental standards, even when several of the assumed conditions of stationarity and independence are severely violated, provided that the analyst is aware of the manner in which these assumptions affect the overall conclusions. This analysis also indicates how the binomial and Poisson models, under certain circumstances, can be used to complement each other, thus strengthening the overall conclusion. The Poisson model also can be used as an approximation to the binomial model in certain cases, providing almost the same results but with a single-parameter rather than a two-parameter model. The choice of which model to apply to a given problem will depend on the physical characteristics of the problem and the perspective of the analyst.

Practical Implications for NAAQS

As indicated in Chapter 4, the results from the above analysis suggest an inconsistency between the statistical goal of the ozone NAAQS and its practical compliance criterion. This inconsistency can be summarized briefly in tabular form (Table 5.3). The statistical goal of the ozone standard seeks to make the expected number of exceedances per year 1.0 or less; that is, $E[K] \leq 1.0$, which naturally implies $E[M] \leq 3.0$. However, actual compliance with the NAAQS is judged by whether the 3-year average is 3.0 or less, implying the event $\{M \leq 3\}$. Even when the statistical goal is satisfied, the compliance criterion will be violated—that is, $\{M \geq 4\}$—in about 35% of the 3-year periods, giving a compliance rate of only 65%. From common experience, a compliance rate of 65% does not seem very high, and it is likely that air pollution control agencies will strive to attain compliance with the standards for more than about two-thirds of the time. If they achieve a higher compliance rate—say, 90% of the 3-year periods—then the expected number of exceedances will be less than 1.0. Using the Poisson model, Javits[2] shows that an expected number of exceedances of 0.5 per year will produce a 3-year compliance rate of 93.4%. If the expected number of exceedances is set to 0.4 per year, the compliance rate will be slightly above 95% (Problem 7).

The apparent inconsistency between the statistical and deterministic parts of the ozone standard shows the advantage of specifying a *compliance rate* with this standard. Is 90% acceptable, or should 95% of the 3-year periods meet the deterministic criterion? Is 65% acceptable? Without a compliance rate, the deterministic part of the NAAQS is incomplete, for the user must guess about the probability with which it is to be attained. As we move increasingly to statistical forms for environmental standards, it will be useful to specify a compliance rate with each standard. The models presented in this chapter and in Chapter 4 provide a simple, practical means for calculating these probabilities, thus allowing the statistical and deterministic parts of the standards to be related to each other in a coherent, logical fashion. Because these models also satisfy the-

Table 5.3. Statistical and Deterministic Components of the Ozone NAAQS

	Statistical Goal	Compliance Criterion
1-Year Period	$E[K] \leq 1.0$	Judged by $\{K \leq 1\}$
3-Year Period	$E[M] \leq 3.0$	Judged by $\{M \leq 3\}$

oretical requirements for their development and application, probabilities computed from them should be sufficiently accurate for enforcement and decision-making purposes.

Five Expected Exceedances

One final air quality example helps illustrate how easily the Poisson model can be used to evaluate the effect of different assumptions. Consider a proposed ambient air quality standard specifying 5.0 expected exceedances per year. If a 3-year deterministic compliance criterion is employed, what will be the probability distribution for the number of exceedances? How often will the deterministic compliance criterion be violated, even though the statistical goal has been attained?

Using the Poisson model, $E[K] = 5.0 = \rho = rt = (r)(365 \text{ days})$. Solving this equation for r gives $r = 5/365 = 0.01370 \text{ days}^{-1}$. If 1-year periods are considered, $\rho = rt = (5/365)(365) = 5$, and the probability distribution for the number of exceedances K will be **P**(5). This distribution is right-skewed (Figure 5.6), although the right-skewness is not as pronounced as for the **P**(1) and **P**(3) distributions (Figures 5.4 and 5.5). If the probabilities for values of K from $K = 0$ to $K = 5$ are added together, the resulting CDF will have the value $P\{K \leq 5\} = F_K(5) = 0.61596$, and therefore $P\{K > 5\} = 1.00000 - 0.61596 = 0.38404$. Thus, for 5.0 expected exceedances per year, there will be more than 5 exceedances in 38.4% of the 1-year periods, giving a 1-year compliance rate of 61.6%.

If 3-year periods are considered, $rt = (5/365 \text{ days}^{-1})(1095 \text{ days}) = 15$, and the probability distribution for M will be **P**(15). The CDF for 3-year periods has the value $P\{M \leq 15\} = F_M(15) = 0.56809$, so the probability of more than 15 exceedances in any 3-year period will be $P\{M > 15\} = 1.00000 - 0.56809 =$

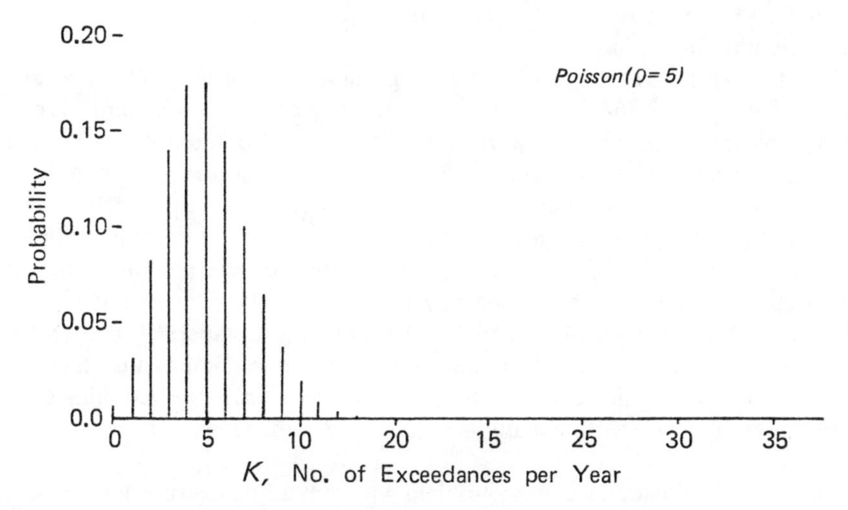

Figure 5.6. Poisson probability distribution **P**(5) for the number of exceedances in 1 year if the expected number of exceedances per year is 5.0.

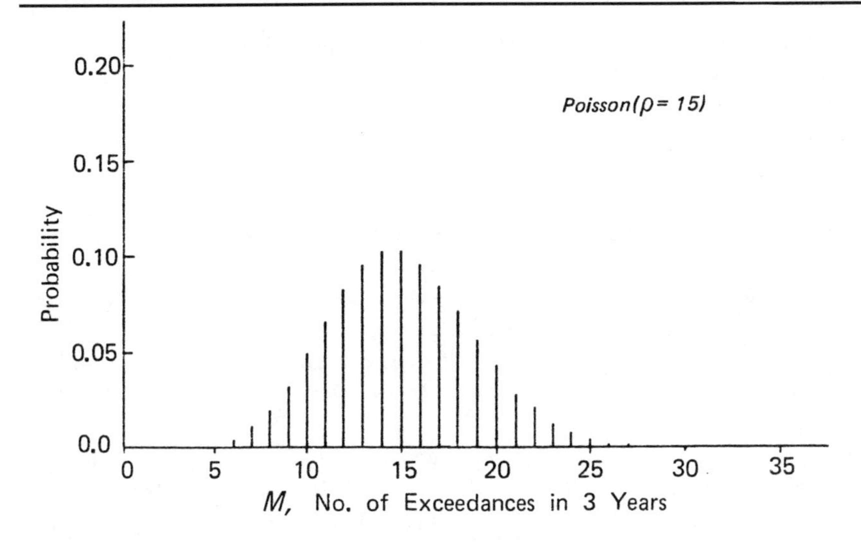

Figure 5.7. Poisson probability distribution **P**(15) for the number of exceedances in 3 years if the expected number of exceedances per year is 5.0.

0.43191. Thus, for 5.0 expected exceedances per year, there will be more than 15 exceedances in 43.2% of the 3-year periods, giving a 3-year deterministic compliance rate of only 56.8%.

The Poisson distribution, like the binomial, rapidly approaches a bell-shaped pattern as the number of events included in the observation period becomes larger, and its envelope increasingly resembles a normal distribution (Chapter 7). This tendency toward symmetry is quite evident if one compares the Poisson distribution **P**(5) in Figure 5.6 with the Poisson distribution **P**(15) in Figure 5.7. Eventually, as ρ becomes very large, this symmetry will cause about half the probability mass to lie to the left of the mean and half the probability mass to the right of the mean.

The 1.0 expected exceedance standard discussed earlier gave a 3-year compliance rate of only 64.7%. The 5.0 expected exceedance standard gives a compliance rate that is even lower: 56.8%. Thus, use of a multiple exceedance standard causes the inconsistency between the statistical goal of the standard and its deterministic compliance criterion to worsen. Eventually, if the expected number of exceedances in the standard is increased further, air pollution control agencies will be out of compliance with the deterministic criterion about 50% of the time. Obviously, they will seek a higher frequency of compliance, and, in the process, they will be attaining a statistical goal that is more stringent than the one specified by the written version of the standard. For example, if they achieve a 95% compliance rate, the expected number of exceedances per year will actually be $E[K] = 3.3$ rather than $E[K] = 5.0$ (Problem 8).

This analysis illustrates a basic problem with a multiple exceedance standard not widely discussed in the literature. Use of multiple exceedances (more than 1.0 expected exceedance per year) increases the inconsistency between the statistical and deterministic parts of the standard originally pointed out by Javits.[2] The Poisson and binomial models provide excellent, effective tools for analyzing these problems and for assessing the impact of alternative forms of air and water quality standards.

COMPUTATION OF P(λt)

Because of the simplicity of the Poisson model, probabilities can be calculated quite easily using a hand calculator or a personal computer. To illustrate the computation, a program was written in BASIC to run on an IBM*-compatible personal computer (Figure 5.8).

In line #150, the program reads in input values for lambda (λ) as L and time t as T, and then it prints these values on the screen. To keep the number of terms printed within limits, the program then reads in the value for N, the highest value of the random variable K to be evaluated (line #170). The PMF is given by P, the probability $P\{K = k\}$, and the CDF is given by Q, which is the sum of P on each successive iteration. The CDF is initialized as Q = 0 in line #190, and the constant S = EXP(–LT) is computed once (line #200) so it does not have to be computed over and over. The main computation loop lies between lines #230 and #280, and K = 0 on the first iteration and is incremented by 1 on each successive iteration of the loop. Statement #230 forms the factorial in the denominator by multiplying by K each time K is incremented. On the first iteration, K = 0, so this step is not carried out, and the divisor remains at D = 1 to

*IBM is the registered trademark of International Business Machines, Inc.

```
100 PRINT
110 PRINT "PROGRAM TO COMPUT TERMS OF POISSON PROBABILITY MASS FUNCTION"
120 PRINT "         (ALL VALUES ARE COMPUTED IN DOUBLE PRECISION)"
130 PRINT
140 DEFDBL A-Z
150 INPUT "Enter Lamda, Observation Time"; L, T
160 PRINT "Lamda = "; L; : PRINT "Time ="; T
170 INPUT "Number of the highest term to be computed"; N
180 PRINT : PRINT
190 Q = 0
200 S = EXP(-1 * L * T)
210 K = 0
220 D = 1
230 IF K <> 0 THEN D = D * K
240 P = S * ((L * T) ^ K) / D
250 Q = Q + P
260 PRINT K, P, Q
270 K = K + 1
280 IF K <= N GOTO 230
290 END
```

Figure 5.8. BASIC computer program for computing the PMF and CDF for the Poisson probability model.

avoid division by 0. The main computation of the PDF is performed on line #240. The terms begin with K = 0, and the test at the end (line #280) assures that they end with K = N on the last iteration, generating a total of N + 1 terms. The program prints a table with three columns: the random variable *K*, the PMF, and the CDF of the Poisson probability model.

PROBLEMS

1. Consider the traffic intersection problem discussed earlier. On one of the roads leading to this intersection, vehicles have a fixed arrival rate of $\lambda = 40$ vehicles per hour. Suppose an important driveway will be blocked on this particular road if more than 5 vehicles arrive while the light is red. If the traffic light normally stays red for 3-minute periods, what is the probability that this driveway will be blocked? [Answer: $P\{N_3 > 5\} = 0.0166$] Plot the probability distribution for the number of vehicles accumulating on this road during the red 3-minute cycle periods. [Answer: **P**(2)]
2. To reduce the likelihood of blockage of the driveway in the problem above, a decision is made to change the duration of the red light. With 3-minute periods, the chance of a blockage is 1.66%, or 1.66 in 100. What will be the chance of a blockage if the duration time is reduced to 2 minutes and 1 minute? [Answers: 2.5 in 1,000; 6.9 in 10,000] Suppose the goal is to reduce the risk of blocking the driveway to just less than 1 blockage in 1,000 cycles of the light. Show that a red light time duration of 1.65 minutes will produce the desired level of risk. Plot the resulting distribution. [Answer: **P**(1.1)]
3. In an hourglass, sand particles fall at an average rate of 1.5 grains per second. In any 2-second period, what is the probability that (a) 0 grains arrive, (b) more than 2 grains arrive, and (c) 9 or fewer grains arrive? [Answers: Since the distribution is **P**(3), the probabilities for *M* given in Table 5.2 can be used: (a) $P\{N_3 = 0\} = 0.04979$, (b) $P\{N_3 > 2\} = 0.35277$, (c) $P\{N_3 \leq 9\} = 0.99890$]
4. In the above problem, what is the probability that there will be at least 1 grain of sand arriving during any observation period of duration *t*? [Answer: $P\{N_t > 0\} = 1 - e^{-1.5t}$] Plot this function. What is the probability of at least 1 grain of sand in 1 second, 3 seconds, 10 seconds? [Answers: 0.6321, 0.9502, 0.9999546]
5. When a water pollutant is discharged continuously into a particular stream, past experience indicates that the number of violations of a water quality standard in the stream can be treated as Poisson throughout a season. If violations occur at an average rate of 8 per month, what is the probability distribution for the number of violations occurring in a week, one-half month, and one month? [Answers: **P**(2), **P**(4), **P**(8)] Plot these distributions and discuss their relative symmetry. What is the probability of 2 or more violations in a week, a 2-week period, and a month. [Answers: 0.5940, 0.9084, 0.9970]
6. In the example given in this chapter about the National Ambient Air Quality Standard for ozone, why is the exceedance rate $r = 1/365$ the same regardless of whether 1-year periods or 3-year periods are considered? Why is the exceedance rate $r = 1/365$ the same for singlets, doublets, and triplets?

7. An air pollution control program seeks to maintain a compliance rate in which the average number of exceedances per year is 3.0 or less in at least 95% of the 3-year periods considered. Show that this objective will be met if the expected number of exceedances is set to 0.4 per year. For this case, what is the probability of 1 or fewer exceedances in 1 year? [Answer: $P\{K \leq 1\} = 0.9384$] What is the probability of 3 or fewer exceedances in 3 years? [Answer: $P\{M \leq 3\} = 0.9662$]
8. An air quality standard is formulated with a statistical goal of 5.0 expected exceedances per year. An air pollution control program implements this standard by controlling sources until 95% of the 3-year periods experience 15 or fewer exceedances. Show that the actual statistical goal will be closer to $E[K] = 3.3$ exceedances per year at that time. For this case, what is the probability of 5 or fewer exceedances in any year? [Answer: $P\{K \leq 5\} = 0.8829$] What is the probability of 15 or fewer exceedances in any 3-year period? [Answer: $P\{M \leq 15\} = 0.9546$]

REFERENCES

1. Cinlar, Erhan, *Introduction to Stochastic Processes* (Prentice-Hall, Inc.: Englewood Cliffs, NJ, 1975).
2. Javits, Jarold S. "Statistical Interdependencies in the Ozone National Ambient Air Quality Standard," *J. Air Poll. Control Assoc.* 30(1):58–59 (January 1980).

6 Diffusion and Dispersion of Pollutants

The two previous chapters dealt with integer random variables. It was shown that certain integer probability distributions such as the binomial and Poisson can arise naturally from relatively simple assumptions that apply approximately to actual situations. Examples were presented of applications of these probability models to environmental problems, such as counts of the number of times that a particular concentration is exceeded. In these applications, the analyst was concerned with the *number of times* that a given standard was violated and not with the *levels observed* themselves. This chapter serves as a transition from consideration of integer to continuous random variables, which are useful for representing actual pollutant concentrations. This chapter also develops the Gaussian plume model from simple analogs.

When a pollutant is released into the environment, many diverse, unrelated forces act on it simultaneously. Because of the complexity of these processes, it is difficult to construct a single model for the movement, transformation, and fate of a pollutant. To gain insight into these phenomena, it is preferable to consider several very simple models of one or more of the important processes at work in the environment. Such models are idealized, but their purpose is to illustrate how the statistical properties of observed environmental concentrations come about. These idealized models are analogs that are intended to illustrate basic processes that operate, with far greater complexity, in actual environmental situations. Many of these concepts are presented for the first time in this book, and it is hoped that readers will evaluate them carefully, and, if warranted, will develop and extend them further.

Diffusion is one of the most important processes that acts upon a pollutant released into the environment. In its simplest form, diffusion occurs when a molecule changes place with an adjacent molecule. Typically, the molecules of the pollutant released into a carrier medium change places with adjacent molecules of the carrier medium. In the absence of outside mechanical forces, such movement will take place naturally due to the constant motion of the molecules of the pollutant and the material comprising the carrier medium. Such natural movements at the molecular level, called Brownian motion, can cause the pollutant to become thoroughly mixed within the carrier medium. Because the molecules will tend to spread out, or become dispersed in the carrier medium, they will also become diluted, and the concentration of the pollutant (number of molecules per unit volume) will be reduced. Because of the importance of *dilution* in generating certain kinds of probability distributions, it is treated in a separate chapter of this book (Chapter 8).

If the carrier medium moves as the diffusion occurs, causing the molecules of the pollutant to exhibit a predominant motion in a particular direction, the process is called diffusion with drift. As diffusion processes become more complex, incorporating the effects of many real factors (winds, temperature changes, turbulence), they sometimes are called dispersion processes. One, two, or three-dimensional diffusion processes can be considered. Although most real environmental diffusion and dispersion processes occur in three dimensions, the one- and two-dimensional analogs embody most of the same fundamental characteristics and are much easier to describe and analyze.

WEDGE MACHINE

Consider a physical system consisting of an array of wedges in uniform rows (Figure 6.1). The wedges might be constructed of pieces of wood, for example, glued to a plywood back panel with a sheet of glass as the face. Suppose that many small particles—say, grains of sand—can be released from a "source" at the very top of the structure and that they eventually fall by gravity through the array to the bottom of the structure. As we shall see, the movement of particles downward through this "wedge machine" is a mechanical analog of the dispersion and diffusion of pollutants in the environment. In the following sections, we examine the probability distributions with respect to space and time that will arise naturally from this mechanical analog.

Distribution with Respect to Space

Consider a single particle that is released from the source at the top of the array and falls downward. When it reaches the upward tip of the wedge in row 1, assume that it has probability $q = 1/2$ of falling down the left side of this wedge (route A_1) and probability $p = 1/2$ of falling down the right side of this wedge (route B_1). Suppose that this particular particle happens to travel down route A_1 (dotted line in Figure 6.1). As it falls, it soon encounters another wedge in row 2, where a similar decision must be made. Once again, it has probability $p = 1/2$ of traveling down the right side of the wedge (route B_2) and probability $q = 1/2$ of traveling down the left side of the wedge (route A_2). The probability that the particle will travel down route A_2, given that it has traveled down route A_1, can be written as the conditional probability $P\{A_2|A_1\} = 1/2$. Then, the probability that a particle initially released from the source actually ends up traveling down route A_2 will be the product of the probabilities at the two wedges: $P\{A_2\} = P\{A_2|A_1\}P\{A_1\} = (1/2)(1/2) = 1/4$. These probabilities are listed on the routes of the wedge machine (Figure 6.1), and the typical path of a particle that begins at the source and ends up in channel C at the bottom is shown (dotted line).

If we now consider route B_2, in the middle of the two wedges in row 2, we see that a particle can reach this point by two alternative routes. It can travel down the left side of the top wedge and then take route B_2, or it can travel down the right side of the top wedge and then take route B_2. Assuming that $p = q = 1/2$ for all wedges in the structure, the probability that the particle actually travels down route B_2 will be given by the union of the two possible paths: $P\{B_2\} = P\{A_2|A_1\}P\{A_1\} + P\{B_2|B_1\}P\{B_1\} = (1/2)(1/2) + (1/2)(1/2) = 1/4 +$

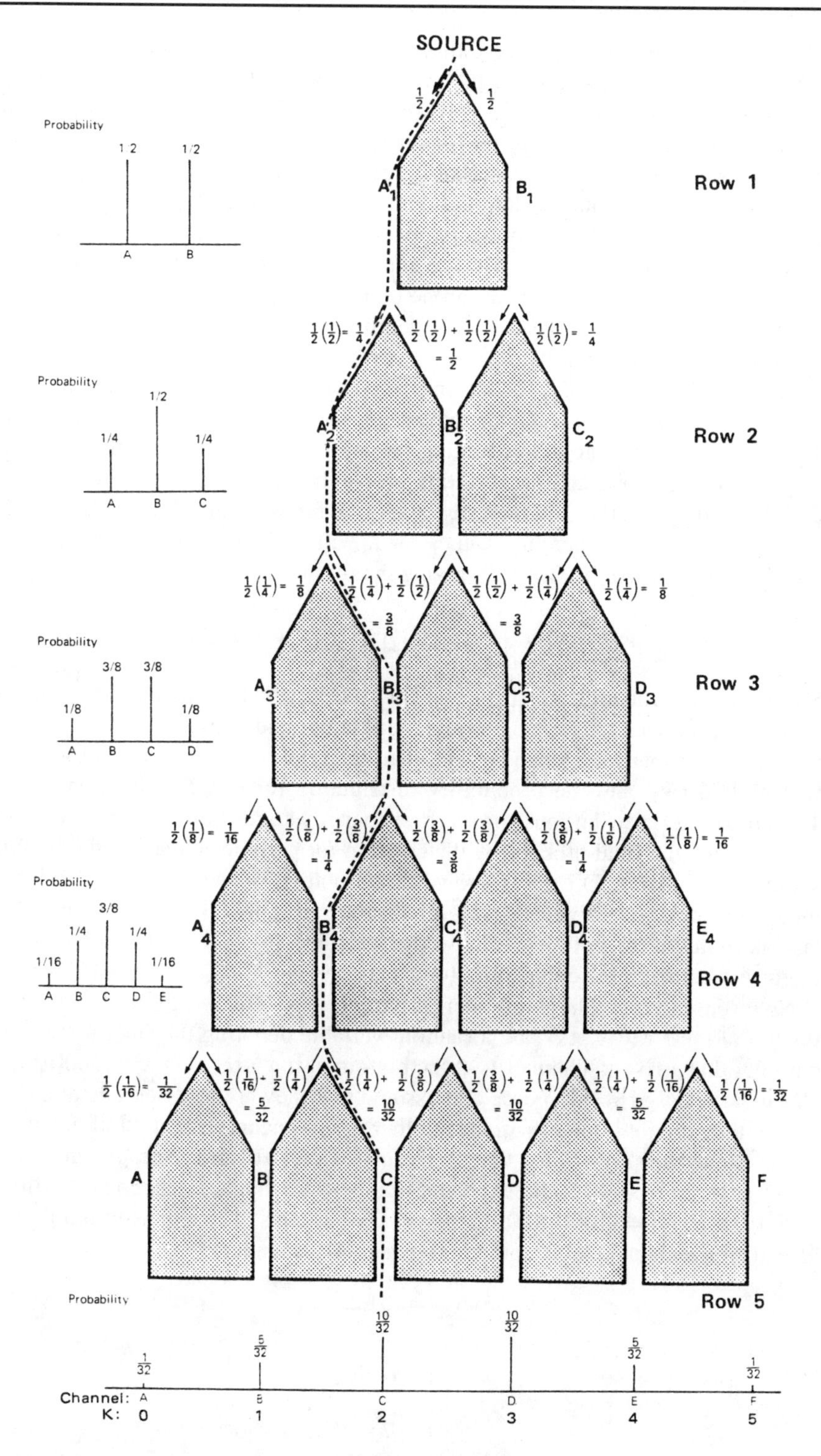

Figure 6.1. Wedge machine: particles released at the source move downward through array in response to gravity, with equal probability of falling to either side of each wedge, thereby generating a binomial distribution at the bottom.

1/4 = 1/2. Finally, the probability that a particle travels down route C_2 will be computed as the intersection of the probabilities for the two relevant decision points: $P\{C_2\} = P\{C_2 | B_1\}P\{B_1\} = (1/2)(1/2) = 1/4$. By this logic, we have computed the probability distribution for all routes in row 2, which comprise the sample space $\{A_2, B_2, C_2\}$. The probabilities for all routes in the array are listed between the wedges on the diagram, and the resulting probability distribution for row 2 is shown to the left of the two wedges (see Figure 6.1).

As the probabilities for each row are computed in the manner described above, it becomes evident that each row can be represented as a Bernoulli process, provided that an index is assigned which identifies the routes from left to right. For the top wedge (row 1), a particle faces two independent choices: route A_1, for which $q = 1/2$, or route B_1, for which $p = 1/2$. Let the random variable K be an index such that $K = 0$ corresponds to route A_1 and $K = 1$ corresponds to route B_1. This situation is analogous to counting the number of heads that will appear on one flip of a fair coin, and the resulting probability distribution will be binomial **B**(1,1/2). In a similar manner, the particle faces three choices in row 2: $\{A_2, B_2, C_2\}$. The corresponding values for K will be {0,1,2}, and the probability distribution for K will be **B**(2,1/2), giving the following PMF:

$$P\{K = k\} = \binom{2}{k} (\tfrac{1}{2})^k (\tfrac{1}{2})^{2-k} \quad \text{for } k = 0, 1, 2.$$

Similarly, the probability distribution for row 3, which contains four possible routes, will be **B**(3,1/2). Using similar reasoning, and considering the possible routes for each particle, it can be seen that the probability distribution for row 4 will be **B**(4,1/2), and the probability distribution for row 5 will be **B**(5,1/2). The probability distributions for the first four rows are shown at the left of the drawing, and the final probability distribution for particle arrivals in the bottom six channels {A,B,C,D,E,F} is shown at the bottom of the drawing. The intermediate calculations giving these probabilities are listed on the drawing. (The last six routes at the bottom of the wedge machine are called "channels" to distinguish them from the connecting routes above.)

Now suppose that a large number of particles are released from the source at the top of the array. Let N_A be a random variable denoting the final number of particles that arrive in channel A at the bottom of the array. Let $n_o = 1000$ particles to be released at the source and assume that they are released one at a time so that they do not interact with each other in any manner as they fall down the array. Since the probability that a given particle will arrive in channel A is $P\{A\} = 1/32 = 0.03125$, the expected number of arrivals in channel A will be $E[N_A] = n_o P\{A\} = (1000)(0.03125) = 31.25$ particles. The expected number of arrivals for each channel is computed in a similar manner:

$$\begin{aligned}
E[N_A] &= \ \ 31.25 \\
E[N_B] &= 156.25 \\
E[N_C] &= 312.50 \\
E[N_D] &= 312.50 \\
E[N_E] &= 156.25 \\
E[N_F] &= \ \ 31.25 \\
\text{Total} &= \overline{1000.00} \text{ particles}
\end{aligned}$$

The distribution of the expected number of particle arrivals is seen to be symmetrical about the midpoint (center line) of the array of wedges.

Suppose that the array of wedges did not end at a row 5 in the example shown but extended well beyond, say to row 19. Then the probability distribution for the number of arrivals in the 20 channels at the bottom of the array would be **B**(19,1/2). In general, if the array extended to any row m, the probability distribution for the number of arrivals would be **B**(m,1/2). As m becomes larger, the heights of the probability bars, when plotted, increasingly resemble the familiar bell-shaped curve of the normal distribution (Chapter 7).

Notice that each row in the wedge machine contains a different number of wedges, giving a "Christmas tree-shaped" pattern. The top row contains only one wedge, and each row has one more wedge than the one above it. If the same number of wedges were present in each row, it would not change the analysis or the final probability distribution. If, for example, there were five wedges in row 1, the paths of the particles still would be determined entirely by a single wedge, the one located immediately below the stream of particles released by the source. Similarly, only two wedges affect the paths of the particles in row 2, and any additional wedges would not change the overall conclusion. Thus, the important characteristic responsible for the binomial distribution at the bottom of the array is the regularity of spacing of the wedges and not the tree-shaped pattern. In general, a regularly spaced array of obstacles in two dimensions will cause a stream of particles passing through it to spread out, generating a bell-shaped distribution that is symmetrical about the centerline of the stream.

The major movements of the particles can be separated into two components: (1) downward motion (drift) due to gravity, and (2) side-by-side movements due to encounters with the wedges. The side-by-side movements are responsible for the binomial distribution. When a particle encounters a wedge, it experiences a displacement either to the left or to the right, with probability $p = q = 1/2$. In any row, its horizontal position is determined by the algebraic *sum* of the left and right horizontal displacements it has experienced.

The horizontal movement of the particle warrants further analysis. Consider an experiment in which some object begins at rest at an origin and experiences "jumps," or fixed displacements of length d, in one dimension until it reaches a final position (Figure 6.2). In the figure, the object begins at position $K = 0$, experiences four displacements, and ends at position $K = 4$. Suppose that a fair coin is flipped on m trials, and the following rule is adopted based on the outcome of each trial: If a "head" appears, the object moves to the right by displacement d; if a "tail" appears, the object remains where it is. In effect, the ob-

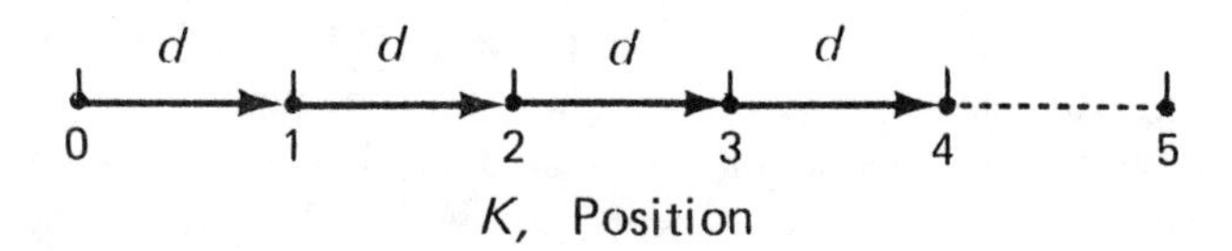

Figure 6.2. Position of an object experiencing four successive, equally-spaced displacements in one direction.

ject moves to the right with probability $p = 1/2$, or it remains stationary with probability $q = 1/2$. After m coin flips, the object can occupy any position $K = \{0,1,2, \ldots , m\}$. Because the coin flipping experiment is a Bernoulli process, the number of displacements also will be a Bernoulli process, and the position of the object will have a binomial distribution $\mathbf{B}(m,1/2)$. The expected value of this distribution will be $E[K] = mp = m/2$, and therefore the distribution will be symmetrical about the midpoint of its range $(0,m)$.

Consider a slightly different case in which the object can move either to the left or right: If a "head" appears, it moves to the right by displacement d; if a "tail" appears, it moves to the left by displacement $-d$. Thus, if, on 5 trials, 3 "heads" and 2 "tails" occur, the object will be at position $Kd = 3d - 2d = d$. In general, the result of allowing rightward-leftward movements is to generate a distribution that is symmetrical about the origin, $K = 0$. For this situation, the binomial distribution is expanded by a factor of two and is translated m units to the left. If J is a random variable with the distribution $\mathbf{B}(m,1/2)$, then the translated random variable $K = 2J - m$ will now describe the position of the object.

The above process is called a "random walk." Suppose a drunken man must walk along a straight line and that he can take a step to the left or a step to the right at any point, beginning at a lamppost that corresponds to position $K = 0$. Suppose that each step is of fixed length $|d|$ and that a step to the right $(+d)$ occurs with probability $p = 1/2$, while a step to the left $(-d)$ occurs with probability $q = 1/2$. After m steps are taken, what will be the position of the man? Here, the random variable $K = 2J - m$, where J is $\mathbf{B}(m,1/2)$, will represent the location of the man in terms of the total number of positive or negative steps from the lamppost. Thus the position of the man relative to the lamppost will be given by Kd, and K will have a translated binomial distribution that is symmetrical about the lamppost.

With the above translation, the random variable K extends from $-m$ to $+m$, and its range is doubled to $2m$, with a consequent increase in its variance. To achieve a simple translation of the binomial distribution to the origin, while preserving its range as m, the random variable K must extend from $-m/2$ to $+m/2$. This shift can be achieved by using the following equation, in which J has the distribution $\mathbf{B}(m/2,1/2)$ and m is even:

$$K = J - m/2$$

Because the resulting translated binomial distribution is centered about the origin like the simplest form of the normal distribution (Chapter 7), and because it resembles the normal distribution for large m, we shall refer to this function as the "normalizing translation."

The physical position of any object is determined by the history of the total number of displacements it has experienced since leaving some starting point. It may have moved smoothly from its starting point to its final position, or it may have experienced many jerky, irregular displacements. Regardless of the type of motion involved, its path always can be broken up into discrete components, or displacements, and its final position will be the sum of these components. Thus, position is an additive variable. In this book, we shall refer to any pure additive process as a *sum process*, or a $\mathbb{S}$-process.

Although the above examples all involve additive variables, they possess an additional attribute: randomness. In each case, the sum is composed not of deterministic increments but of fixed increments that behave as independent random variables. Whenever a "head" appears in the coin-flipping experiment, the sum (number of "heads") is incremented by one. The drunken man takes equally likely right or left steps that are of fixed length, and the sum (his position relative to the lamppost) is the total of all the step lengths. Similarly, the object experiences right-left displacements with probability $p = q = 1/2$, and the sum (the object's position relative to the origin) reflects the total number of right and left displacements. Because the sums all are formed by adding independent random variables, we shall call this result a *random sum process*, or a $\mathfrak{RS}$-process. If the increments are of equal size, any $\mathfrak{RS}$-process also will be a Bernoulli process.

In the above $\mathfrak{RS}$-process examples, the result has been a bell-shaped probability distribution that is symmetrical with respect to space (e.g., horizontal distance from the origin). As we shall see, the above examples are analogous to the position of a molecule subjected to dispersion and diffusion in an environmental carrier medium.

Distribution with Respect to Time

In the wedge machine discussed above, n_o total particles were released, one at a time, above the midpoint of a uniform array of wedges. As the "stream" of particles fell downward through the array due to gravity, it spread out horizontally. The probability distribution of arrivals in any horizontal row was seen to be binomial with $p = 1/2$ and its mean centered on the midpoint of the row. Such a distribution is symmetrical and appears increasingly smooth and bell-shaped (like the normal distribution, Chapter 7) as more and more channels are encountered in the lower rows. The resulting distribution describes arrival probabilities as a function of *space* (horizontal distance from the midpoint of each row).

Using this probability distribution as a function of space, the expected number of arrivals for each channel {A,B,C,D,E,F} at the bottom of the array was computed. Although the expected number of arrivals for every channel has been obtained, we have yet to develop a distribution for particle arrivals within a channel. That is, if the expected number of arrivals in channel A is known to be $E[N_A] = n_o(1/32)$, then what is the probability distribution of N_A? What are the probability distributions of N_B, N_C, N_D, N_E, and N_F?

For each of the bottom channels, the probability that a particle arrives is fixed, and arrivals are assumed independent. Thus, N_A is the sum of n_o possible events, each occurring with fixed probability $P\{A\} = 1/32$, and the result is a Bernoulli process. The probability distribution for the number of arrivals in channel A will be given by the binomial distribution $\mathbf{B}(n_o,1/32)$:

$$P\{N_A = j\} = \binom{n_o}{j}\left(\frac{1}{32}\right)^{n_o}\left(\frac{31}{32}\right)^{n_o - j}$$

Similarly, the probability distributions for the remaining bottom channels will be $\mathbf{B}(n_o,5/32)$, $\mathbf{B}(n_o,10/32)$, $\mathbf{B}(n_o,10/32)$,$\mathbf{B}(n_o,5/32)$, and $\mathbf{B}(n_o,1/32)$.

As discussed earlier in this book (Chapter 4), any binomial distribution for which $p < 1/2$ will be right-skewed, even though the right-skewness will not be very evident to the eye if n_o is large (e.g., $n_o > 1000$). This right-skewness can be illustrated by considering the effect of different values of n_o on the distribution of particle arrivals in channel A. For small $n_o = 10$, $E[N_A] = (10)(1/32) = 0.3125$ particles, and the distribution of particle arrivals is extremely right-skewed, with its mode at the origin (Figure 6.3). For $n_o = 100$, $E[N_A] = (100)(1.32) = 3.125$ particles, and the mode (3.0 particles) shifts to the right,

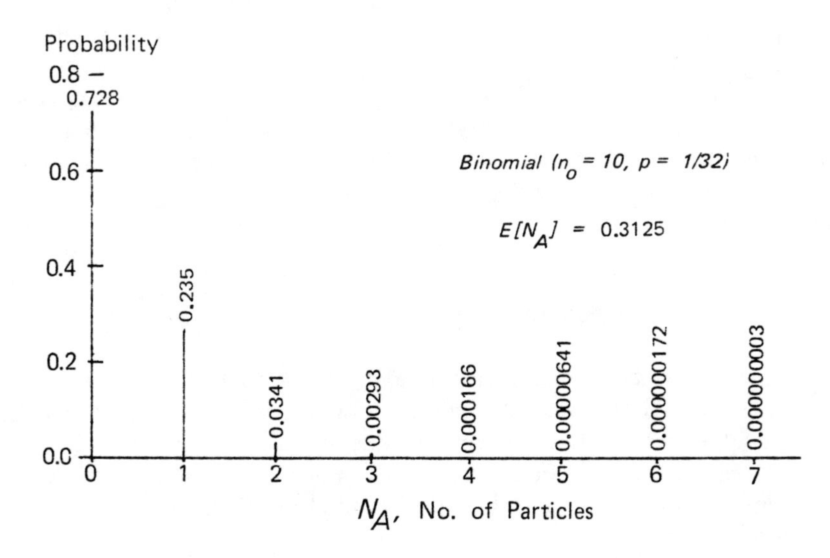

Figure 6.3. Probability distribution for the number of particles arriving in channel A of the wedge machine for the case in which $n_o = 10$, showing extreme right-skewness.

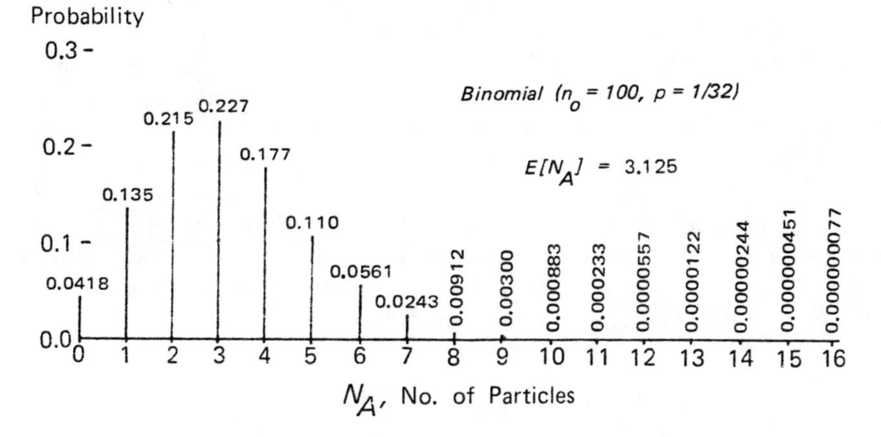

Figure 6.4. Probability distribution for the number of particles arriving in channel A of the wedge machine for the case in which $n_o = 100$, showing considerable right-skewness.

even though the right-skewness is still very striking (Figure 6.4). For $n_o = 100$, $E[N_A] = (1000)(1/32) = 31.25$ particles (with a mode of 31 particles) and the right-skewness no longer is very obvious (Figure 6.5). Comparison of these three figures reveals that the mean (expected value) is greater than the mode in all cases, a characteristic of unimodal, right-skewed distributions (Chapter 3: "Variance, Kurtosis, and Skewness," page 41). The right-skewness is strikingly apparent only when the number n_o of particles initially released is relatively small (several hundred or less). As the number of particles released becomes greater, the distribution increasingly becomes bell-shaped, more and more resembling the normal distribution. Careful examination of Figure 6.5 shows that, for $n_o = 1000$, the difference between the mode (31.00) and the mean (31.25) is small. As n_o increases further, the mode will approach the mean, and the heights of the probability bars will more and more resemble the density curve of the normal distribution (Chapter 7), for which the mode and mean coincide.

The right-skewness of these distributions also can be related to time. Because the particles are assumed to be released at a fixed rate, the 100-particle case will taken 10 times as long as the 10-particle case; similarly, the 1000-particle case will take 10 times as long as the 100-particle case. Actually, the 1000-particle case may be viewed as consisting of ten 100-particle cases occurring consecutively one after another in time.

Suppose that particles are released from the source continuously, one at a time, at a rate of $r = 1$ particle per minute. Instead of allowing particles to accumulate at the bottom of the wedge array, suppose that, after they are counted, they can continue falling freely. Suppose that this process has been operating for some time, so that transients associated with start-up conditions have subsided, and an equilibrium "flow" of particles through the array is occurring. Then the number of particles released from the source during any time period t will be $n_o(t) = rt$. If 10-minute periods are considered, the number of particles

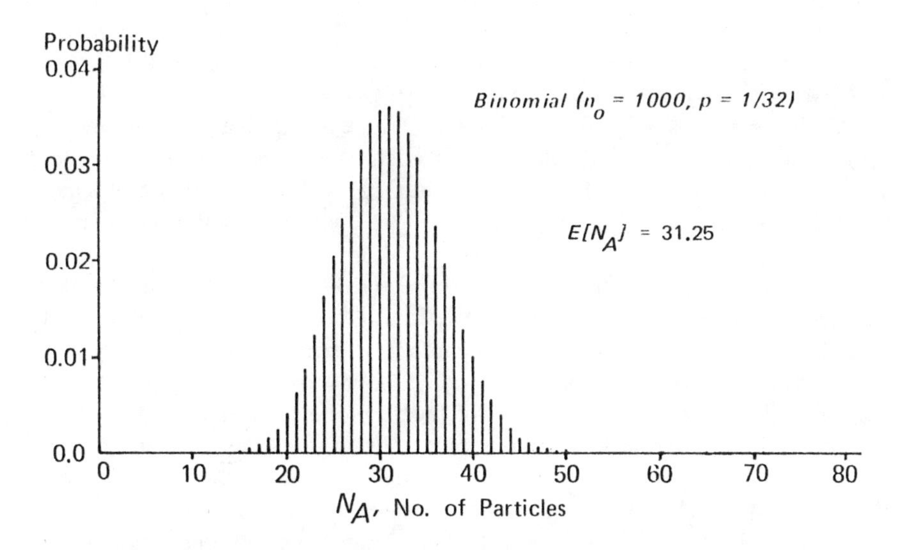

Figure 6.5. Probability distribution for the number of particles arriving in channel A of the wedge machine for the case in which $n_o = 1000$, revealing similarity to a symmetrical, bell-shaped (normal) curve.

released will be n_o = 10 particles, and the expected number of arrivals in channel A during any 10-minute period will be $10P\{A\} = 10(1/32) = 0.3125$. The probability distribution of arrivals during 10-minute periods will be approximately the same as in Figure 6.3. On the average, we would expect 72.8% of the 10-minute periods to contain no particles; approximately 23.5% would contain one particle; approximately 3.41% would contain two particles. Because these three situations account for 72.8 + 23.5 + 3.4 = 99.7% of the probability mass, only 0.3% of the 10-minute periods will contain three or more particles. The probability distribution is extremely right-skewed, because the 10-minute observation period is so brief (see Problem 4).

If observation periods lasting 100 minutes are considered, the expected number of arrivals in channel A will be $E[N_A] = 3.125$ particles, and the probability distribution of arrivals during 100-minute periods will be the same as in Figure 6.4. When compared with the 10-particle case, the mode will shift to the right, and we can expect more 100-minute periods to contain 3 particles (22.7%) than any other count. Like the 10-minute case (Problem 4), the distribution will be right-skewed, but the right-skewness will be slightly less pronounced.

Finally, if 1000-minute time periods are considered, the expected number of arrivals in channel A will be $E[N_A] = 31.25$ particles, and the probability distribution will have the bell-shaped pattern given by Figure 6.5. Comparison of these three figures illustrates the tendency of the probability distribution of particle arrivals to become less right-skewed as the observation period becomes longer.

Suppose that the concentration of a pollutant in channel A is related to the *average number* of particles X_A counted there. During any observation period of duration t, the average number of particles arriving in channel A per unit time will be given by $X_A = N_A/t$. For 10-minute time periods, for example, the expected value of the average will be given by the expected number of arrivals divided by time, or $E[X_A] = E[N_A]/t = 0.3125/10 = 0.03125$ particles per minute. For 100-minute periods, the expected value of the average will be $E[X_A] = 3.125/100 = 0.03125$ particles per minute, or the same result. Similarly, the expected value of the average for 1000-minute periods will be $E[X_A] = 31.25/1000 = 0.03125$ particles per minute. The expected value of the average number of particle arrivals is the same in each case, because it is equal to the particle generation rate at the source r multiplied by the probability that a particle arrives at channel A; that is, $E[X_A] = rP\{A\} = (1)(1/32) = 0.03125$ particles per minute.

All of the above cases give the same (expected) particle arrival rate for channel A, $r_A = 0.03125$ particles per minute, a basic property of channel A for the conditions specified. However, if the number of particles arriving in the three cases described above were divided by 10, 100, and 1000, respectively, to form the normalized averages, the shapes of the resulting distributions of averages would be similar to the distributions plotted in Figures 6.3, 6.4, and 6.5. All cases would have the same expected value, but right-skewness would be greatest for the 10-minute averages, intermediate for the 100-minute averages, and least for the 1000-minute averages. If the averaging period were made longer than 1000 minutes, the distribution of averages would appear increasingly bell-shaped, and the right-skewness would be very difficult to detect with the eye.

The observed concentration of a pollutant in the ambient atmosphere, when averaged over very short time periods, usually gives a frequency distribution that is unimodal and quite right-skewed. For example, if observed atmospheric carbon monoxide (CO) concentrations are averaged over successive 10-minute periods, the resulting histogram of 10-minute averages usually will be unimodal and very right-skewed. If the same CO concentration time series is divided into 100-minute averages, however, the overall average will be the same, but the right-skewness will be less pronounced. Similar distributional characteristics with respect to time are predicted for water pollutant concentrations in streams (Chapter 8). An important question arises: Do the statistical characteristics of real observations result from the same basic processes that produce unimodal, right-skewed distributions in the wedge machine?

In the wedge machine, the right-skewness with respect to time is a result of a binomial distribution with small p and large n. Over the averaging periods considered, only small numbers of particles (for example, $E[N_A] = (100)(1/32) = 3.125$ particles for 100-minute periods) are expected in channel A. Because the number of particle arrivals usually is small in the outer channels of the wedge machine (and cannot be negative in any channel), it is intuitively understandable why excursions involving large numbers might occur occasionally, giving right-skewed distributions.

Before answering the above question, we consider how the distribution of the average number of particles over time varies spatially from channel to channel within a row. In channel A of row 5, the arrival probability is $P\{A\} = 1/32$, while the arrival probabilities in the innermost channels C and D are $P\{C\} = P\{D\} = 10/32$. The arrival probabilities in any row must add horizontally to one, and the arrival probabilities for all channels of the wedge machine below those in the top row (and the middle channel of row 2) are less than 1/2. Thus, for almost all channels, the probability distribution of arrivals with respect to time will be right-skewed. The right-skewness of these binomial distributions will be most pronounced in those channels located farthest from the midpoint of the row, becoming least evident in the middle channels.

It may seem odd at first that the distribution of particle arrivals can be right-skewed for all channels in the same row, indicating that all channels can exhibit excursions involving large numbers of particles. Such excursions do not, however, occur in all channels at the same time on a given realization. On any realization, one or two channels may experience a large excess, while many others will make up the difference by experiencing numbers that are slightly less than normal. Nevertheless, the probability distribution of arrivals will be right-skewed in all channels.

Actual measurements of a pollutant in the environment commonly give concentrations of several parts per million or parts per billion by volume. Although such quantities are very small, they represent billions or trillions of molecules of the pollutant in a cubic centimeter. The time averages of these concentrations yield frequency distributions that often are unimodal and strikingly right-skewed. Because the right-skewness found in the wedge machine time series occurs primarily in channels with only a few particle arrivals, which would correspond to only a few molecules, it is doubtful that the right-skewed distributions observed in environmental processes result from the same phenomena at work in the wedge machine. Furthermore, the binomial distribution's property that its variance (npq) cannot exceed its mean (np) is very restrictive, and it seems unlikely that real observations will meet this requirement, except in rare situations. To explain the right-skewness with respect to time observed for actual environmental concentrations, a stochastic process that will generate right-skewed distributions under less restrictive conditions must be found. The Theory of Successive Random Dilutions (Chapter 8) offers such an explanation.

Summary and Discussion

The above analysis shows that a very simple mechanical model, the wedge machine, can generate probability distributions that are both symmetrical and right-skewed, depending on which variable is examined and how it is examined. If one considers the probability distribution of particle arrivals with respect to space (distance from the midpoint of any row), the result is a symmetrical binomial distribution ($p = 1/2$), causing the expected number of particle arrivals to be symmetrical also. If one considers the number of arrivals with respect to time (counts within observation periods of fixed duration, or counts on successive realizations of the process), the distribution also is binomial, but, because $p < 1/2$ in all but the top two rows, the binomial distribution will be right-skewed. The right-skewness will be most evident in channels at the outer edges of the wedge machine, where p is smallest. If the number of particles released during the observation period is large, however, the right-skewness of particle arrivals within a channel will not be very apparent, because of the binomial distribution's tendency to approach a symmetrical, bell-shaped distribution for large n, regardless of the value of p.

Although the distribution of particle arrivals with respect to time will be unimodal and right-skewed for nearly all channels, the right-skewness will not be very pronounced, except for those channels at the outer edges where few particles ordinarily arrive. Therefore, the pronounced right-skewness usually found in actual environmental concentrations, which result from trillions of molecules, cannot be explained by the processes at work in the wedge machine. A better explanation can be found in the Theory of Successive Random Dilutions presented later in this book (Chapter 8). The wedge machine does, however, provide an explanation of the way in which symmetrical, bell-shaped distributions of concentrations with respect to space may be generated naturally by a real physical system that is analogous to a diffusion process. It is an important analog for the processes at work in the particle frame machine discussed next and the Gaussian plume model presented at the end of this chapter.

PARTICLE FRAME MACHINE

As indicated in the literature, various mechanical devices have been constructed in order to generate binomial and other related distributions. Aitchison and Brown[1] mention the apparatus of Galton designed to produce binomial and normal distributions, and a U.S. company[2] offers a wooden frame filled with particles that is suitable for classroom demonstrations.

The wooden particle frame (Figure 6.6) is approximately 13″ (33 cm) high,

Figure 6.6. Particle frame analog of diffusion process showing reservoir with plunger released (top), particles spreading out as they fall through pin array (middle), and equally spaced collection columns with bell-shaped distribution emerging (bottom).

11″ (30 cm) wide, and 1 1/4″ (3 cm) deep. It contains a large number of particles, each of which is about the size of a grain of coarse sand but lighter in weight. The front of the frame has a clear plastic window, allowing the particle movements to be observed. The upper part of the frame is a compartment with a V-shaped floor, which acts as a reservoir capable of holding the entire population of particles. The floor has a hole which is plugged by a vertical plunger. When the plunger is lifted, as shown in the figure, the particles are released and fall down through the hole in a steady stream.

The middle section of the particle frame consists of a triangular array of uniformly spaced metal pins. As the particles fall downward through the pins, they collide with each other and with the individual pins and spread out horizontally. Because each pin is round and smooth, any particle can be assumed to have probability $p = 1/2$ of moving to the right of a given pin after hitting it, or probability $q = 1/2$ of moving to the left of the pin. Right-left movements are not as restricted as they are in the wedge machine, where only certain channels are available to the particles, but the process is similar: a stream of particles drifting downward through a uniform array experiences random (equally likely) right-left displacements.

The bottom section consists of 20 equally spaced vertical channels designed to collect the particles. Because each channel is the same width, the height of the column of particles in each channel is proportional to the number collected. After all the particles have fallen through the array and been collected in the appropriate channels, the entire frame can be turned upside down and the process repeated to give another realization.

To examine the spatial distributions generated, we conducted a simple experiment on an actual particle frame machine. A single realization was undertaken: the particles were collected in the upper reservoir, the stopper was released, and the particles fell downward and accumulated in the 20 channels. By first measuring the vertical heights of the columns in a few channels, and then counting, one by one, the number of particles, it was determined that each column contained approximately 300 particles/cm. By measuring the height of each column and multiplying by 300, the number of particles found in each column was estimated. The resulting counts, when plotted, yielded a symmetrical distribution (Figure 6.7).

If, as before, we allow K to be an index representing the position of the channel so that $K = 0$ denotes the first channel on the left, and $K = 19$ denotes the last channel on the right, then the probability distribution of arrivals will be **B**(19,1/2), using the same logic developed from the wedge machine.* When all the particle arrivals observed above were added, the frame was estimated to contain 17,900 particles. The expected number of arrivals in each channel could be computed by multiplying the total number of particles by the arrival probabilities. The resulting expected values gave a bell-shaped distribution that showed reasonably good agreement with the observed counts (Figure 6.7).

Actually, agreement between expected and observed numbers was reason-

---

*Because there is open space between the pins, the particles are dispersed more widely as they collide with the pins than in a wedge machine, which restricts the movement to specific channels. Thus, the important parameter of the particle frame is not the 10 pins in the bottom row but the 20 collecting channels at the bottom of the frame.

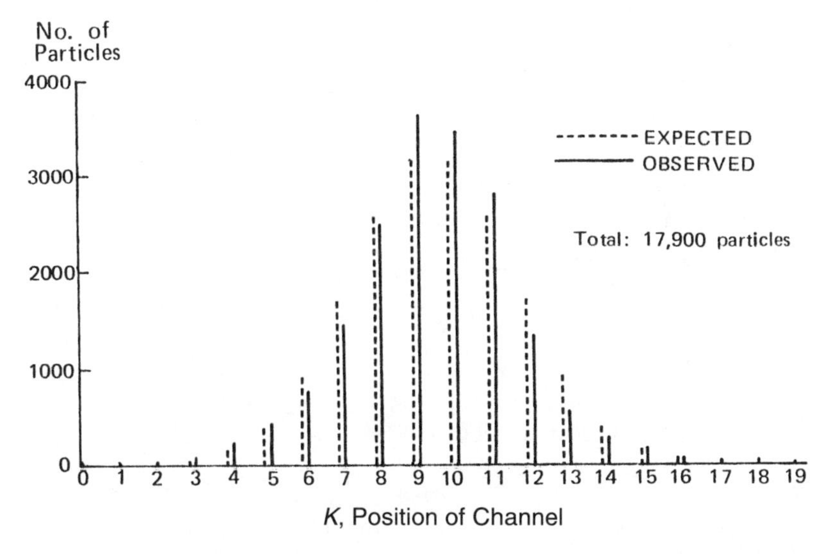

Figure 6.7. Counts of the number of particles observed in each channel of the particle frame, compared with the expected number of particles predicted by the binomial distribution **B**(19,1/2).

ably good for channels located near the midpoint of the frame, but observed counts were much higher than expected values for channels located at the extreme outer edges, a fact not evident in Figure 6.7. For example, the expected number of arrivals predicted by the binomial distribution **B**(19,1/2) was $E[N] = 0.03$ particles for the first channel ($K = 0$) and the last channel ($K = 19$), but 6 particles arrived in the first channel and 9 particles arrived in the last channel. Although expected values seldom will agree with the results from just one trial, these differences may be due partly to the manner in which the particle frame was constructed. The array of pins occupies only a limited area in this machine, causing the movements of particles to be less restricted than they were in the channels of the wedge machine. In the photograph (Figure 6.6), one actually can see particles "bouncing" to the left and to the right, outside the area occupied by the pin array. These free horizontal movements cause channels at the ends to receive more particles than they otherwise would if the pin array were more extensive, and the particles also collide with each other.

To gain insight into the probability distribution with respect to time generated by this mechanical particle frame system, all particles were allowed to fall 52 times, giving 52 realizations for every channel. On each trial, the particles arriving in the end channels were counted very carefully. Of the 52 realizations, the most frequent number of arrivals in the first channel ($K = 0$) was 5 particles (8 occurrences) and 6 particles (8 occurrences), giving modes of 5 and 6 (Figure 6.8). The smallest count was 1 particle (1 occurrence), and the largest count was 46 particles (1 occurrence). The observed number of arrivals, when plotted, gave a distribution that appeared right-skewed to the eye (solid vertical bars in the figure), even if the two highest values (24 and 46 particles) are excluded. To compare the results with an actual right-skewed probability distribu-

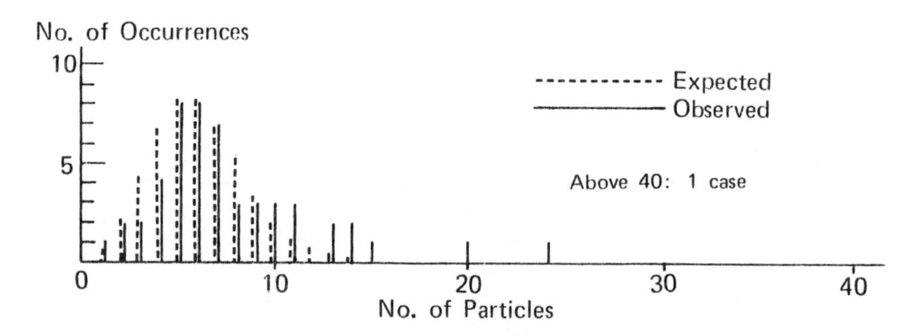

Figure 6.8. Counts of the number of particles observed in the first channel of the particle frame on 52 experimental trials, compared with the number of particles predicted by the Poisson distribution **P**(6).

tion, the Poisson probability model was chosen (dotted vertical bars in the figure). With its parameters selected at $\lambda t = 6$, the **P**(6) distribution has two modes, $N = 5$ and $N = 6$, and the resulting probability bars are similar in appearance to the actual counts. Examination of counts in nearby channels, such as at positions $K = 1$ and $K = 18$, revealed distributions with greater symmetry, although some right-skewness was evident. Channels closer to the midpoint gave distributions about the expected value that were nearly symmetrical.

The results indicated that the mechanical particle frame, despite the imperfections of any mechanical system, generated probability distributions similar to those predicted theoretically for the wedge machine. The observed distribution of the number of particles with respect to space was similar to the expected number of particles predicted for the wedge machine: a symmetrical, bell-shaped distribution. The observed distribution of counts on successive realizations (distribution with respect to time), like that predicted for the wedge machine, was right-skewed for channels at the outer edges, becoming more symmetrical for channels closer in. The results illustrate how a two-dimensional physical system can generate distributions that are both symmetrical and right-skewed, depending on one's perspective, although the overriding tendency, if many particles are involved, is toward symmetry.

PLUME MODEL

Suppose that the particles (or molecules) of a pollutant are released from a point source as a continuous stream into a moving carrier medium. Instead of gravity, which was responsible for the drift in the wedge machine and particle frame machine, assume that the drift now is caused by the predominant motion of the carrier medium at a constant speed and direction. An air pollution example is the discharge of particles (or gases) from a smokestack into a wind moving with laminar flow in a constant direction. A water pollution example is the discharge of an effluent from a pipe into a large stream with steady current. Such processes occur in three dimensions, but one can imagine a simplified process in two dimensions (for example, the projection of the particles' position on a two-dimensional surface). What will happen to the particles released into this moving carrier medium?

The path of a particle will be affected by many sub-microscopic collisions with molecules or particles (clusters of molecules) present in the mixture. These collisions are analogous to collisions of the particles in the wedge machine with the array of pins, except that the molecules are in motion. Although the drift of the carrier medium will cause the particles to move gradually in the direction of flow, each particle's actual path will be quite erratic. Because of these random collisions, the particles of the pollutant will disperse, occupying ever larger physical volume as they are carried along, thus creating a "plume." Assuming that the carrier medium is homogeneous and free of interfering boundaries, the plume will be symmetrical, with its center line extending from the source parallel to the direction of flow.

Consider a two-dimensional version of this process. As in the particle frame analog, each particle experiences longitudinal drift after leaving the source, but it is subject to numerous collisions as it moves along. On the *i*th collision, the particle will experience horizontal displacement d_i (see Figure 6.9). If $y(m)$ denotes the horizontal position of the particle after m collisions, then $y(0) = 0$ is the initial position at the source, and the final position of the particle will be the sum of all the horizontal displacements it has experienced since it left the source:

$$y(m) = d_1 + d_2 + \ldots + d_m$$

Let Y be a random variable denoting the horizontal position of the particle measured by its positive or negative horizontal distance from the center line. Let D be a random variable denoting its horizontal displacement on each suc-

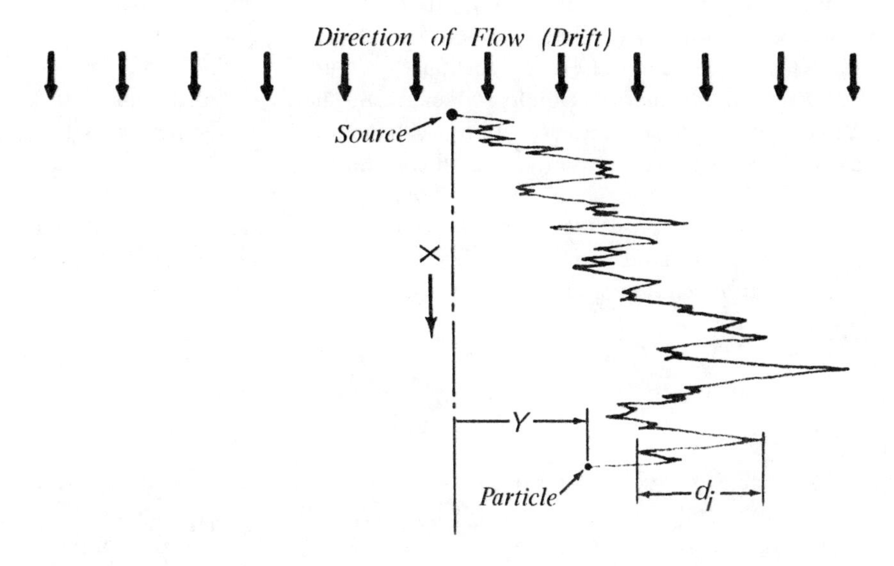

Figure 6.9. Path of a particle in two dimensions subject to drift and random collisions.

cessive collision. Then Y for an individual particle will be the sum of the realizations $\{d_1, d_2, \ldots, d_m\}$ obtained from the random variable D. Assume that the particles are displaced to the left or to the right with equal probability and that the expected value of the displacement is zero: $E[D] = 0$. Because the displacements are independent, the expected value of Y will be the sum of the expected values of the displacements by the rules for combining expected values (Chapter 3, pages 39–40):

$$E[Y] = E[d_1] + E[d_2] + \ldots + E[d_m] = mE[D] = 0$$

Similarly, the variance of Y will be the sum of the variances of the individual displacements by the rules for combining variances (Chapter 3, page 42):

$$Var(Y) = Var(d_1) + Var(d_2) + \ldots + Var(d_m) = mVar(D)$$

Because Y consists of the sum of independent random variables with finite variance, it asymptotically approaches a normal distribution in accordance with the Central Limit Theorem (Chapter 7, page 170).

This analysis indicates that the final position Y of the particle (or of many similar particles) after m collisions will asymptotically approach a normal distribution whose mean is at the center line and variance is proportional to m. The probability density function of the normal distribution with mean 0 and variance $\sigma^2 = m$ is as follows:

$$f_Y(y) = \frac{1}{\sigma\sqrt{2\pi}} e^{-\frac{1}{2}\left(\frac{y^2}{\sigma^2}\right)}$$

As time passes, more collisions will occur. Thus, it is reasonable to assume that the number of collisions is directly proportional to the elapsed time. If we let $m = \alpha t$, where α is a constant of proportionality, then the variance $\sigma^2 = m = \alpha t$ of the resulting normal distribution. Because the particle is drifting in the X-direction, the variance of its probability distribution in the Y-direction will increase with successive times $t_1, t_2, \ldots$, which also will correspond to successive distances $x_1, x_2, \ldots$.

From physical principles, Einstein showed that the probability distribution of the position of particles subjected to Brownian motion, as they are here, must satisfy the *diffusion equation*,[3] where B is the diffusion coefficient and f is the normal density:

$$\frac{\partial f}{\partial t} = B\frac{\partial^2 f}{\partial y^2}$$

The value of B is given by the formula $B = 2RT/Ns$, where R is the gas constant, T is temperature, N is Avogadro's number, and s is the coefficient of friction. By choosing the proper scale, we may take $B = 1/2$, and the following probability density function will satisfy the diffusion equation:

$$f_Y(y) = \frac{1}{\sqrt{2\pi t}} e^{-\frac{1}{2}\left(\frac{y^2}{t}\right)}$$

This result can be verified by taking the partial derivative of $f_Y(y)$ with respect to t and comparing the answer with the second partial derivative of $f_Y(y)$ with respect to y (Problem 5). The same equation is obtained in both cases.

From the above analysis, one can visualize the position of a drifting particle as described by a bell-shaped probability distribution, symmetrical about the center line, which spreads out as the particle drifts along. As the number of collisions increases, the distribution asymptotically becomes normal with its variance proportional to time. Figures 6.9 and 6.10 both show a two-dimensional system in which the probability distribution of the horizontal (Y-direction) distance of the particle is considered at successive longitudinal (X-direction) distances from the center line. If the probability distribution of the horizontal component of the position of the particle were plotted as a function of elapsed time in Figure 6.10 (instead of distance), the resulting plot would show a series of normal distributions with their standard deviations increasing with time as $\sqrt{t}$. Thus, if a particle is released from the source as time $t = 0$ and snapshots are taken at times $t = t_1, t_2, t_3$, and t_4, its horizontal position will be asymptotically normally distributed with mean $\overline{Y} = 0$ and standard deviation $\sigma_y = \sqrt{t_1}$, $\sqrt{t_2}$, $\sqrt{t_3}$, and $\sqrt{t_4}$. One can imagine the time plot by replacing x_1, x_2, x_3, and x_4 with $\sqrt{t_1}$, $\sqrt{t_2}$, $\sqrt{t_3}$, and $\sqrt{t_4}$, in Figure 6.10.

If many particles were released at once from the source, then their "expected arrival density" (expected number of particles per unit length) will follow similar normal distributions, except that the quantities will be multiplied by the number of particles released. The expected number of particles arriving in a given segment of the Y-axis will be given by the area under the normal curve

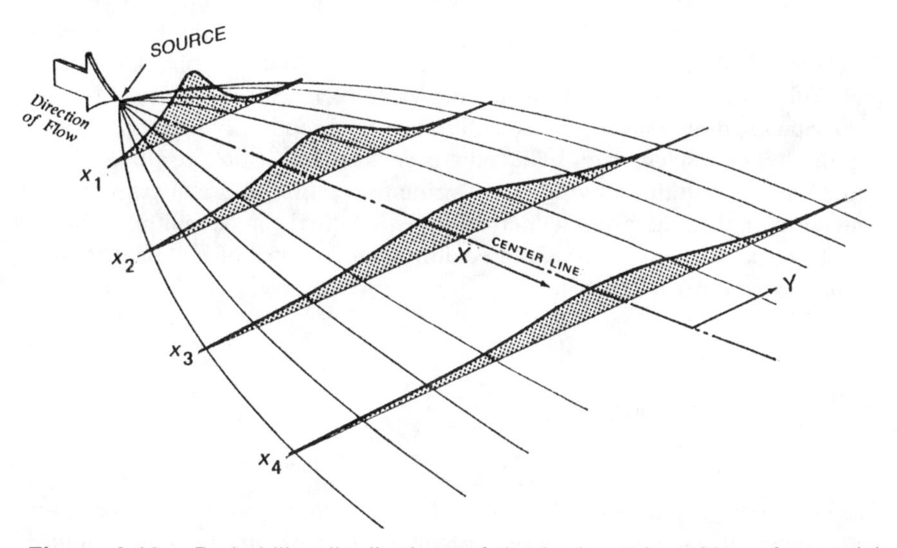

Figure 6.10. Probability distributions of the horizontal position of a particle subject to Brownian motion and drift. The vertical axis denotes probability, and the normal distributions occur at four different distances (and times) from the source.

for that segment. If n_o particles are released at time $t = 0$, then their expected particle densities at times $t = t_1, t_2, t_3$, and t_4 will be the product of n_o and normal distributions with variances $\sigma_y^2 = t_1, t_2, t_3$, and t_4. Any process in which particles behave in this fashion—a normal distribution with respect to space whose variance is proportional to time—satisfies the diffusion equation and is a diffusion process.

For a three-dimensional model, each particle has a horizontal (Y-direction) coordinate and a vertical (Z-direction) coordinate, and both coordinates will be normally distributed. In a real plume, the standard deviations of these two normal distributions will not necessarily be the same, because mixing characteristics of the carrier medium may differ in the horizontal and vertical directions. As time passes, more collisions will occur, and the standard deviations $\sigma_y(t)$ and $\sigma_z(t)$ will be functions of time. The position of the particle will have a bivariate normal probability distribution whose standard deviations are a function of time:

$$f_{YZ}(y,z) = \frac{1}{2\pi\sigma_y(t)\sigma_z(t)} e^{-\frac{1}{2}\left(\frac{y^2}{\sigma_y^2(t)} + \frac{z^2}{\sigma_z^2(t)}\right)}$$

This result is obtained by multiplying the PDFs for the two normal distributions, each with their respective variances, together.

For practical applications of this theory, we usually consider a constant plume, such as air pollution from a smokestack, in which the source is emitting a pollutant continuously. Drift is caused by physical movement of the carrier medium, such as wind traveling with a constant speed and direction. In this situation, the standard deviations of the probability distributions of the coordinates of each particle increase with time, as above. As distance from the source in the X-direction increases, the elapsed time for each particle also becomes greater. Thus, the standard deviations $\sigma_y(x)$ and $\sigma_z(x)$ also can be written as functions of distance in the X-direction. The result will be a bivariate normal probability density function in which the variance in both the vertical (Z) and horizontal (Y) directions increases with source-to-receptor (x) distance.

If the source discharges its pollutant at a constant emission rate Q[mass/time] into a carrier medium moving in a constant direction at a constant speed u [length/time], then the bivariate normal PDF [length^{-2}] can be multiplied by the ratio Q/u [mass/length] to give a three-dimensional model of pollutant concentration $C(x,y,z)$ [mass/length^3]:

$$C(x,y,z) = \frac{Q}{2\pi u\sigma_y(x)\sigma_z(x)} e^{-\frac{1}{2}\left(\frac{y^2}{\sigma_y^2(x)} + \frac{z^2}{\sigma_z^2(x)}\right)}$$

This equation, known as the *Gaussian plume equation*, has wide applicability in the fields of air pollution and meteorology. In most practical air pollution problems, the equations used for $\sigma_y(x)$ and $\sigma_z(x)$ have been obtained empirically from field experiments and depend on the atmospheric stability class.

A considerable body of literature is available on the Gaussian plume model.

For example, Turner[4] has prepared a workbook on air pollution dispersion estimates which contains many practical examples. Noll and Miller[5] have reviewed this model and other related versions used for making air quality predictions for point, line, and area sources. Benarie[6] has summarized the field of air pollution modeling in a book. Many additional papers on this topic appear in the periodical research literature, such as the *Journal of the Air and Waste Management Association* and *Atmospheric Environment*. The reader wishing to pursue this topic further is referred to the research literature. Our goal has been to illustrate how the basis for this model could be developed from relatively few assumptions and from analogs that have intuitive clarity.

Although the plume model usually is applied to air pollution problems, one can imagine similar applications to plumes created by water pollutants discharged into streams or lakes. One also can imagine the model applied approximately to the dispersion of chemicals as they seep downward through porous soils. This model is of general applicability whenever one chemical is transported by a carrier medium, experiencing random collisions on a molecular level as it drifts.

SUMMARY AND CONCLUSIONS

This chapter began by describing an idealized mechanical system, the wedge machine, which was intended as a simple mechanical analog of physical diffusion and dispersion processes in the environment. In the wedge machine, particles are released from a source at the top and fall by gravity through a uniform array of wedges. Whenever a particle encounters a wedge, it has probability $p = q = 1/2$ of falling to the left or right of the wedge. If K denotes the number of the channel in row m, then it is shown that the probability that a particle arrives in channel K is distributed as $\mathbf{B}(m,1/2)$. If n_o particles are released, and $p = P\{K = k\}$ denotes the probability that a particle arrives in a particular channel, then the expected number of arrivals will be given by $n_o p$. Thus, the expected number of arrivals will have a symmetrical, bell-shaped (binomial) distribution about the center line of the wedge machine.

It also was shown that the probability distribution of the number of arrivals in any channel on successive realizations is $\mathbf{B}(n_o,p)$, where p is the arrival probability computed from the binomial distribution $\mathbf{B}(m,1/2)$ given above. Because the arrival probabilities in each channel, which sum horizontally to 1, are virtually always less than $p = 1/2$, the distribution of arrivals in individual channels will be right-skewed. However, this right-skewness will not be very evident in most channels—except those at the extreme ends—because of the tendency of the binomial distribution to approach the normal distribution in appearance if n_o is large. If the particles are emitted from the source at a fixed rate with respect to time, then counts of the number of particles arriving in any channel (except those at the very top) in successive time intervals will be right-skewed, with expected value $n_o p$. In the author's view, this intrinsic right-skewness with respect to time is not sufficient to explain the right-skewness usually encountered in observed environmental concentrations, which often are believed to be lognormally distributed. Concentrations observed in the environment usually involve billions or trillions of molecules, and it is difficult to see how the diffusion processes of the plume model could, by their own stochastic properties, naturally give rise to lognormal distributions. The Theory of Successive Random Dilutions (Chapter 8) offers, I believe, a more plausible mechanism for generating right-skewed (asymptotically lognormal) distributions than do diffusion processes, even when random components are included.

This chapter also illustrates that an actual mechanical device, the particle frame machine, gives results very similar to those predicted theoretically for the wedge machine. In both machines, the horizontal position of the particle, like the drunken man engaged in random walks, is the random sum of linear increments, or a $\mathcal{RS}$-process. The horizontal position of particles colliding with molecules in a carrier medium, which creates a plume in the environment, also is a $\mathcal{RS}$-process. Both kinds of processes—the wedge machine and the plume model—asymptotically give rise to normal distributions, but by slightly different reasoning. In the wedge machine, like the drunken man taking steps of fixed size, the changes in position occur with probability $p = 1/2$ and are of fixed size. The final position is the result of adding these fixed increments, and it can be called a fixed increment random sum ($\mathcal{FIRS}$-process). All $\mathcal{FIRS}$-processes generate binomial distributions. If $p = 1/2$, as is the case here, the binomial distribution approaches the normal curve as n becomes large. The plume model also is a $\mathcal{RS}$-process, but the horizontal increments are not fixed and can take on any finite continuous values $d_1, d_2, \ldots$, etc. The final position also is a $\mathcal{RS}$-process, but, because the increments are variable, it can be called a variable increment random sum process ($\mathcal{VIRS}$-process). By the Central Limit Theorem, all $\mathcal{VIRS}$-processes generate probability distributions that are asymptotically normal (Chapter 7).

PROBLEMS

1. Suppose that a wedge machine is constructed in a manner similar to the one described in this chapter, except that it has one more row of wedges at the bottom. What will be the probability distribution of particle arrivals for the 7 channels comprising row 6? [Answer: **B**(6,1/2)] If $n_o = 1000$ particles are released from the source during every realization, what will be the expected number of arrivals in each channel at the bottom of the machine? [Answers: 15.625, 93.750, 234.375, 312.5, 234.375, 93.750, 15.625 particles]
2. In the wedge machine in Problem 1, the expected number of particles arriving in the first channel of row 6 will be given by $E[N] = (1000)(1/2)^6 =$ 15.625 particles. What will be the probability distribution of the number of arrivals in this channel on successive realizations of this process. [Answer: **B**(1000, 0.015625)] If the number of particles released from the source is reduced to $n_o = 100$ and $n_o = 10$, what will be the probability distributions for the number of arrivals on successive realizations in the first channel? [Answers: **B**(100, 0.015625), **B**(10, 0.015625)] Plot the first few terms of these three distributions and discuss their differences in shape.
3. A drunken man is asked to walk along a straight line, beginning at a lamppost. He takes a step to the left with probability $q = 1/2$ and a step to the

right with probability $p = 1/2$, and every step is of constant length $\pm d$. If $Y = Kd$ represents the final position relative to the lamppost, and if the man takes m steps, what will be the probability distribution of Y? [Answer: $Y = (2J - m)d$, where J is distributed as $\mathbf{B}(m,1/2)$] If this man takes exactly 3 steps, and each step is of length $d = 1$ ft, what is the probability that he ends up 3 ft to the left of the lamppost? [Answer: $P\{Y = -3 \text{ ft}\} = 0.125$] By tracing all possible routes that this man can take, show that positions $Y = \pm 2$ ft and $Y = 0$ ft are impossible and that all other positions have probabilities given by the translated binomial distribution.

4. Suppose that particles in the wedge machine described in this chapter are released at a constant rate of $r = 1$ particle per minute. Because arrivals are rare in channels at the outer edges of the wedge machine, their distribution can be approximated by the Poisson probability model. If $t = 10$ minutes, what will be the resulting distribution of particle arrivals in channel A of row 5? [Answer: $\mathbf{P}(rt) = \mathbf{P}(0.3125)$] What will be the probabilities associated with $N_A = 0,1,2$, and 3 arrivals in this channel? [Answers: 0.7316, 0.2286, 0.03572, 0.00372]
5. Consider a normal PDF with mean 0 and variance t describing the one-dimensional distance Y of a particle from the center line, as discussed in this chapter. Demonstrate that this density satisfies the diffusion equation. First, differentiate this PDF with respect to y to obtain the following result:

$$\frac{\partial f}{\partial y} = -\frac{y}{t^{3/2}\sqrt{2\pi}} e^{-\frac{1}{2}\left(\frac{y^2}{t}\right)}$$

Next, differentiate this result with respect to y once again. Finally, differentiate the original PDF with respect to t and compare it with the result obtained in the previous step. These two independent paths of analysis both should give the same following answer:

$$\frac{\partial f}{\partial t} = \frac{1}{2}\frac{\partial^2 f}{\partial y^2} = \left(\frac{y^2}{t^{5/2}} - \frac{1}{t^{3/2}}\right)\frac{1}{\sqrt{2\pi}} e^{-\frac{1}{2}\left(\frac{y^2}{t}\right)}$$

Obtaining this identical result in both cases shows that the Gaussian PDF satisfies the diffusion equation.

REFERENCES

1. Aitchison, J., and J.A.C. Brown, *The Lognormal Distribution* (London: Cambridge University Press, 1973).
2. "Statistics and Quality Control Products," Catalog No. 8, TECHNOVATE, Inc., 910 Southwest 12th Avenue, Pompano Beach, FL 33060, 1976.
3. Karlin, Samuel, and Howard M. Taylor, *A First Course in Stochastic Processes*, Chapter 7 (New York, NY: Academic Press, 1975).

4. Turner, D. Bruce, "Workbook of Atmospheric Dispersion Estimates," U.S. Department of Health, Education, and Welfare, Public Health Service, National Center for Air Pollution Control, Public Health Service Publication No. 999-AP-26, Cincinnati, OH, 1967.
5. Noll, Kenneth E., and Terry L. Miller, *Air Monitoring Survey Design*, Chapter II (Ann Arbor, MI: Ann Arbor Science Publishers, 1977).
6. Benarie, Michael M., *Urban Air Pollution Modelling* (Cambridge, MA: The MIT Press, 1980).

7 Normal Processes

Normal processes are those that give rise to normally distributed random variables. The normal, or Gaussian, distribution—a symmetrical, bell-shaped curve—is one of the most important probability models in statistics. Normally distributed random variables tend to arise naturally when many continuous, independent random variables are added together. As discussed in the previous chapter, a person taking randomly sized steps ends up at distances from the starting point that tend to be normally distributed. Also, many of the distributions presented in this book can become normal, or nearly normal, when their parameters approach certain limiting values.

Examples of common processes that lead to distributions that are approximately normal include the following:

1. A housewife seeks to save $20 per week, placing the amount in a cookie jar. Although she tries very hard to put aside exactly $20 each week, sometimes she saves a little more; sometimes a little less. At the end of the year, she expects to have a total amount of (52 weeks)($20/week) = $1,040, which she removes and spends. In some years, she has more than $1,040; in some years she has less than $1,040. If this same process occurs on a number of years (or if a number of different housewives behave in a similar manner), what will the distribution of the yearly sums in the cookie jar (or the housewives' cookie jars) look like?
2. In a particular tracer study, a particle starts at some point in space (the origin). Considering only movement in one dimension, the particle can move arbitrarily sized increments either to the right (positive) or to the left (negative). After a thousand of these small arbitrary displacements, the location of the particle relative to the origin is measured. What will the probability distribution of the particle's position look like? If the mean value of these incremental displacements is zero and their standard deviation is 10 microns, what will be the distribution of the particle's final position?
3. In a stream-monitoring study, 30 chloride samples are collected in a particular location. Past history indicates that the standard deviation of chloride concentrations in this area is 25 mg/l. If a sample of 30 readings has a mean value of 560 mg/l, how confident can we be that the average represents the true average of the stream? Can we say that there is 95% chance, or even a 99% chance, that the true mean lies within ± 10 mg/l of the observed mean?

4. In a personnel monitoring field study, 100 people work in a large room (such as a warehouse) where carbon monoxide (CO) is present. Suppose 12 people are selected at random on a particular date, and their CO exposure is measured over 8 hours. The resulting 12 samples give a mean value of 10.5 ppm and a standard deviation of 1.9 ppm. Can we say that the true mean (if more people were sampled) lies within 10.5 ± 1 ppm with 95% assurance?

In all of these situations, the final variable of interest—amount of money saved, distance of a particle from the origin, formation of the arithmetic mean—is produced by summing many observations.

CONDITIONS FOR NORMAL PROCESS

Many variables found in nature result from the summing of numerous unrelated components. When the individual components are sufficiently unrelated and complex, then the resulting sum tends toward normality as the number of components comprising the sum becomes increasingly large. Two important conditions for a normal process are: (1) summation of many continuous random variables, and (2) independence of these random variables. In summary, ...

> A normal process results when a number of unrelated, continuous random variables are added together.

A normal process is a random-sum process ($\mathcal{RS}$-process), as discussed in Chapter 6. The $\mathcal{RS}$-process is a form that is complementary to the random product process ($\mathcal{RP}$-process) introduced in Chapters 8 and 9.

DEVELOPMENT OF MODEL

Consider an idealized model in which $X_1, X_2, \ldots, X_n$ are mutually independent random variables with a common distribution in which the mean $\mu_o = 0$ and the variance $\sigma_o^2 = 1$. Suppose we form the normalized sum Z of these n random variables:

$$Z = \frac{X_1 + X_2 + \ldots + X_n}{\sqrt{n}}$$

By the Central Limit Theorem[1], Z tends to be normally distributed in the limit as n becomes large. The standardized normal distribution has the following PDF:

$$f_Z(z) = \frac{1}{\sqrt{2\pi}} e^{-\frac{1}{2}z^2} \qquad -\infty < z < \infty$$

This PDF is bell-shaped and symmetrical about zero (Figure 7.1). Its CDF, $F_Z(a)$,

$$F_Z(a) = \int_{-\infty}^{a} \frac{1}{\sqrt{2\pi}} e^{-\frac{1}{2}z^2} dz$$

cannot be evaluated analytically, but tables are available for its evaluation.[2] Another approach for obtaining values for the CDF is by using an approximation on a hand calculator or a personal computer (see last section of this chapter, pages 185–186). Because of its importance, the standardized normal CDF, $F_Z(a)$ often is represented in the literature by the standardized notation $\Phi(a)$.

Several useful values of the CDF are listed on the intercepts of Figure 7.1. Because $F_Z(-1) = 0.1587$ and $F_Z(+1) = 0.8413$, the probability that Z lies within the range $-1 \leq Z < +1$ is $0.8413 - 0.1587 = 0.6826$. Thus, 68.26% of the distribution lies within the range $Z = \pm 1$. Similarly, the probability that Z lies within the range $-2 \leq Z < +2$ is given by $F_Z(+2) - F_Z(-2) = 0.97725 - 0.02275 = 0.9545$. Thus, 95.45% of the distribution lies within the range $Z = \pm 2$. The range of Z that accounts for exactly 95% of the distribution is $Z = \pm 1.96$, a result that often is useful for calculating the 95% confidence interval for a normally distributed random variable.

From Figure 7.1, the probability that Z lies within the range $-3 \leq Z < +3$ is given by $F_Z(+3) - F_Z(-3) = 0.99865 - 0.00135 = 0.9973$. Thus, 99.73% of the distribution lies within the range $Z = \pm 3$. The range of Z that accounts for exactly 99% of the distribution is $Z = \pm 2.58$. This result is useful for calculating the 99% confidence interval for a normally distributed random variable.

Usually, it is of interest to consider normally distributed random variables

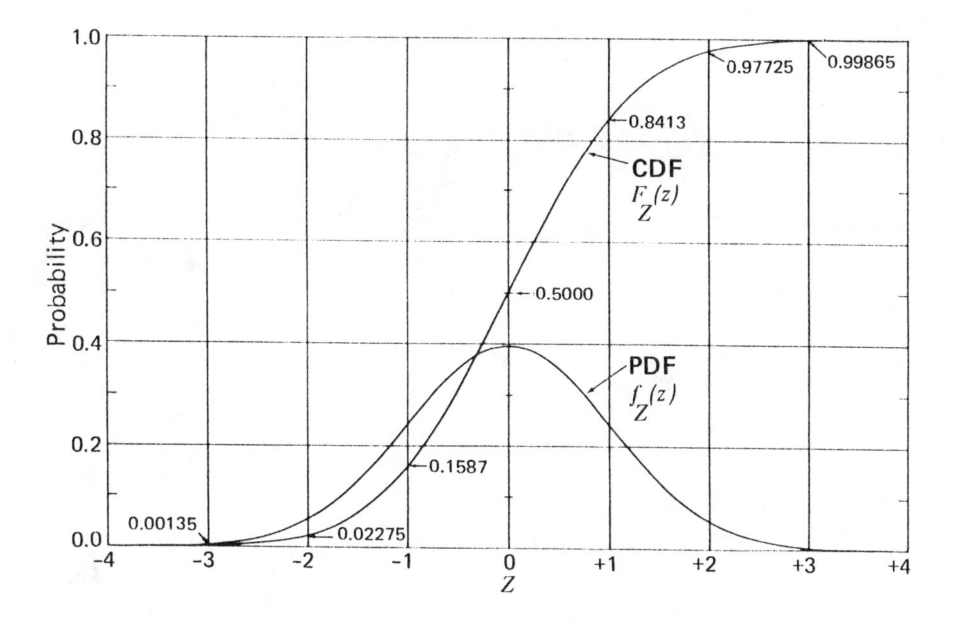

Figure 7.1. PDF and CDF of the standardized normal distribution.

that are translated such that they are symmetrical about values other than the origin. To make this translation, we employ the following transform:

$$z = \frac{y-\mu}{\sigma}$$

Differentiating this transform to obtain $dz/dy = 1/\sigma$, we solve for dz to obtain $dz = dy/\sigma$. The resulting PDF for the transformed random variable Y is written as follows:

$$f_Y(y) = \frac{1}{\sigma\sqrt{2\pi}} e^{-\frac{1}{2}\left(\frac{y-\mu}{\sigma}\right)^2}$$

This PDF has the same characteristic bell shape as in Figure 7.1, except that it is centered about μ and its spread is determined by σ. Thus, μ is its location parameter, and σ is its dispersion, or scale, parameter. By this transformation, $y = \sigma z + \mu$, so the CDF of the standardized normal distribution $F_Z(z)$ can be used to compute the CDF of the transformed normally distributed random variable $F_Y(y)$. We denote this transformed normally distributed random variable Y as $\mathbf{N}(\mu,\sigma)$ (see Table 7.1). The standardized normal distribution $f_Z(z)$ may be viewed as the special case $\mathbf{N}(0,1)$.

The transformed normally distributed random variable Y is extremely useful in a variety of problems. As indicated in the following section, summing processes tend naturally to produce normally distributed random variables. Be-

Table 7.1. The Normal Distribution $\mathbf{N}(\mu,\sigma)$

Probability Density Function:

$$f_Y(y) = \frac{1}{\sigma\sqrt{2\pi}} e^{-\frac{1}{2}\left(\frac{y-\mu}{\sigma}\right)^2} \qquad -\infty < y < \infty$$

Expected Value:

$$E[Y] = \mu$$

Variance:

$$Var(Y) = \sigma^2$$

Coefficient of Skewness:

$$\beta_1 = 0$$

Coefficient of Kurtosis:

$$\sqrt{\beta_2} = 3$$

cause formation of the arithmetic mean of a set of observations is a summing process, the distribution of the sample mean about the true population mean tends to be normal. This result is useful for calculating confidence intervals about the arithmetic mean.

As indicated in Table 7.1, the normal distribution $\mathbf{N}(\mu,\sigma)$ is fully specified by its expected value μ and its variance σ^2. Because the distribution is symmetrical, its coefficient of skewness is zero: $\sqrt{\beta_1} = 0$. Because its characteristic shape remains unchanged even though it is translated by changes in μ and its spread is altered by changes in σ, its coefficient of kurtosis remains unchanged: $\beta_2 = 3$.

Summing Processes

Consider the mutually independent random variables $X_1, X_2, \ldots, X_n$ with means $\mu_1, \mu_2, \ldots, \mu_n$ and variances $\sigma_1^2, \sigma_2^2, \ldots, \sigma_n^2$. Suppose we form the simple sum of these random variables:

$$Y = X_1 + X_2 + \ldots + X_n$$

By the rules of expected values (Chapter 3, pages 38–41), the expected value of this sum $\mu = E[Y]$ will be the sum of the individual expected values:

$$\mu = E[Y] = E[X_1] + E[X_2] + \ldots + E[X_n] = \mu_1 + \mu_2 + \ldots + \mu_n$$

Similarly, by the rules for combining variances (Chapter 3), the variances of this sum $\sigma^2 = Var(Y)$ will be the sum of the individual variances:

$$\sigma^2 = Var(Y) = Var(X_1) + Var(X_2) + \ldots + Var(X_n) = \sigma_1^2 + \sigma_2^2 + \ldots + \sigma_n^2$$

Consider the special case in which each of the n random variables comprising the sum has the same mean $\mu_o = \mu_1 = \mu_2 = \ldots = \mu_n$ and the same variance $\sigma_o^2 = \sigma_1^2 = \sigma_2^2 = \ldots = \sigma_n^2$. Then the mean and variance of the sum Y will be:

$$\mu = n\mu_o \qquad \sigma^2 = n\sigma_o^2$$

Thus, the standard deviation of Y will be given by $\sigma = \sqrt{n}\,\sigma_o$ and its mean will be given by $\mu = n\mu_o$.

The sum Y characteristically will approach the normal distribution $\mathbf{N}(\mu,\sigma)$. Here we are able to predict the distribution of Y even though we do not know the distributional forms of the individual random variables comprising the sum. The summing process brings about this characteristic distributional form, the normal distribution, even though the distributions of the variables making up the sum may not be normal. Indeed, these variables may have the same distribution or they may have different distributions. Furthermore, once we know the means and variances of the individual variables within the sum, we can predict the two parameters of the resulting normally distributed sum. Thus, we can fully specify the distributional form of the random variable of interest Y.

To apply these concepts to the earlier problem of the housewife who is trying to save \$20 per week, we observe that the total amount accumulated at the end of each year is the *sum* of her 52 weeks of savings. Thus, the yearly total will tend toward a normal distribution. What are the parameters of this distribution? Because the housewife seeks to save \$20 per week, we can assume the weekly expected value will be μ_o = \$20. Thus, the expected value of the yearly total will be $\mu = n\mu_o$ = (52)(\$20) = \$1,040. If, from past experience with the housewife's saving habits, we can assume that the standard deviation of the weekly savings is σ_o = \$2, then the standard deviation of the yearly savings will be $\sigma = \sqrt{n}\sigma_o = \sqrt{52}\,(2)$ = \$14.42. Then the PDF of the total amount of money Y at the end of the year will be **N**(1040,14.42), which is written as follows:

$$f_Y(y) = \frac{1}{14.42\sqrt{2\pi}} e^{-\frac{1}{2}\left(\frac{y - 1040}{14.42}\right)^2}$$

Because 95% of the standardized normal distribution occurs between $Z = \pm 1.96$, 95% of the transformed normal distribution will lie within $\mu \pm 1.96\sigma$. Applying this result to the housewife's problem, her yearly total will be within \$1,040 ± (1.96)(\$14.42) = \$1,040 ± \$28.26 with probability $P = 0.95$. That is, there is a 95% chance that the yearly total lies between \$1,011.74 and \$1,068.26. Similarly, 99% of the standardized normal distribution lies within $Z = \pm 2.58$, so 99% of the transformed normal distribution will lie within $\mu \pm 2.58$. Using this fact, there is a 99% chance that the yearly total will lie within the limits \$1,040 ± (2.58)(\$14.42) = \$1,040 ± \$37.20. The ranges for other probabilities can be computed using the computer program for the CDF of the standardized normal distribution given at the end of this chapter (Figure 7.8, page 186). In general, as the probabilities become higher, the ranges become larger.

Notice that the mean of the sum is proportional to the number of terms n in the sum while the standard deviation increases as $\sqrt{n}$. Since the coefficient of variation v is the ratio of the standard deviation to the mean (see Chapter 3, page 44), v will be a function of $1/\sqrt{n}$. For the housewife's weekly savings, for example, v = \$2/\$10 = 0.10, or 10%. By comparison, the coefficient of variation for the yearly total is v = \$14.42/\$1,040 = 0.0139, or about 1.4%. Thus, the relative variation, as measured by v, shows a significant reduction due to the summing process. This result agrees with common sense: as more values are added together, the positive and negative weekly deviations about the mean tend to offset each other, causing the relative variation to become smaller. At the same time, and perhaps less obvious, a bell-shaped curve emerges from these sums.

The particle problem at the beginning of this chapter is similar to the housewife problem, except that the mean is zero. Like the particle examples in the previous chapter, the final position of the particle is the sum of all the displacements it has experienced. Because the means of the individual displacements are $\mu_o = 0$ (zero drift), the overall mean will be the sum of these means, or $\mu = 0$. Since there are 1,000 displacements, and the standard deviation of the individual displacements is σ_o = 10 microns, the standard deviation of the overall distribution of the particle's position will be $\sigma_o\sqrt{n} = 10\sqrt{1000} = 31.6$ microns. Thus, the distribution of the particle's final position will be **N**(0,31.6).

Averaging Processes

When we compute the mean of a set of n observations, we perform a summing process, except that we then divide the sum by n. Suppose we collect n observations from some process, representing them as $X_1, X_2, \ldots, X_n$. Then we form the sample mean $\overline{X}$ of these observations as follows:

$$\overline{X} = \frac{X_1 + X_2 + \ldots + X_n}{n} = \frac{X_1}{n} + \frac{X_2}{n} + \ldots + \frac{X_n}{n}$$

Since the samples $X_1, X_2, \ldots, X_n$ are random variables, the sample mean $\overline{\mathrm{X}}$ is a function of these random variables and is itself a random variable. Suppose the individual observations have means $\mu_1, \mu_2, \ldots, \mu_n$ and variances $\sigma_1^2, \sigma_2^2, \ldots, \sigma_n^2$.

How does the expected value of $\overline{X}$, $E[\overline{X}] = \mu$, relate to the observational means $\mu_1, \mu_2, \ldots, \mu_n$? Typically, the observations are collected in some scheme to make them as representative of the process as possible. In random sampling strategies, for example, each member of the sampled population customarily has an equally likely chance of being selected. If the resulting observations are independent, we can apply the general rules for expected values given in Chapter 3 (pages 38–41):

$$\mu = E[\overline{X}] = E[X_1/n] + E[X_2/n] + \ldots + E[X_n/n]$$

$$= \frac{1}{n}(E[X_1] + E[X_2] + \ldots + E[X_n]) = \frac{1}{n}(\mu_1 + \mu_2 + \ldots + \mu_n)$$

How does the sample mean's variance, $Var(\overline{X}) = \sigma^2$, relate to the observational variances $\sigma_1^2, \sigma_2^2, \ldots, \sigma_n^2$? For independent observations, the rules for combining variances given in Chapter 3 (pages 41–42) yield the following:

$$\sigma^2 = Var(\overline{X}) = Var(X_1/n) + Var(X_2/n) + \ldots + Var(X_n/n)$$

$$= \frac{1}{n^2} Var(X_1) + \frac{1}{n^2} Var(X_2) + \ldots + \frac{1}{n^2} Var(X_n) = \frac{1}{n^2}(\sigma_1^2 + \sigma_2^2 + \ldots + \sigma_n^2)$$

Now consider the case in which the process is stationary during the period in which the samples are collected. Thus, all the n observations have the same mean $\mu_o = \mu_1 = \mu_2 = \ldots = \mu_n$ and all have the same variance $\sigma_o^2 = \sigma_1^2 = \sigma_2^2 = \ldots = \sigma_n^2$. If we substitute these conditions into the above expressions for μ and σ^2, we obtain the following results:

$$\mu = \frac{1}{n}(n\mu_o) = \mu_o \qquad \sigma^2 = \frac{1}{n^2}(n\sigma_o^2) = \frac{\sigma_o^2}{n}$$

These results are important and useful for many applications. These equations show that the expected value of the sample average $E[\overline{X}]$ will be identical to the population mean μ_o. Thus, the observed sample mean $\overline{X}$ will tend to approach the population mean μ_o. How rapidly will $\overline{X}$ approach μ_o? The variance σ^2 of $\overline{X}$ will decrease as n increases; as indicated by the above expression, the variance will be proportional to $1/n$. Thus, the standard deviation of the sample mean is given by $\sigma = \sigma_o/\sqrt{n}$, and the standard deviation will approach zero as n becomes increasingly large. As this happens, the sample mean approaches the population mean, or $\bar{X} \rightarrow \mu_o$ as n becomes increasingly large. This conclusion agrees with common sense: as more samples are averaged, their mean approaches the true population mean and their variability about the true mean becomes smaller.

The quantity $\sigma = \sigma_o/\sqrt{n}$ is called the *standard error of the mean*, sometimes denoted as $SE(\bar{X})$. That is,

$$SE(\bar{X}) = \sigma = \sigma_o / \sqrt{n}$$

The quantity $SE(\bar{X})$ is a measure of variation just like σ_o; however, it is a measure of the variation of the sample mean rather than the variation of the individual observations. It is an important statistical quantity, because it gives an indication of the "sampling error." Sampling error differs from other kinds of error, in that it often can be reduced by the person collecting the data. Increasing the sample size n reduces the sampling error, but it also generally increases the data collection costs. Thus, there is a trade-off between reducing sampling error and increasing the cost of collecting the data.

Finally, what is the distribution of the observed average $\bar{X}$? Because forming the average is a summing process, the sample mean $\bar{X}$ will be approximately normally distributed with mean μ_o. The above analysis gives us not only one parameter of this normal distribution, its mean, but also its second parameter, its standard deviation $\sigma = \sigma_o/\sqrt{n}$. Thus, the distribution of the random variable $\bar{X}$ is fully specified to be $\mathbf{N}(\mu_o, \sigma_o/\sqrt{n})$, approximately.

The tendency for the observed sample mean to be normally distributed rests on the Central Limit Theorem (CLT). Proof of the CLT is not a suitable subject for this book; however, it is discussed in detail by Feller.[1] A simple approach for convincing one of the Theorem's truthfulness is to generate random numbers on a computer, form their sums, and examine the resulting frequency distributions. Kuzma[2] samples from distributions of unusual shapes, forms the means, and plots the distribution of the means. Even when the distributions are uniform, bimodal, or strongly right-skewed, the distribution of the means appears symmetrical for sample sizes as small as $n = 5$. The characteristic bell-shaped form of the normal distribution emerges for sample sizes above $n = 30$.

It is instructive to compare the resulting equations for summing processes with the equations for averaging processes (Table 7.2):

Table 7.2. Parameters of the Normally Distributed Random Variable Resulting from Two Common Normal Processes

	Mean	Variance	Standard Deviation
Summing Process:	$\mu = n\mu_o$	$\sigma^2 = n\sigma_o^2$	$\sigma = \sigma_o\sqrt{n}$
Averaging Process:	$\mu = \mu_o$	$\sigma^2 = \sigma_o^2/n$	$\sigma = \sigma_o/\sqrt{n}$

For a summing process, the mean increases linearly with n while the standard deviation increases with $\sqrt{n}$. For an averaging process, the mean approaches the population mean as n increases, with the standard deviation decreasing in inverse proportion to $\sqrt{n}$.

To summarize, the following important conclusions are valid for averaging processes and apply to sample means calculated from a set of observations:

1. The mean μ of the distribution of sample means approaches the population mean μ_o.
2. The standard deviation of the sample mean σ is computed from the standard deviation of the population σ_o by using the relationship $\sigma = \sigma_o/\sqrt{n}$.
3. The distribution of the sample means is approximately normal.

These conclusions, derived theoretically from the CLT, can be applied to a surprisingly large number of different situations.

CONFIDENCE INTERVALS

By extending these concepts further, we can obtain a general procedure with many practical applications. The above concepts are of limited usefulness in that they require us to know the mean and variance of the original population from which the samples were obtained; with this information, we then can compute the probability that the observed sample mean will lie within some specified range of the population mean. That is, if we know μ_o and σ_o, then we can use the standardized normal distribution (Figure 7.1) to determine the probability that $\overline{X}$ will lie within some particular range of μ_o. For example, if we happen to choose the range of two standard deviations, or $\mu_o \pm 2\sigma_o/\sqrt{n}$, then we have:

$$P\left\{\mu_o - 2\frac{\sigma_o}{\sqrt{n}} \leq \overline{X} \leq \mu_o + 2\frac{\sigma_o}{\sqrt{n}}\right\} = 0.9545$$

It is common custom to consider the range of values that corresponds to a probability of 0.95 exactly. Because $F_Z(+1.96) - F_Z(-1.96) = 0.95$ from the previous discussion of the standardized normal distribution, the range $\mu_o \pm 1.96\sigma_o/\sqrt{n}$ will correspond to a probability of 0.95 exactly:

$$P\left\{\mu_o - 1.96\frac{\sigma_o}{\sqrt{n}} \leq \overline{X} \leq \mu_o + 1.96\frac{\sigma_o}{\sqrt{n}}\right\} = 0.95$$

Although this result is interesting, it has limited direct usefulness, because we seldom know μ_o or σ_o. Rather, we ordinarily have only a set of n observations; from these, we can calculate an observed sample mean $\overline{x}$, and we then would like to know how close the true population mean μ_o lies to the sample mean.

How can the above expression be modified to give the range in which the true mean lies with known probability, given the observed sample mean as a reference point?

Suppose we begin by subtracting μ_o from all terms in the above inequality, giving the following expression:

$$P\left\{-1.96\frac{\sigma_o}{\sqrt{n}} \le \overline{X} - \mu_o \le 1.96\frac{\sigma_o}{\sqrt{n}}\right\} = 0.95$$

Next, we change all the signs in the equation and reverse the direction of inequality:

$$P\left\{1.96\frac{\sigma_o}{\sqrt{n}} \ge -\overline{X} + \mu_o \ge -1.96\frac{\sigma_o}{\sqrt{n}}\right\} = 0.95$$

Finally, we add $\overline{X}$ to all terms in the equation and rearrange:

$$P\left\{\overline{X} - 1.96\frac{\sigma_o}{\sqrt{n}} \le \mu_o \le \overline{X} + 1.96\frac{\sigma_o}{\sqrt{n}}\right\} = 0.95$$

From this result, we can state that the true mean μ_o lies within the range $\overline{X} \pm 1.96\sigma_o/\sqrt{n}$ with probability 0.95. Thus, if we calculate a sample mean $\overline{x}$ from a set of n observations, we can use this expression to predict the range in which the true mean can be found with probability 0.95.

The above equation is of the general form:

$$P\{\overline{X} - d \le \mu_o \le \overline{X} + d\} = p$$

The probability p is called the *confidence level*, sometimes denoted as CL, and usually expressed as a percentage. That is, CL = $100p$. The resulting range $(\overline{X} - d, \overline{X} + d)$ is called the *confidence interval*. CL values of 90%, 95%, and 99% are quite popular. For the case in which σ_o is known, the values of d for these CLs, consistent with the CLT, are summarized in Table 7.3.

Table 7.3. Confidence Levels and Corresponding Confidence Intervals for the Case in which σ_o is Known, as Derived from the CLT

Confidence Level (CL) $100p$	Confidence Interval $\pm d$
90%	$\pm 1.65\frac{\sigma_o}{\sqrt{n}}$
95%	$\pm 1.96\frac{\sigma_o}{\sqrt{n}}$
99%	$\pm 2.58\frac{\sigma_o}{\sqrt{n}}$

Example 3 at the beginning of this chapter (page 163) helps to illustrate these concepts. Here, 30 observations have been collected in a stream-monitoring study, and their mean is 560 mg/l. From the past history of measurements under the same conditions at this location, we conclude that the standard deviation is 25 mg/l. The investigator wants to know how close the true mean may be to the observed mean.

Is the true mean within ± 10 mg/l of the observed mean with 95% confidence? Using the expression for the 95% confidence interval in Table 7.3, we obtain the following:

$$\bar{x} \pm 1.96\frac{\sigma_o}{\sqrt{n}} = 560 \pm (1.96)\frac{25}{\sqrt{30}}$$

$$= 560 \pm (1.96)(5.48) = 560 \pm 8.9$$

Thus, the investigator can be 95% certain that the true mean lies within the range 560 – 8.9 = 551.1 mg/l and 560 + 8.9 = 568.9 mg/l. Because the 95% confidence interval is ± 8.9 mg/l, the true mean lies within a range that is narrower than ± 10 mg/l with at least 95% assurance.

Is the true mean within ± 10 mg/l of the observed mean with 99% confidence? Using the expression for the 99% confidence interval in Table 7.3, we obtain the following:

$$\bar{x} \pm 2.58\frac{\sigma_o}{\sqrt{n}} = 560 \pm (2.58)\frac{25}{\sqrt{30}}$$

$$= 560 \pm (2.58)(5.48) = 560 \pm 14.1$$

With the same data, the investigator can be 99% certain that the true mean lies within the range above 560 – 14.1 = 545.9 mg/l and below 560 + 14.1 = 574.1 mg/l; therefore, the answer is "No," and the true mean does not lie within the desired ± 1 mg/l interval with 99% assurance. Thus, in this problem, the confidence interval lies within the desired ± 10 mg/l range of the observed mean at the 95% confidence level but not at the 99% confidence level.

In real environmental situations, we ordinarily can calculate the sample mean from a set of observations quite easily, but we seldom know the population standard deviation σ_o. Unless extensive historical data are available from which to infer σ_o, some other approach must be found for obtaining a value for this parameter to be used in the relationships of Table 7.3. Can the sample standard deviation s computed from the set of observations be used as a substitute for σ_o?

Chapter 3 describes ways to calculate an estimate of the sample variance s^2 from a set of empirical observations. The following relationship from Chapter 3 (page 49) gives an unbiased estimate of the sample standard deviation s:

$$s = \sqrt{\frac{\sum_{i=1}^{n}(x_i - \bar{x})^2}{n-1}}$$

Chapter 3 discusses both one-pass and two-pass techniques for calculating s from a set of observations. If we substitute s for σ_o, additional error will be introduced in the confidence interval calculation process, for we now will have a different distribution. This distribution was discovered in 1906 and published in 1908 by William S. Gossett,[3] an English chemist and statistician employed by the Guinness Brewer in Dublin. As noted by Kuzma,[2] the Brewery did not encourage publications that might reveal trade secrets, so Gossett published it under the pseudonym "Student." Thus, his distribution is commonly called the Student's *t*-distribution.[4] In 1925, Gossett published values for this distribution,[5] and tables can be found in most statistical reference books.[6] This distribution is "bell shaped" like the normal distribution, but it has greater spread.[4] As n becomes large (above $n = 100$), values from the Student's *t*-distribution are virtually identical to those obtained from the normal distribution.

To deal with the larger uncertainty that is introduced when s is used as an estimate of σ_o, especially for small sample sizes, the confidence interval about the mean should be calculated using t values from the Student's *t*-distribution in place of the CDF of the normal distribution:

$$P\left\{\overline{X} - t\frac{s}{\sqrt{n}} \le \mu_o \le \overline{X} + t\frac{s}{\sqrt{n}}\right\} = p$$

Tables of t values for particular confidence levels, CL = $100p$, and various degrees of freedom $n - 1$ are readily available in reference sources.[6] As an alternative to using a table, t values also can be read from a nomogram or a graph (Figure 7.2). Using this graph, it is possible to read t values to two significant figures, which is adequate for many calculations. For sample sizes less than 10, the t values increase very rapidly as n decreases, reflecting the large uncertainty caused by very small sample sizes. As n approaches 100 for the 95% confidence level, the t value approaches 1.96, the same value as for the normal distribution.

To illustrate use of the Student's *t*-distribution to calculate confidence intervals, consider Example 4 at the beginning of this chapter. In this personnel monitoring study, carbon monoxide exposure readings were made on 12 people, yielding an average of $\overline{x} = 10.5$ ppm and an observed standard deviation of $s = 1.6$ ppm. What is the 95% confidence interval of the observed mean about the true mean? From the curve for the 95% confidence level with $n - 1 = 12 - 1 = 11$ degrees of freedom (Figure 7.2), we obtain $t = 2.2$. Substituting the observed sample mean of 10.5 ppm into the above equation for the confidence interval with the Student's *t*-distribution, along with these values for s, t, and n, we obtain the following result:

$$P\left\{10.5 - 2.2\frac{1.9}{\sqrt{12}} \le \mu_o \le 10.5 + 2.2\frac{1.9}{\sqrt{12}}\right\} = 0.95$$

$$P\{10.5 - 1.2 \le \mu_o \le 10.5 + 1.2\} = 0.95$$

From this result, the true mean lies within the range $10.5 - 1.2 = 9.3$ ppm and $10.5 + 1.2 = 11.7$ ppm with probability 0.95. If the investigator compares this result with a 9 ppm health related standard, he can have 95% assurance that the average of all 100 persons in the room is above the standard, a matter that may be of concern to him.

Since the resulting confidence interval is ± 1.2 ppm, it is wider than the ± 1 ppm confidence interval that the investigator originally sought. Thus, he

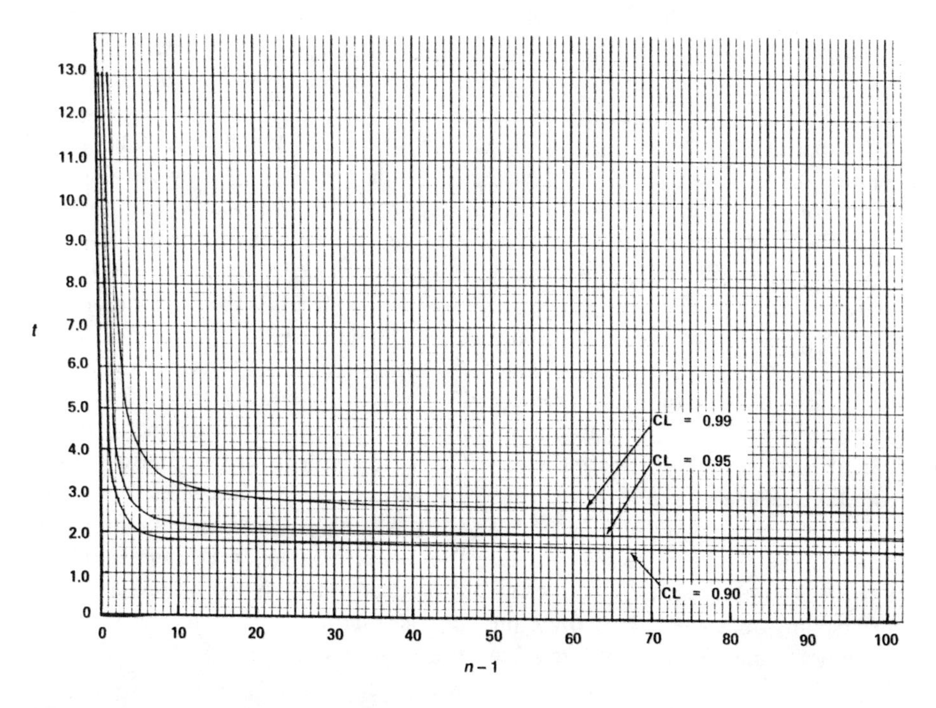

Figure 7.2. Plot of the factor t as a function of degrees of freedom $n - 1$ from the Student's t-distribution.

may reject this finding as not being sufficiently precise. Increasing the sample size will reduce the width of the confidence interval, but we can't say exactly how many people should be sampled, because we don't yet know the value of the standard deviation that will be observed with the larger sample. A good practical approach is to use the present standard deviation to calculate the number of samples that should be collected, add a few extra samples, and repeat the experiment. The 90% and 99% confidence intervals can be calculated for this problem using the same expression as above, except that the t value is read from the other curves in Figure 7.2.

APPLICATIONS TO ENVIRONMENTAL PROBLEMS

Calculating confidence intervals from a set of observations using the methods just described is straightforward. However, environmental quality data may exhibit certain pecularities that cause the analyst to question the assumptions on which these techniques are based. For example, hourly readings of water quality or ambient air quality usually have strong serial dependencies. Thus, the concentration observed in a particular hour is not independent of the concentration observed in the next hour. Because of such serial autocorrelation, convergence toward the mean usually tends to be more gradual than the above expressions predict. How does such autocorrelation alter the confidence intervals calculated from these environmental data?

Ott and Mage[7] examined the problem of calculating confidence intervals from ambient air quality data that are known to be dependent. They obtained a year's observations of ambient carbon monoxide (CO) concentrations measured at the San Jose, CA, air monitoring station (the 12-month period from June 1, 1970, to May 31, 1971). The histogram of these same data appears in Chapter 3 (Figure 3.13, page 67) of this book. Although a year ordinarily consists of 8,760 hours, 187 observations were missing, giving 8,573 hourly average CO readings (see discussion in Chapter 3). Missing observations generally are due to instrument calibration and maintenance; however, this data set was fairly complete with 97.9% of the year covered. Furthermore, the 187 missing observations seemed to be evenly distributed throughout the year.

The histogram of this year of CO observations (Figure 3.13) exhibits a single mode and considerable right-skewness. When the cumulative frequencies obtained from this histogram are plotted on logarithmic-probability paper (Figure 7.3), the curve appears relatively straight, indicative of a lognormal distribution. Further analysis of this data set by the authors revealed considerable autocorrelation. If we treat the year of observations as the original population from which a sample is to be drawn, then the population mean is calculated as $\mu_o = 3.537$ ppm, and the population standard deviation is calculated as $\sigma_o = 2.271$ ppm. Because μ_o is known, one can use Monte Carlo sampling to draw random samples of various sizes from this population and see how close the mean of these samples lies to the true mean μ_o. Because σ_o is known, one can see how well the standard deviation of the observed means agrees with σ predicted by confidence interval formulas based on the assumption of independence.

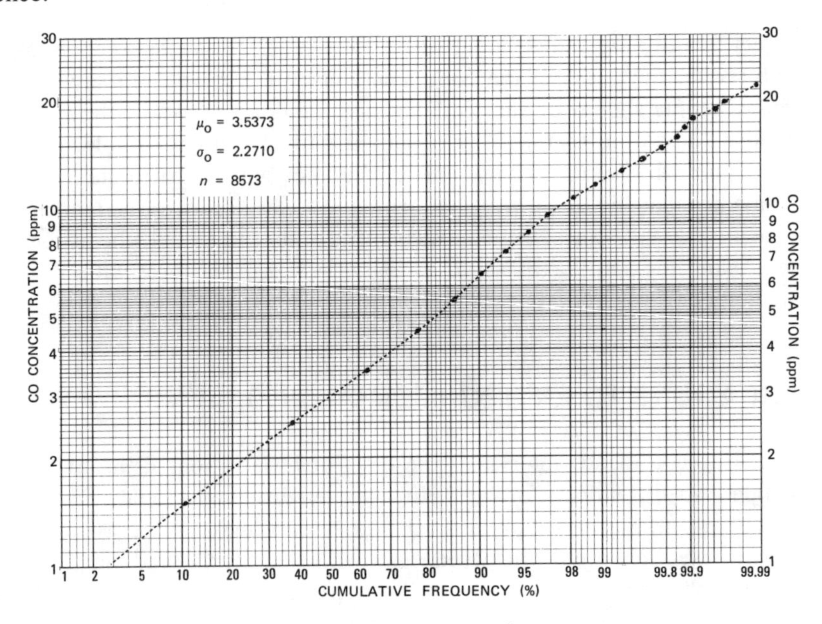

Figure 7.3. Logarithmic-probability plot of hourly CO concentrations observed for one-year period at the San Jose, CA, air monitoring station.

To make these comparisons, Ott and Mage[7] wrote a computer program that first selected the day of the year at random and next selected the hour within that day at random. By this approach, the CO reading for each hour of the year had an equally likely chance of being selected. Once a particular hour was selected, it could not be selected again, thus sampling "without replacement." They tested four different sample sizes—$n = 9$, $n = 36$, $n = 144$, and $n = 576$—obtaining 1,000 groups of each size. For each group of samples, they calculated

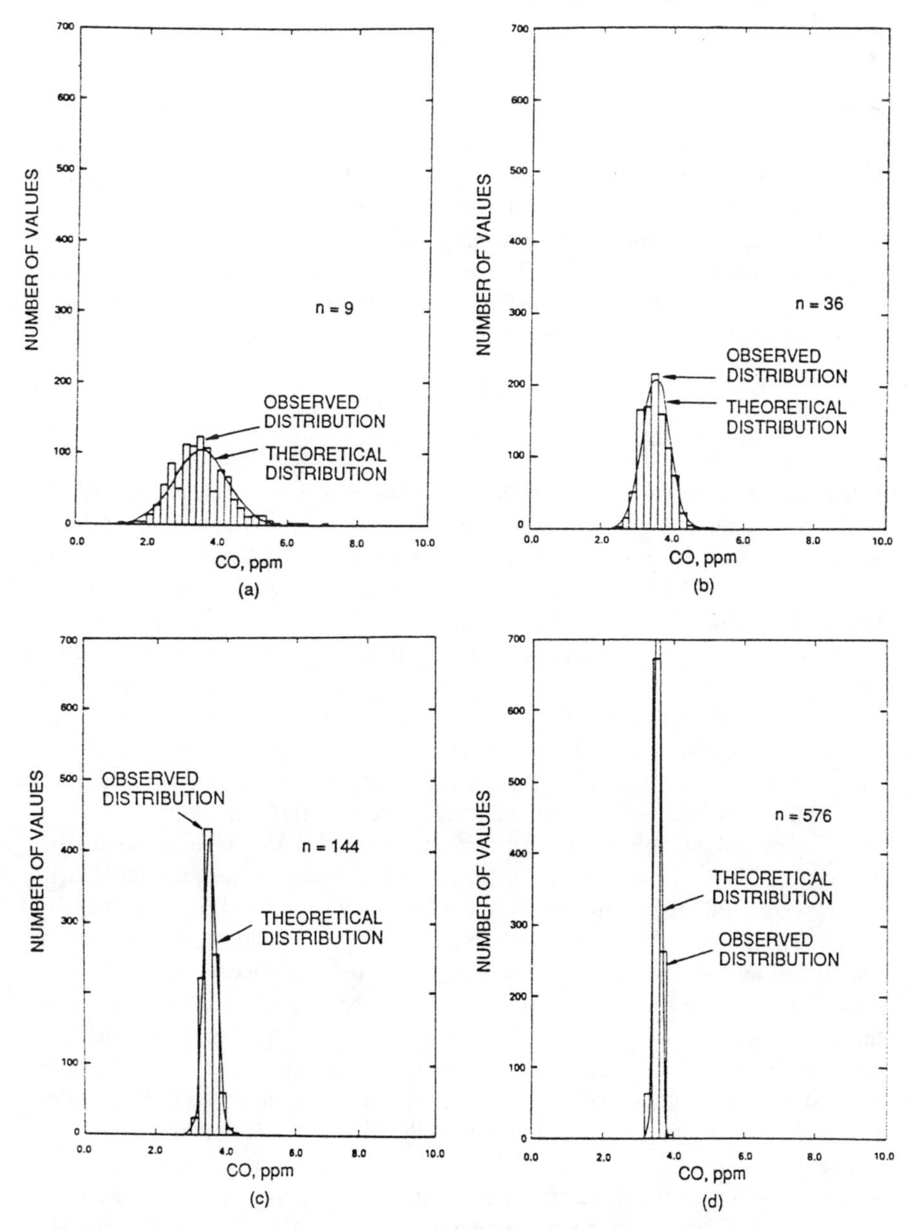

Figure 7.4. Histogram of the mean values of 1,000 groups of samples of size *n* randomly selected from the San Jose CO data set, along with the normal distribution: (a) $n = 9$; (b) $n = 36$; (c) $n = 144$; (d) $n = 576$ (Source: Ott and Mage[7]).

Table 7.4. Comparison of Predicted and Observed Parameters for a Year of CO Concentrations Observed in San Jose[7]

	Mean μ(ppm)		Standard Deviation σ(ppm)	
n	Population	Observed	Predicted	Observed
9	3.537	3.512	0.757	0.761
36	3.537	3.505	0.379	0.371
144	3.537	3.529	0.189	0.184
576	3.537	3.538	0.0946	0.0920

the arithmetic mean, thus giving 1,000 means for the groups of size 9; 1,000 means for the groups of size 36; and so on. When they plotted the histograms of these observed means (Figure 7.4), no right-skewness was evident, and the distributions became narrower as the sample size increased.

The average of all the means in each group (Table 7.4) was very close to the overall population mean of 3.537 ppm, as predicted by the CLT equations for averaging processes (Table 7.2). If the observations actually were independent, these equations predict the standard deviation of the means as follows:

$$\sigma = \frac{\sigma_o}{\sqrt{n}} = \frac{2.271}{\sqrt{n}}$$

When the standard deviation predicted by this equation is compared with the observed standard deviation of the means (Table 7.4), extremely good agreement is evident. For groups of size 9, for example, the CLT predicts a standard deviation of 0.757 ppm, and the observed standard deviation of 1,000 means was 0.761 ppm.

A graphical way to evaluate how closely two variables agree is to plot them on the horizontal and vertical axes of standard linear graph paper. If the resulting points lie close to a 45° line on this paper, then they are nearly equal; that is, $x = y$. If the standard deviations of Ott and Mage's random sampling experiment (y-axis) are plotted versus the CLT theoretical predictions (x-axis), the points lie very close to the 45° line and agree very well (Figure 7.5).

In Figure 7.4, smooth curves for the normal distribution have been superimposed over the four histograms. The mean and variance of these normal distributions were predicted by theory from the CLT in Table 7.4. These results show that, even though these air quality data are not independent and are approximately lognormal, the mean values drawn from these observations using representative sampling for sizes from 9 to 576 are approximately normal and obey the same formulas as for the CLT, to be a good approximation. The important practical implication is that, for small-to-moderate sample sizes, the formulas derived for independent observations can be used to compute confidence intervals for these air quality data, even though they are not independent.

The good agreement between CLT predictions and observed results provides a solid basis for calculating confidence intervals about the mean for the case in which the mean and variance are known. Using the San Jose air monitoring station's variance, we can calculate the confidence intervals for different values of n (Table 7.5) using the expressions in Table 7.3 (page 172). For $n = 9$, the observed mean will be within ± 1.5 ppm (actually within ± 1.48 ppm) of the true

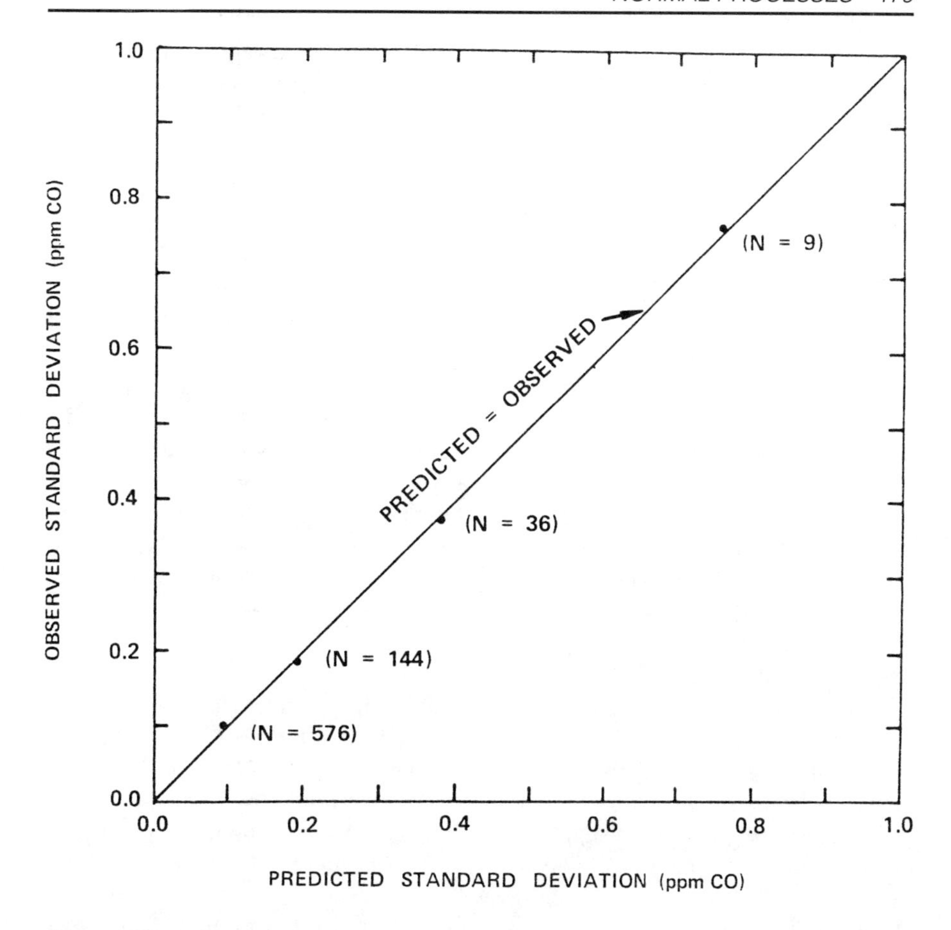

Figure 7.5. Standard deviation, as predicted by the Central Limit Theorem (CLT), versus standard deviation of the mean values of 1,000 randomly selected groups of samples of size *n*, from Ott and Mage.[7]

population mean with $p = 0.95$. Using only 9 samples distributed evenly over the year will give a reading that is approximately ± 2 ppm (actually ± 1.94 ppm) of the actual mean with 99% assurance. For $n = 36$, the sample mean will be within ± 1 ppm (actually within ± 0.97 ppm) of the true mean with 99% assurance. A sample of size 144 assures that the resulting mean is ± 1/2 ppm (actually ± 0.49 ppm) of the true mean with 99% assurance.

Table 7.5. Confidence Intervals for Mean CO Concentrations (ppm) from Sample Groups of Different Size, Based on the San Jose Data[7]

	Confidence Level		
n	0.90	0.95	0.99
9	±1.25	±1.48	±1.94
36	±0.62	±0.74	±0.97
144	±0.31	±0.37	±0.49
576	±0.16	±0.19	±0.24

These results show that the number of samples required to provide a precise estimate of the mean, based on the San Jose data, is surprisingly small. This high precision from small sample sizes is caused by the relatively small standard deviation of the San Jose data set. To what degree do these findings apply to other cities?

Generalizing to Other Cities

One way to evaluate whether these narrow confidence intervals apply to other cities is to compare San Jose's CO data with data sets from other U.S. cities. To make this comparison, Ott and Mage[7] plotted the annual means and standard deviations of 1-hour average CO concentrations from 83 air monitoring stations in 74 cities on a histogram (Figure 7.6). The annual mean CO concentration ranged from 0.9 ppm in Magna, UT, to 17.5 ppm in New York City, NY, with an overall average value for all cities of 3.66 ppm. The corresponding standard deviation ranged from 1.0 ppm in Magna to 9.5 ppm in New York City, with an average standard deviation of 2.48 ppm. The mean and standard deviation of 1-hour average CO concentrations at the San Jose air monitoring station ($\bar{x} = 3.54$ ppm; $s = 2.27$ ppm) therefore are similar to the mean and average standard deviation of all U.S. monitoring sites.

San Jose's data must be compared not only with U.S. averages but also with extreme values. The histogram (Figure 7.6) shows that the ranges of the means and standard deviations across the U.S. are relatively small. If one ranks the 83 standard deviations from these monitoring stations from highest to lowest, 94% are less than 5 ppm and 84% are less than 4 ppm. Using 5 ppm as the 94-percentile value and 4 ppm as the 84-percentile value, we can apply the CLT formulas (Table 7.3) to calculate confidence intervals for these "extreme" cases.

Suppose 144 samples are collected at each of these 83 sites. In 86% of these cases, the confidence interval will be equal to or less than the following result:

$$\bar{X} \pm 1.96\frac{\sigma_0}{\sqrt{144}} = \bar{X} \pm 1.96\frac{4}{\sqrt{144}} = \bar{X} \pm 0.65 \text{ ppm}$$

Similarly, in 94% of the U.S. sites, the confidence interval will be equal to or less than the following result:

$$\bar{X} \pm 1.96\frac{\sigma_0}{\sqrt{144}} = \bar{X} \pm 1.96\frac{5}{\sqrt{144}} = \bar{X} \pm 0.82 \text{ ppm}$$

Thus, with 100 samples, 94% of these U.S. monitoring sites will have a confidence interval about the mean equal to or less than ± 1 ppm. Confidence intervals for several other sample sizes are computed in Table 7.6.

From these findings, there appears to be justification for using a very small number of samples to determine the arithmetic mean. For the monitoring stations considered above, only 100 samples would have been adequate in at least 94% of the cases to determine the true annual mean concentration with 95% assurance of obtaining the precision of ± 1 ppm (actually, ± 0.98 ppm). This high precision for small sample sizes occurs because the standard deviations at these

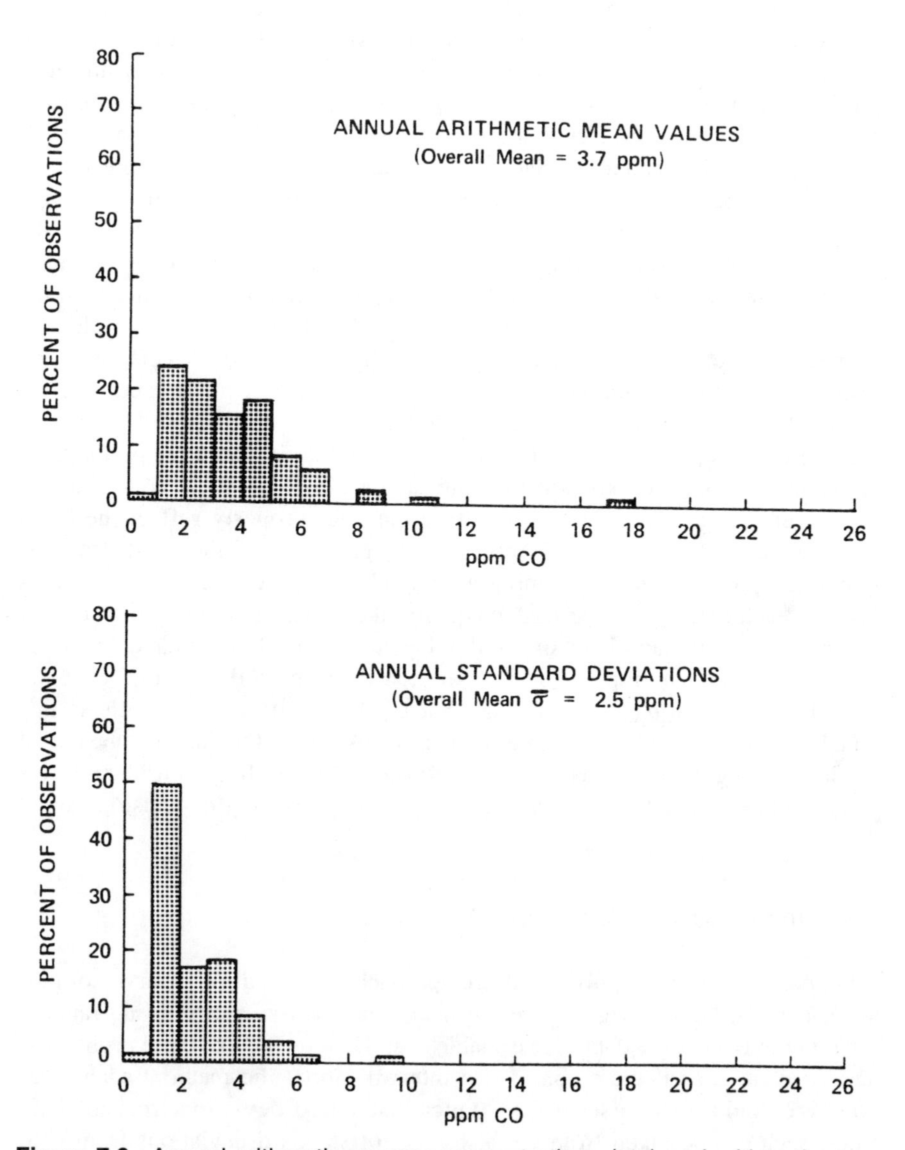

Figure 7.6. Annual arithmetic average concentrations (top) and arithmetic standard deviations (bottom) for carbon monoxide in 83 U.S. cities in 1972.

Table 7.6. Extreme 95% Confidence Intervals for 83 U.S. Monitoring Sites[7]

n	86% of sites (ppm CO)	94% of sites (ppm CO)
9	± 2.61	± 3.26
36	± 1.31	± 1.63
100	± 0.78	± 0.98
144	± 0.65	± 0.82
576	± 0.33	± 0.41

monitoring stations are so small. These results are important for the trade-off between collecting many observations at a single site or collecting only a few observations at a large number of sites. If many sites can be monitored at sufficient precision with only a few samples, then greater spatial information on pollutant concentrations can be collected at the same cost. If other pollutants besides CO have small hourly standard deviations, as usually is the case, then a great many locations can be sampled at modest cost and with adequate precision.

Because the adverse health effects of CO ordinarily are associated with the higher concentrations occurring over short time periods, interest often is directed toward the highest 8-hour or 1-hour average concentrations rather than the annual mean. However, for studying air quality trends—for example, the changes in air quality before and after a highway is constructed—the long-term (6-month average or annual average) may be of greater importance than the highest 1-hour or 8-hour average concentration.

Simpson[8] has pointed out that samples collected during the time period from 8:00 a.m. to 5:00 p.m. may not show the same behavior as those collected during the entire daily period and therefore may not properly reflect the long-term average concentration. As noted by Mage and Ott,[9] the most desirable sampling approach for determining an annual average concentration requires that "... each day in the year has an equally likely chance of being selected for measurement, and each hour of the day has an equally likely chance of being chosen." If the sampling period is restricted to from 8:00 a.m. to 5:00 p.m., then the investigator can claim only that the long-term average reflects that time segment. Even when the time segment is restricted, the data may be useful for determining trends, provided that the investigator carefully considers the restricted time segment when making statements or drawing conclusions about the results.

Random Sampling Field Surveys

Suppose we wish to apply the above approaches to a real field survey of pollutant concentrations over a large physical area, such as a city, using random sampling to collect the data at reasonable cost. How do we use these probabilistic techniques to calculate confidence intervals about the mean at each location? We could, of course, use a "typical" standard deviation from all U.S. cities, such as one taken from the histogram of standard deviations from U.S. cities (Figure 7.6). A better approach is to apply the Student's t-distribution as described above, allowing the observations themselves to provide the basis for the standard deviation at each site.

To apply the Student's t-distribution, we use Gossett's formula[3] in the same manner as was described above for independent observations:

$$P\left\{\overline{X} - t\frac{s}{\sqrt{n}} \leq \mu_o \leq \overline{X} + t\frac{s}{\sqrt{n}}\right\} = p$$

Here, s is the sample standard deviation calculated from n observations at each site, and t is obtained from the Student's t-distribution (Figure 7.2) for $n - 1$ degrees of freedom.

To illustrate this approach, Ott[10–11] measured CO concentrations at each of 9 locations in San Jose, CA, by traveling to each location and filling a sampling bag with a small, portable pump. The sampling locations (R1 to R9) were widely distributed over a rectangular grid covering a 13-square mile area of the city, including the downtown area and the local air monitoring station (Figure 7.7). Because filling each bag took about 5 minutes, the resulting observations were 5-minute averages, and the sample variance therefore is slightly greater than would be obtained for 1-hour averages. The samples—about 50 from each location—were analyzed for CO using standard techniques (Nondispersive Infrared Absorption Spectrometry) on the same day they were collected.

At location R3, for example, 49 samples were collected with a mean value of

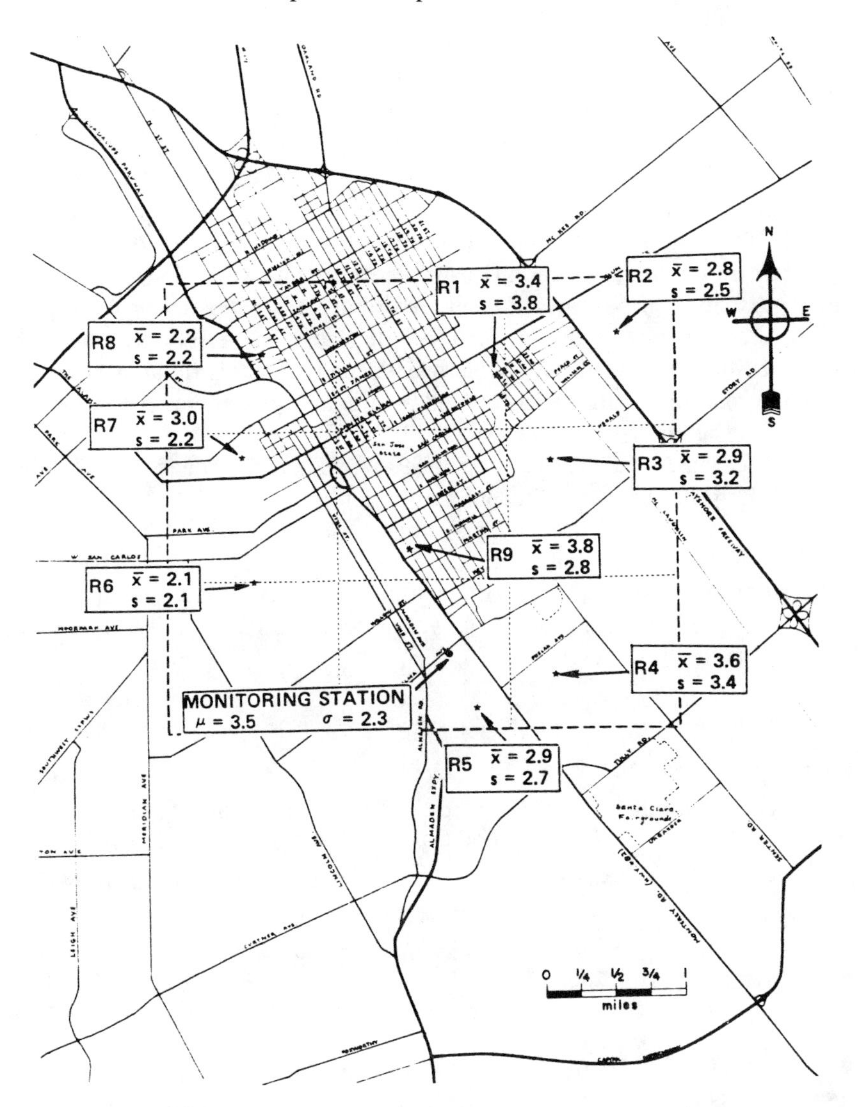

Figure 7.7. Map of San Jose, California, showing sites for random sampling and monitoring station, along with CO means and standard deviations (Source: Ott and Mage[7]).

$\bar{x} = 2.9$ ppm and an observed standard deviation of $s = 3.2$ ppm. By referring to the earlier graph of t-values (Figure 7.2, page 175) for $n = 49$, we read the curve's intercept for $n - 1 = 48$ as approximately 2.7. Using this result and the observed standard deviation, we calculate the 99% confidence interval as follows:

$$\bar{x} \pm t\frac{s}{\sqrt{n}} = 2.9 \pm 2.7\frac{3.2}{\sqrt{49}} = 2.9 \pm 1.23 \text{ ppm}$$

If the normal distribution had been used instead of the Student's t-distribution, the factor 2.58 (Table 7.3) would be used instead of 2.7 with $\sigma_o = s$, giving a narrower (underestimated) confidence interval:

$$\bar{x} \pm 2.58\frac{\sigma_o}{\sqrt{n}} = 2.9 \pm 2.58\frac{3.2}{\sqrt{49}} = 2.9 \pm 1.18 \text{ ppm}$$

If both these confidence intervals are rounded to one decimal place, the answers are the same, 2.9 ± 1.2 ppm. However, as n becomes smaller, the difference between confidence intervals using the normal distribution and the Student's t-distribution becomes greater, reflecting a greater uncertainty about using s as an estimate of σ_o for small sample sizes. Conversely, as n becomes larger, the Student's t-distribution approaches the normal distribution so closely that the normal distribution can be used as a good approximation for the Student's t-distribution.

Using the San Jose data, Ott and Mage[7] calculated confidence intervals for all 9 sampling sites using the Student's t-distribution (Table 7.7). The 99% confidence interval was less than ± 1 ppm for 4 cases (R2, R6, R7, and R8). The 95% confidence interval was less than ± 1 ppm for 6 cases. Thus, there was 95% assurance that the true mean was within ± 1 ppm of the observed mean at all but 3 sites (R1, R3, and R4). Finally, the 90% confidence interval was less than ± 1 ppm for all 9 sites.

Table 7.7. Confidence Intervals for CO Concentrations Measured at 9 Locations in San Jose, CA, Calculated Using the Student's t-Distribution[7]

		Mean	s	Confidence Interval (ppm)		
Site	n	(ppm)	(ppm)	90%	95%	99%
R1	54	3.4	3.8	±0.85	±1.20	±1.40
R2	53	2.8	2.5	±0.59	±0.76	±0.91
R3	49	2.9	3.2	±0.80	±1.00	±1.20
R4	48	3.6	3.4	±0.81	±1.10	±1.30
R5	46	2.9	2.7	±0.70	±0.90	±1.10
R6	48	2.1	2.1	±0.53	±0.68	±0.81
R7	48	3.0	2.2	±0.56	±0.71	±0.85
R8	46	2.2	2.2	±0.57	±0.73	±0.87
R9	48	3.8	2.8	±0.71	±0.91	±1.10

COMPUTATION OF N(μ,σ)

Traditionally, books on probability and statistics have included tables of the CDF of the normal distribution. With the widespread availability of programmable calculators and personal computers, generally it is more convenient to program a mathematical approximation into these machines and calculate the CDF for any arbitrary input value. Hastings' book,[12] for example, contains a number of convenient approximations for the CDF of the normal distribution, and their precision is adequate for most applications.

If y is the variate of interest, we can employ the transform $z = (y - \mu)/\sigma$ and then compute the standardized normal CDF, $F_Z(z)$. A very precise approximation for the normal CDF from the mathematical handbook by Abramowitz and Stegun[13] first requires calculation of the PDF of the normal distribution:

$$f_Z(z) = \frac{1}{\sqrt{2\pi}} e^{-\frac{z^2}{2}}$$

Once $f_Z(z)$ is computed, it is substituted into the following expression:

$$F_Z(z) = 1 + f_Z(z)\left\{b_1 t + b_2 t^2 + b_3 t^3 + b_4 t^4 + b_5 t^5\right\}$$

where $t = \dfrac{1}{1 + a|z|}$ with $a = 0.2316419$

and

$b_1 = 0.319381530$ $\quad b_4 = -1.821255978$

$b_2 = -0.356563782$ $\quad b_5 = 1.330274429$

$b_3 = 1.781477937$

The result of this approximation is valid only for the lower half of the standardized normal CDF: that is, for negative values of Z: $-\infty < z < 0$. To compute the upper half, we use the same approximation but note that $F_Z(z) = 1 - F_Z(-z)$. The overall error—the difference between the approximation and the true CDF—is an oscillating function whose maximum value is less than 7.5×10^{-8}. Thus, one generally can rely on this result to about 7 places to the right of the decimal.

A computer program has been written for this book in Microsoft BASIC™ to compute this approximation (Figure 7.8). This double-precision program first reads the 5 constants—b_1,b_2,b_3,b_4, and b_5—into the vector B(5) using a FOR loop (see lines #140 to #200). The IF statement (line #230) tests whether Z is nonzero; if so, the main computation is carried out (lines #270 to #380). If not, the Gaussian CDF, represented by GAUCDF, is set to 0.5 because Z = 0, and the program branches to the end (lines #390 to #420). In the main computation section (lines #270 to #380), the variable MEMORY keeps track of the sign of Z. Initially, MEMORY = 0 (line #270), and, if Z is negative, MEMORY remains unchanged. If Z is positive, then MEMORY = 1 (line #280). The normal PDF, represented by PDFN, is computed in line #290 using the constant

```
100 REM Program to Compute Gaussian CDF
110 PRINT "PROGRAM TO COMPUTE GAUSSIAN PDF AND CDF"
120 PRINT : PRINT
130 DEFDBL B, G, P, S-T, V, Z
140 DIM B(5)
150 DATA 0.31938153,-0.356563782,1.781477937
160 DATA -1.821255978,1.330274429
170 REM Read in the Constants
180 FOR I = 1 TO 5
190 READ B(I)
200 NEXT I
210 REM Read In the Variate
220 INPUT "VARIATE"; Z
230 IF (Z <> 0) GOTO 270
240 REM If Z=0, set GAUCDF=0.5
250 GAUCDF = .5#
260 GOTO 390
270 MEMORY = 0
280 IF (Z > 0) THEN MEMORY = 1
290 PDFN = .39894228# * EXP(-.5 * Z * Z)
300 T = 1# / (1# + .2316419 * ABS(Z))
310 SUM = 0#
320 V = 1#
330 FOR I = 1 TO 5
340 V = T * V
350 SUM = SUM + B(I) * V
360 NEXT I
370 GAUCDF = SUM * PDFN
380 IF (MEMORY = 1) THEN GAUCDF = 1# - GAUCDF
390 PRINT "For Z =", Z; ",       The Normal PDF =", PDFN
400 PRINT "                    and the Normal CDF =", GAUCDF
410 GOTO 220
420 END
```

Figure 7.8. BASIC computer program for computing the normal CDF $F_Z(z)$.

$1/\sqrt{\pi} = 0.398942280$. Because $f_Z(z) = f_Z(-z)$, it makes no difference whether Z is positive or negative in calculating PDFN (line #290). A FOR loop (lines #330 to #360) computes the main equation. The value for t (T in line #300) is raised to the power 1 and is multiplied by b_1 on the first iteration on this loop. On the second iteration, t is raised to the power 2 and is multiplied by b_2. Successive iterations raise t to increased powers and store the total in SUM. Finally, if MEMORY = 1 (indicating that Z > 0), then GAUCDF = 1.0 – GAUCDF. In lines #390 and #400, the normal PDF and CDF computed by this approximation are printed out.

The computation above gives the value of $F_Z(z)$ for any value of $Z = z$. Suppose $F_Z(z)$ is known, but we do not know z and wish to find it. For example, we may know the probability associated with a particular normally distributed random variable and may wish to find the corresponding value of the variate. In such cases, it is necessary to find the inverse value of the normal CDF, or $z = F_Z^{-1}(z)$. A useful approximation for this purpose is given by Hastings[12] and also included in the reference book by Abramowitz and Stegun.[13] If we let $Q = 1 - F_Z(z)$ in this approximation, then, for the case in which $0 < Q \leq 0.5$, we compute z as follows:

$$z = \eta - \frac{a_1 + a_2\eta + a_3\eta^2}{1 + b_1\eta + b_2\eta^2 + b_3\eta^3}$$

where $$\eta = \sqrt{\ln \frac{1}{Q^2}}$$

$a_1 = 2.515517$ $\qquad$ $b_1 = 1.432788$

$a_2 = 0.802853$ $\qquad$ $b_2 = 0.189269$

$a_3 = 0.010328$ $\qquad$ $b_3 = 0.001308$

For the case in which $0.5 < Q \leq 1$, we know that $0 < F_Z(z) < 0.5$, since $F_Z(z) = 1 - Q$. In that case, we simply substitute $F_Z(z)$ for Q into the above expression; however, we must remember to make the resulting value of z negative by multiplying it by -1.

A double-precision computer program for computing the inverse normal CDF (Figure 7.9) first reads the constants in the numerator—a_1, a_2, and a_3—

```
100 REM Program to Compute Inverse Gaussian CDF
110 PRINT "PROGRAM TO COMPUTE INVERSE GAUSSIAN PDF AND CDF"
120 PRINT : PRINT
130 DEFDBL A-B, E, G, P-Q, X
140 DIM A(3), B(3)
150 DATA 2.515517,0.802853,0.010328
160 DATA 1.432788,0.189269,0.001308
170 REM Read in the Constants
180 FOR I = 1 TO 3
190 READ A(I)
200 NEXT I
210 FOR I = 1 TO 3
220 READ B(I)
230 NEXT I
240 MEMORY = 0
250 INPUT "CDF"; P
260 Q = 1# - P
270 IF P = 0# THEN GOTO 240
280 IF P = 1# THEN GOTO 240
290 IF P = .5# THEN Z = 0#: GOTO 430
300 IF Q > .5 THEN MEMORY = 1
310 IF MEMORY = 1 THEN Q = P
320 AIDA = SQR(LOG(1# / (Q * Q)))
330 XNUM = 0#
340 DENOM = 1#
350 G = 1#
360 FOR I = 1 TO 3
370 XNUM = XNUM + A(I) * G
380 G = G * AIDA
390 DENOM = DENOM + B(I) * G
400 NEXT I
410 Z = AIDA - XNUM / DENOM
420 IF MEMORY = 1 THEN Z = -1# * Z
430 PRINT "FOR CDF =", P; ",        The Inverse Normal PDF =", Z
440 GOTO 240
450 END
```

Figure 7.9. BASIC computer program for computing the inverse normal CDF: $z = F_Z^{-1}(z)$.

into the vector A(3) (lines #150 and #180 to #200). It then reads the constants in the denominator into the vector B(3) (lines #160 and #210 to #230). In this program, the CDF $F_Z(z)$ is input as P (line #250), and then Q = 1 – P is computed (line #260). (The notation "1#" in line #260 denotes double precision values.) The variable MEMORY "remembers" whether Q is less than (or equal to) or greater than 0.5. Initially, MEMORY = 0 (line #240). If Q is greater than 0.5, then MEMORY = 1 (line #300). In this case, the computation proceeds using P instead of Q (line #310), but the program makes the computed value of Z negative at the end by multiplying it by –1.0 (line #420).

Lines #270, #280, and #290 take care of special cases of P that cannot be handled by the approximation. If P = 0 (line #270) or P = 1 (line #280), the program returns to line #240 to clear the memory and to line #250 to obtain another input CDF value. If P = 0.5, the correct result should be $F_Z^{-1}(0.5) = 0$ exactly, but the approximation gives a very small value that is close to 0. Thus, the result is set to Z = 0 exactly (line #290), and this result is printed out (line #430). The value of η (AIDA) is computed in line #320, and the computations of the above equation are carried out in a FOR loop (lines #360 to #400). With this approximation, the error between the true variate and the calculated variate is less than 4.5×10^{-4}, which is satisfactory for many applications.

PROBLEMS

1. The members of a charitable society all agree to donate approximately $10 every week for a year to a particular cause. If 1,000 people make gifts every week and the average individual gift is $10, what is the expected value of the weekly receipts? [Answer: $10,000] Suppose the managers make a histogram of the weekly total contributions; discuss why the resulting frequency distribution should appear normal. Suppose past history shows that the standard deviation of each person's gift-giving activity is $1; what will be the standard deviation of the weekly receipts? [Answer: $31.62] Assuming that weekly contributions continue for 52 weeks per year, what is an appropriate distribution for the year's receipts and what parameters should it have? [Answer: Normal with mean μ = $520,000 and standard deviation σ = $228] What is the probability that the total annual receipts will lie within the narrow (3σ) range of $520,000 ± $684? [Answer: 0.99865 – 0.00135 = 0.9973]
2. A group of small particles are released from a point source in a large chamber, and the movement of these particles is carefully monitored. For a particle's displacement in the horizontal dimension, assume that each particle is equally likely to move to the left (negative) or right (positive) relative to the source (the origin). If each particle experiences independent incremental displacements of arbitrary size, then the particle's horizontal distance from the origin will be the sum of all its displacements. If the standard deviation of the displacements is 1.5 cm, and a particle experiences 10,000 displacements, what will be the probability distribution of its horizontal distance from the origin. [Answer: **N**(0,1.5m)]

3. A water quality specialist wants to determine the average chloride concentrations over a 15-day period in a stream. The investigator collects a sample in the stream each day, making sure the time of day selected is evenly distributed over the 24-hour period to make the samples as representative as possible. The resulting average of the 15 samples is 460 mg/l with a standard deviation of 30 mg/l. In his final report, the investigator would like to say that "The true chloride mean during the 15-day period is 460 mg/l $\pm x$ with 99% assurance." Using the Student's *t*-distribution (Figure 7.2), find x. [Answer: 23 mg/l]
4. Hourly average SO_2 measurements are made at a particular location such that the hours are evenly distributed over a 3-month period. Suppose 75 hourly averages are taken in this fashion, yielding a calculated mean value of 29 $\mu g/m^3$ and a standard deviation of 11.4 $\mu g/m^3$. Using the Student's *t*-distribution, calculate the 99% confidence interval about the observed mean of 29 $\mu g/m^3$. [Answer: $\pm$ 3.5 $\mu g/m^3$]
5. One would like to measure the true monthly mean total suspended particulate concentrations in an indoor setting, but it will not be possible to monitor every day. The investigator can collect only twelve 24-hour average observations, so he uses a table of random numbers to space the sample collection dates in a representative fashion over the 31-day period. The result is an observed mean of 80 $\mu g/m^3$ with an observed standard deviation of 19 $\mu g/m^3$. He would like to report how close the sample mean is to the true mean, with a stated level of confidence. [Answer: The true mean lies within the range 70.1 $\mu g/m^3$ and 89.9 $\mu g/m^3$ with probability 0.90. It lies within the range 67.9 $\mu g/m^3$ and 92.1 $\mu g/m^3$ with probability 0.95.]
6. CO concentrations are measured next to an urban arterial highway on weekdays during a 6-month period by collecting 100 observations that are evenly distributed over the hours of the day and the weekdays of each month. The mean of these 100 hourly observations is 12.5 ppm with a standard deviation of 1.5 ppm. What are the 95% and 99% confidence intervals about the weekday mean CO concentrations? [Answers: $\pm$0.29 ppm, $\pm$0.39 ppm] Repeat the calculation by substituting the 95% and 99% confidence intervals for the normal distribution (Table 7.3), and discuss the differences, if any.
7. Hourly NO_2 readings are made by randomly sampling at a monitoring point for 2 weeks, yielding 49 readings with a mean value of 20.6 $\mu g/m^3$ and a standard deviation of 18.9 $\mu g/m^3$. What is the 99% confidence interval about the mean? [Answer: $\pm$ 7.2 $\mu g/m^3$]

REFERENCES

1. Feller, William, *An Introduction to Probability Theory and Its Applications*, Volume II (New York: John Wiley and Sons, 1970).
2. Kuzma, Jan W., *Basic Statistics for the Health Sciences* (Palo Alto, CA: Mayfield Publishing Company, 1984).
3. Student, "On the Probable Error of the Mean," *Biometrika*, 6:1–25 (1908).
4. Johnson, Norman L., and Samuel Kotz, *Continuous Univariate Distributions-2* (New York: John Wiley and Sons, 1970).

5. Student, "New Tables for Testing the Significance of Observations," *Metron*, 5:105–108, 114–120 (1925).
6. Beyer, William H., 2nd ed., *Handbook of Tables for Probability and Statistics* (Boca Raton, FL: CRC Press, Inc., 1968).
7. Ott, Wayne R., and David T. Mage, "Measuring Air Quality Levels Inexpensively at Multiple Locations by Random Sampling," *J. Air Poll. Control Assoc.*, 31(4):365–369 (April 1981).
8. Simpson, R. W., "Comment on 'Measuring Air Quality Levels Inexpensively at Multiple Locations by Random Sampling,' " *J. Air Poll. Control Assoc.*, 34(9):952–953 (September 1984).
9. Mage, David T., and Wayne R. Ott, "Authors' Reply," *J. Air Poll. Control Assoc.*, 34(9):953 (September 1984).
10. Ott, W., "An Urban Survey Technique for Measuring the Spatial Variation of Carbon Monoxide Concentrations in Cities," Ph.D. dissertation, Stanford University, Department of Civil Engineering, Stanford, CA (1971).
11. Ott, W. R., and R. Eliassen, "A Survey Technique for Determining the Representativeness of Urban Air Monitoring Stations with Respect to Carbon Monoxide," *J. Air Poll. Control Assoc.*, 23(8):685–690 (August 1973).
12. Hastings, Cecil, Jr., *Approximations for Digital Computers* (Princeton, NJ: Princeton University Press, 1955).
13. Abramowitz, Milton, and Irene A. Stegun, eds., *Handbook of Mathematical Functions with Formulas, Graphs, and Mathematical Tables*, Applied Mathematics Series No. 55 (Washington, DC: National Bureau of Standards, 1972).

8 Dilution of Pollutants

Polluters discharge chemicals into the environment to dispose of them. Environmental pollution becomes a problem to society because such chemicals sometimes do not disappear. In some cases, such as the pesticide DDT, they are sufficiently inert, or persistent, to remain in the environment for many years. However, discharging a chemical into the environment usually reduces its adverse effects by reducing its concentration (quantity of chemical per unit volume). Concentration reduction is achieved whenever a particular quantity of a chemical (the pollutant) mixes with some other material (the carrier medium), thereby becoming *diluted*. Dilution is the process by which the molecules of a substance become distributed over a larger physical volume than they formerly occupied.

Dilution always accompanies the formal process of diffusion (Chapter 6), because the molecules in a diffusion process tend to spread out and occupy an ever-increasing volume. However, dilution also can occur in processes that do not primarily involve diffusion, such as the simple mechanical mixing of one chemical liquid with another. For example, if two chemically unreactive liquids are combined together in one container and shaken until they are uniformly mixed, the volume of the resulting mixture will be the sum of the volumes of the two original constituents, and each original constituent will become diluted. Dilution occurs in a large number of common everyday examples: introduction of syrup flavoring when one prepares a milkshake; addition of pigments to raw base paint to give it color; insertion of drops of medicine into a glass of water. Dilution also occurs when concentrated soap is added to a bucket of water for scrubbing the floor. It occurs when table salt is dissolved in hot water prior to cooking rice. Recipes for most foods prepared in the kitchen call for the mixing together of diverse ingredients, thereby causing each ingredient to become diluted.

Dilution also appears in a variety of environmental phenomena. An effluent discharging into a stream becomes diluted as it enters the receiving waters. Pesticides and other chemicals sprayed on crops become diluted as they are carried away to streams as surface runoff. Chemical wastes dumped at sea become diluted as they are mixed and dispersed by ocean currents. Toxic chemicals stored in drums or underground tanks become diluted as they leak out and leach through the soil. Acids and other substances present in rain become diluted as the rainwater falls to earth, ultimately passing to receiving waters. Water pollutants become diluted as they pass through porous soils. Air pollutants discharged from smokestacks become diluted as they are diffused and transported by winds. Even food additives, such as nitrates (which can ultimately form carcinogenic nitrosamines under certain conditions), become diluted in the foods to which they are added. Poisons and other chemicals absorbed by the body become diluted as they enter the blood and other tissue. Thus, dilution is an extremely ubiquitous process, occurring in a great variety of phenomena in our bodies, our daily lives, and the environment about us.

Dilution can be represented by at least two successive states in time: before and after. Before dilution occurs, some finite quantity of material is present, occupying a particular volume. In the simplest case, a diluting agent, or "diluent," is added. After dilution has occurred, the final mixture—original material plus diluent—occupies a greater volume than before. The resulting mixture can be uniform or nonuniform. To simplify the analysis, we assume that the final mixture is uniform and that the materials do not chemically react. A dilution can occur in either the gas, liquid, or solid phase, and we shall assume that the phase is the same before and after dilution. In environmental examples, the original material is assumed to contain a pollutant at some defined concentration. After dilution, the pollutant concentration (quantity per unit volume) is reduced because the volume is increased. In this context, dilution is defined as follows:

> Dilution is the process by which one constituent is mixed with others, causing the ingredients to spread out over a larger volume than before, with a consequent reduction in concentration.

DETERMINISTIC DILUTION

Consider a pollutant present in some constituent prior to dilution. If v_o denotes the initial volume of the constituent with pollutant prior to dilution and c_o denotes the initial concentration of the pollutant, then the total quantity of pollutant present initially will be $q_o = c_o v_o$. If the units of c_o are in grams/cubic meter, for example, then the units of q_o will be in (grams/cubic meter)(cubic meters) = grams. Assuming that no chemical reactions take place, conservation of matter requires that the total quantity of pollutant present after dilution be equal to the total quantity of pollutant present before dilution. If v_1 denotes the volume of the mixture after dilution, and c_1 denotes the concentration of the pollutant in this mixture, then the quantity of pollutant present both before and after dilution must be the same:

$$c_o v_o = q_o = c_1 v_1$$

Solving this equation for c_1, we find that the new concentration is proportional to the ratio of the initial volume v_o to the final volume v_1:

$$c_1 = \frac{v_o}{v_1} c_o$$

To simplify the analysis, we introduce a dilution factor $\alpha_1 = v_o/v_1$, which is a dimensionless number between 0 and 1. In a deterministic dilution process, the dilution factor will be a constant and not a random variable.

Successive Deterministic Dilution

Suppose that the dilution process has ended, and the pollutant concentration has become uniform because the constituents are completely mixed. Now suppose another dilution takes place with this mixture. By adding more diluent, the volume of the mixture changes from v_1 to v_2, and the concentration of the pollutant decreases further from c_1 to some value c_2. By the above reasoning, the resulting concentration c_2 again will be inversely proportional to the ratio of the two mixing volumes, v_2:v_1, and a second dilution factor will be $\alpha_2 = v_1/v_2$:

$$c_2 = \frac{v_1}{v_2} c_1 = \alpha_2 c_1$$

Because the first dilution stage above gave the concentration c_1, this result can be substituted into the above equation, giving the following expression for c_2:

$$c_2 = \alpha_1 \alpha_2 c_o$$

It is seen that the final concentration c_2 after two successive dilutions can be written as the product of the two dilution factors α_1 and α_2 and the initial concentration c_o.

If a third dilution now occurs, the resulting concentration c_3 will be the product of a third dilution factor multiplied times c_2, or $c_3 = \alpha_1 \alpha_2 \alpha_3 c_o$. It is evident that a pattern emerges as successive dilutions are carried out. In general, if a succession of n dilutions occurs, then the resulting concentration c_n will be the product of n individual dilution factors and the original concentration c_o:

$$c_n = \alpha_1 \alpha_2 \ldots \alpha_n c_o = c_o \prod_{i=1}^{n} \alpha_i$$

Here the Greek capital letter pi is a shorthand notation representing the process of multiplying n dilution factors together. This product occurs because the ending mixture at the completion of one dilution becomes the beginning mixture for the next dilution. In this book, we shall describe any multiplicative process of this kind as a *product process*, or a $\mathcal{P}$-process. It also can be described as a "proportionate process," because the concentration in each successive state is some constant proportion of the concentration in any previous state. Because the proportions are always constant, the process is deterministic. That is, if the entire set of n dilutions is repeated, all the dilution factors will be the same, and exactly the same concentration c_n will result.

Example

Suppose that five beakers are available (Figure 8.1). The first beaker contains 250 milliliters (ml) of water with a dissolved chemical pollutant at a concentra-

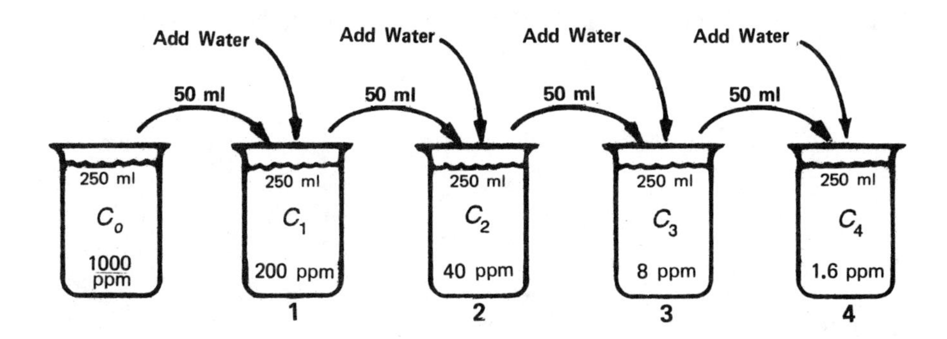

Figure 8.1. Deterministic successive dilution process in which four beakers each have the same dilution factor $\alpha_1 = \alpha_2 = \alpha_3 = \alpha_4 = 1/5$ (Source: Ott[11]).

tion of c_o = 1000 parts-per-million (ppm) by volume. A person carefully measures out 50 ml of this solution and pours it into the next beaker (beaker 1). Then clean water is added to this beaker to bring its volume up to 250 ml, and the resulting solution is thoroughly mixed. By this action, the concentration of the pollutant in the original solution is diluted by a ratio of $v_o/v_1 = 50/250 = \alpha_1 = 1/5$, and the resulting concentration in beaker 1 will be c_1 = (1/5)(1000 ppm) = 200 ppm. Now suppose this process is repeated for beaker 2: one-fifth of the quantity (50 ml) in beaker 1 is poured into beaker 2, and clean water is added to bring the level up to 250 ml. Again, the concentration is diluted by a ratio of 1:5, and the final concentration in beaker 2 will be c_2 = (1/5)(200 ppm) = 40 ppm. If the same process is repeated in the two remaining beakers, the final concentration in beaker 4 will be the product of the original concentration c_o and the four dilution factors, $\alpha_1 = \alpha_2 = \alpha_3 = \alpha_4 = 1/5$, giving the following result:

$$c_4 = \alpha_1 \alpha_2 \alpha_3 \alpha_4 c_o = (1/5)^4 (1000 \text{ ppm}) = 1.6 \text{ ppm}$$

For this example, it is assumed that exactly 50 ml of the contents of a beaker is transferred on each pouring. Thus, if the entire process of four pourings were repeated again, the concentration in the final beaker again would be exactly 1.6 ppm. Obviously, such a precise result is highly idealized, because it assumes that each of the four pourings is measured so exactly that no variation occurs in the final result. In any real process, the individual pourings sometimes will be above 50 ml and sometimes below 50 ml, because the person will not always be able to measure the correct amount exactly. What effect will these pouring errors have on the final answer? To answer this question, we must move from the deterministic to the stochastic version of this process.

STOCHASTIC DILUTION AND THE THEORY OF SUCCESSIVE RANDOM DILUTIONS (SRD)

As indicated above, the example assumed that the person poured out exactly 50 ml of the mixture—never more nor less—from each beaker into the next beaker. In an actual experiment, errors typically will occur in the measurement of liquid quantities. Even though the experimenter tries very hard to measure out exactly 50.000 ml, for example, the first beaker actually may receive 50.145 ml of solution, and the next beaker actually may receive 49.823 ml of solution. Because the experimenter has no reason to show a bias with regard to these errors, it is reasonable to expect the quantities actually poured to be either above or below 50 ml with equal probability. If the entire experiment—four pourings—is repeated many times, it is obvious that these different pouring errors will cause the final concentrations observed in beaker 4 to vary in some manner about 1.6 ppm. Because of the similarity of this process to events actually occurring in nature, it is important to explore the manner in which these errors affect the distribution of the final concentration. Will a characteristic distribution result?

A simple calculation offers a clue about the properties of the distribution. Suppose the entire experiment is repeated many times, each time beginning with c_o and pouring approximately 50 ml of mixed solution into each successive, clean beaker. On one occasion, suppose that the investigator happens to shortchange every beaker by exactly $\Delta c = -1$ ml, causing 49 ml instead of 50 ml to be transferred. (He makes up the shortage by adding an extra 1 ml of water.) For this set of 4 pourings, $\alpha_1 = \alpha_2 = \alpha_3 = \alpha_4 = 49/250 = 0.196$, and the final beaker will end up with $c_4 = (0.196)^4(1000 \text{ ppm}) = 1.476$ ppm. For this case, the final beaker will contain 1.600 – 1.476 = 0.124 ppm less concentration than it should.

In contrast, suppose another pouring cycle occurs in which each of the 4 beakers receives an excess of $\Delta c = + 1$ ml, causing 51 ml instead of 50 ml to be transferred. For this second case, $\alpha_1 = \alpha_2 = \alpha_3 = \alpha_4 = 51/250 = 0.204$. Here, the final beaker will end up with concentration $c_4 = (0.204)^4(1000 \text{ ppm}) = 1.732$ ppm. Now the final beaker contains 1.732 – 1.600 = 0.132 ppm more concentration than it should. Notice an interesting result: although the error of the individual pourings varied as –1 ppm, +1 ppm about 50 ppm for these two cases, the error of the final result varied *asymmetrically* as –0.124 ppm, +0.132 ppm about the correct answer of 1.600 ppm. Because the magnitude of the excursion below 1.6 ppm was less than the magnitude of the excursion above 1.6 ppm, the result suggests a tendency toward right-skewness. Thus, additive symmetrical excursions in the individual pourings cause multiplicative asymmetric, right-skewed excursions in the final answer. This concept is developed further in the following section.

Successive Random Dilutions: Multiple Beaker Example

In the above example, only two possible pouring errors were allowed, permitting only two different quantities to be transferred from beaker to beaker: (a) 50 – 1 = 49 ppm, and (b) 50 + 1 = 51 ppm. In a real experiment, the quantities could take on a continuum of values, and the investigator would try to minimize the error as much as possible. Assuming that an accurate laboratory measuring device is available, such as a graduated cylinder, the quantity transferred from each beaker to the next one will be very close to the desired value of 50 ml, and large excursions away from 50 ml will be unlikely. The error is equally likely to be above 50 ml as below 50 ml, causing the distribution of transferred quantities to be symmetrical about 50 ml. However, the asymmetry of the distribution of concentrations in the final beaker is of particular interest. Does the distribution of concentration in the final beaker have a single mode? Is this distribution right-skewed? Does any particular probability distribution, or family of distributions, describe its shape better than others? Is there a theoretical basis for this result?

An example illustrates a SRD process with multiple beakers. If the experimenter very carefully uses a graduated cylinder to measure the amount poured from one beaker to the next, there will be little deviation about 50 ml, and the concentration in beaker 4 always will be very close to 1.6 ppm. Because little deviation from 1.6 ppm will be evident, the characteristics of the distribution of the final concentration will be difficult to detect due to its small variability. An experimenter with poor eyesight, or one performing the experiment very carelessly, would cause greater variation in the final concentration, and the distribution of the final concentration on repeated trials then could be examined more easily.

One can artificially increase the error, or randomness, in the pouring process by physically modifying the graduated cylinder. A common laboratory graduated cylinder is a clear glass container with a linear scale painted or etched on its surface. If the top marking corresponds to 100 ml, then the cylinder will contain 50 ml when the level of the liquid is at the midpoint. Suppose an opaque section of paper is taped over a portion of the cylinder, thus covering up a portion of its scale (Figure 8.2). If this masked cylinder is used to measure the quantities transferred from beaker to beaker, the experimenter will be unable to determine the amount being transferred in the range covered by the mask. Suppose that the mask is taped around the middle portion of the cylinder, covering up the scale from 30 ml to 70 ml. The experimenter will be able to determine if the level is *below* 30 ml or *above* 70 ml, but he will have no information about the true level *within* the range from 30 ml to 70 ml. If the experimenter is told to ignore any other clues (weight of the liquid, etc.), then it is possible to imagine that all levels within this range are equally likely, so that all concentration values will occur with equal probability within this range. As a result, the quan-

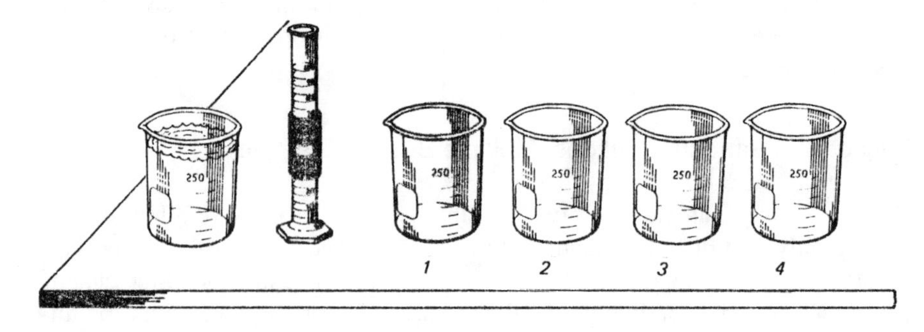

Figure 8.2. By covering the graduated cylinder with a mask between 30 ml and 70 ml, the experimenter cannot see the level of the liquid, and the quantity transferred from beaker to beaker will be a uniformly distributed random variable in the range 30–70 ml (Source: Ott[11]).

tities actually measured will be uniformly distributed as **U**(30,70) with an arithmetic mean of 50 ml.

For each quantity transferred in the range from 30 ml to 70 ml, there will be a corresponding dilution factor, and the minimum, mean, and maximum dilution factors will be as follows:

$$\textit{Minimum}: \ \alpha_{\min} = 30/250 = 0.12$$

$$\textit{Mean}: \ \bar{\alpha} = 50/250 = 0.20$$

$$\textit{Maximum}: \ \alpha_{\max} = 70/250 = 0.28$$

Because the dilution factors are a linear function of the quantities poured, each dilution factor resulting from pouring with a masked cylinder also will be a uniformly distributed random variable (Figure 8.3). Its distribution will be **U**(0.12,0.28) with a mean of 0.20.

Suppose this complete experiment were repeated 1,000 times. Beginning with a fresh beaker with concentration c_o and carrying out the four individual pourings 1,000 times in any real setting would be a tedious project indeed. However, we can employ computer simulation to replicate this experiment 1,000 times. Therefore, a BASIC computer program was written (Figure 8.4) and run on the IBM-PC* to generate four uniformly distributed random variables **U**(0.12,0.28), multiply them together, and generate a histogram of the re-

*IBM is a registered trademark of International Business Machines, Inc. It should be noted that different versions of BASIC have different random number seeds, giving different realizations than the one in Figure 8.5, even though the same program given in Figure 8.4 is used.

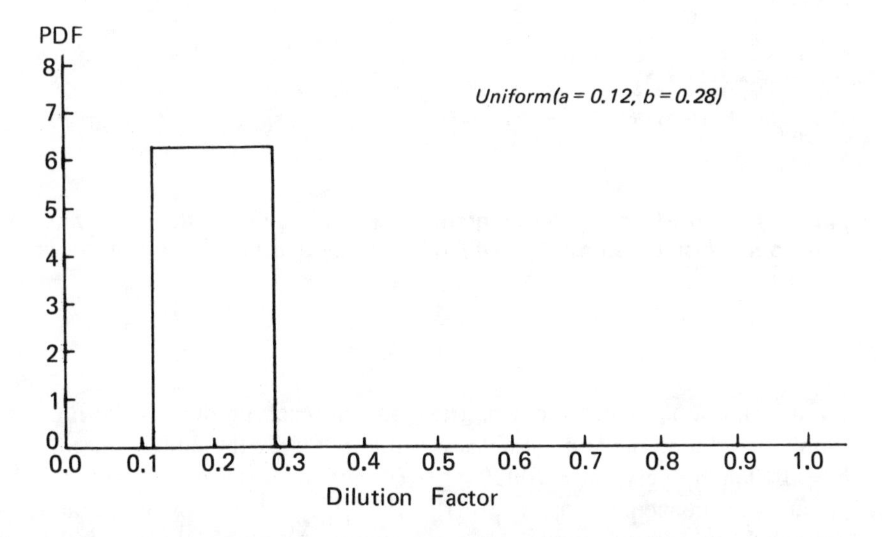

Figure 8.3. The mask covering the graduated cylinder's markings between 30 ml and 70 ml causes the dilution factor to be uniformly distributed as **U**(0.12, 0.28).

```
100 PRINT "PROGRAM TO GENERATE PRODUCT OF UNIFORMLY DISTRIBUTED RANDOM VARIABLES"
110 PRINT : PRINT
120 KEY OFF
130 INPUT "Enter the Histogram Interval Width, Delta"; DELTA
140 PRINT
150 INPUT "Enter the Number of Intervals (Not More than 100)"; K
160 DIM NUM(100)
170 REM Initialize Values
180 SUM = 0
190 VSUM = 0
200 VSUM2 = 0
210 REM Create a uniformly distributed random variable U(A,B)
220 PRINT
230 INPUT "Constant Multiplier"; CONSTANT
240 INPUT "Number of times uniform random variable is multiplied by itself"; M
250 INPUT "Scale Values (A,B)"; A, B
260 SCALE = B - A
270 PRINT
280 INPUT "Number of random values to be generated"; N
290 PRINT
300 FOR J = 1 TO N
310 C = CONSTANT
320 FOR I = 1 TO M
330 U = A + (SCALE) * RND
340 C = C * U
350 NEXT I
360 PRINT J, C
370 REM Compute Statistics
380 VSUM = VSUM + C
390 VSUM2 = VSUM2 + C * C
400 REM Sort result into intervals.
410 SUM = 0!
420 FOR I = 1 TO K
430 SUM = SUM + DELTA
440 IF C < SUM THEN NUM(I) = NUM(I) + 1
450 IF C < SUM THEN GOTO 470
460 NEXT I
470 NEXT J
480 REM Print Results.
490 AVE = VSUM / N
500 STD = SQR(VSUM2 / (N - 1) - AVE * AVE * (N / (N - 1)))
510 LPRINT "Mean ="; AVE; " and Standard Deviation ="; STD
520 REM ...Print Histogram Intervals
530 NCUM = 0
540 PSUM = 0
550 DSUM = 0
560 FOR I = 1 TO K
570 DSUM = DSUM + DELTA
580 NCUM = NCUM + NUM(I)
590 PER = NUM(I) / N
600 PSUM = PER + PSUM
610 LPRINT I, DSUM, NUM(I), NCUM, PSUM
620 NEXT I
630 END
```

Figure 8.4. BASIC computer program for generating the product of any number of uniformly distributed random variables **U**(A,B) and producing a histogram of the result.

sulting values. To match the actual experiment, each product of the four uniformly distributed random variables was multiplied by the initial concentration $c_o = 1000$ ppm to give $C_4 = 1000U_1U_2U_3U_4$, and the series of 1,000 realizations produces the concentrations $(c_4)_1, (c_4)_2, \ldots, (c_4)_{1000}$. The histogram of the 1,000 simulated concentrations in beaker 4, when plotted with an interval width of $\Delta c = 0.5$ ppm, has a single mode and distinct right-skewness (Figure 8.5). The histogram's mode (311 values in this example) lies in the interval

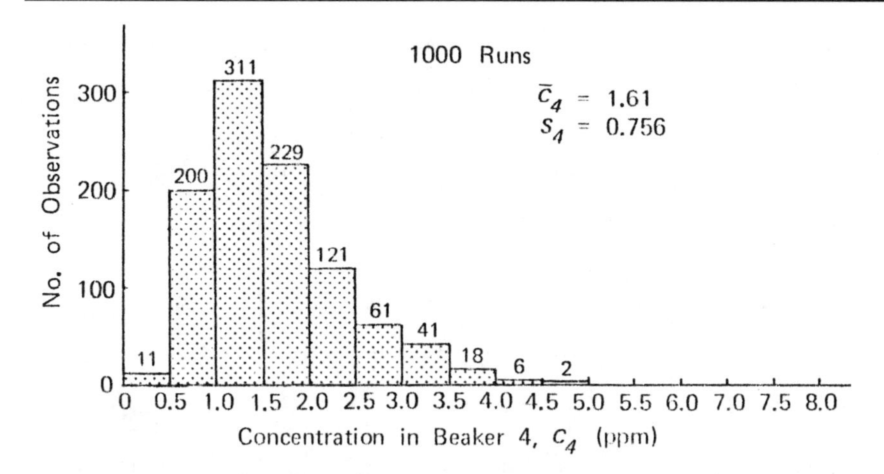

Figure 8.5. Histogram of the final concentration in beaker 4 from computer simulation of 1,000 successive random dilution experiments. The computer simulates each of 4 pourings in an experiment to be uniformly distributed between 30 ml and 70 ml, causing the individual dilution factors to be distributed as **U**(0.12,0.28).

$0.5 \le c_4 < 1.5$. The highest interval containing observations, $4.5 \le c_4 < 5.0$, lies considerably further to the right of the mode than the lowest interval containing observations, $0 \le c_4 < 0.5$, lies to the left of the mode. To the left of the mode, there are two intervals containing 11 + 200 = 211 observations; to the right of the mode, there are 7 intervals containing 229 + 121 + 61 + 41 + 18 + 6 + 2 = 478 observations. Thus, in this example, more than twice as much probability mass lies to the right as to the left of the histogram's mode, showing a pronounced right-skewness.

If this set of 1,000 observations were obtained from actual environmental measurements of a pollutant, the investigator, after examining the right-skewed histogram, probably would decide to plot the cumulative frequencies of this distribution on logarithmic-probability paper. The result (Figure 8.6) is a nearly straight line exhibiting very slight downward curvature, just like so many of the frequency distributions observed in real ambient environmental measurements.[1] This line has been drawn by hand through the points and has no particular theoretical significance.

Using the methods of Chapter 3 (page 38), the expected value of this distribution is the product of the initial concentration and the expected values of the individual dilution factors:

$$E[C_4] = E[c_o]E[\alpha_1]E[\alpha_2]E[\alpha_3]E[\alpha_4] = c_o(\bar{\alpha})^4 = (1000)(0.20)^4 = 1.600 \text{ ppm}$$

The observed sample mean of $\overline{c_4}$ = 1.6126 ppm agrees well with the theoretical mean of 1.600 ppm, and the theoretical and observed means probably would lie closer together if more than 1,000 runs were made.

It is of interest to describe briefly the computer simulation program (Figure 8.4) used to generate the histogram of simulated concentrations in beaker 4

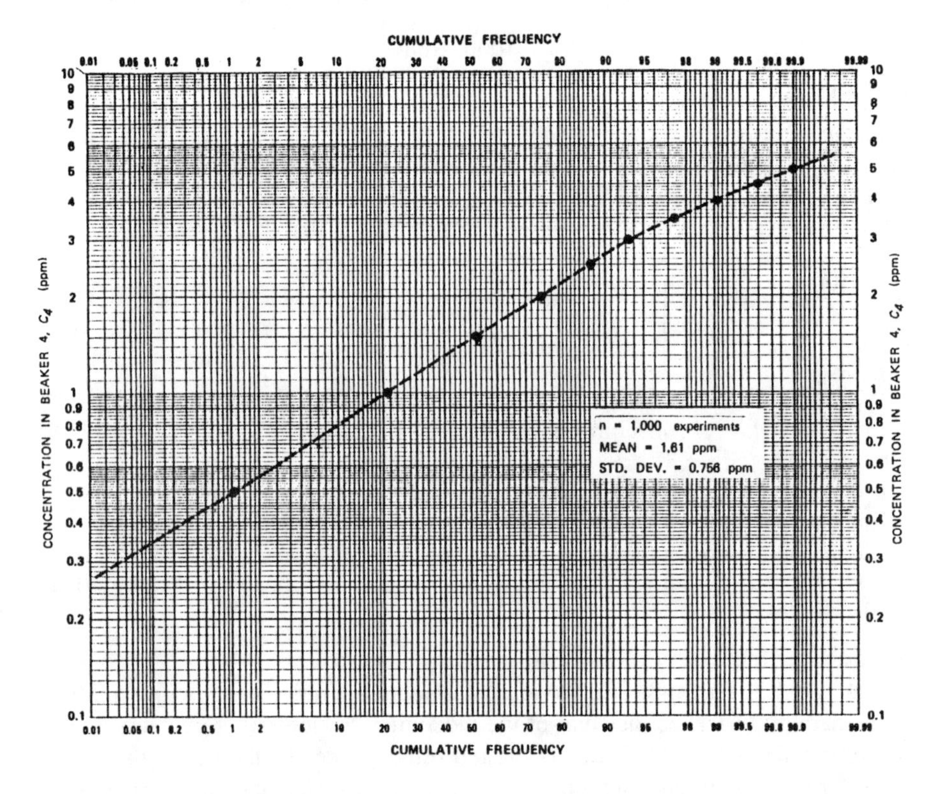

Figure 8.6. Logarithmic-probability plot of the cumulative frequencies resulting from multiplying 4 uniformly distributed random variables **U**(0.12,0.28) to simulate an experiment in which a 1000 ppm source is successively diluted 4 times using a masked measuring cylinder to increase pouring error. The relative straightness of the line approximates lognormality, even though the distribution is not truly lognormal because the asymptotic conditions have not been met.

(Figure 8.5) and the cumulative frequency plot (Figure 8.6). In this BASIC program, the statement RND generates a uniformly distributed random variable **U**(0,1). This random variable then is scaled linearly using the following equation, in which "*" denotes multiplication, to give the desired uniformly distributed random variable **U**(A,B), represented in the program by U:

$$U = A + (B - A) * RND$$

In the program, the constant SCALE = B – A is computed first and substituted into this equation to avoid computing it repeatedly. Notice that this formula can be checked very easily: substituting RND = 0 gives U = A, and substituting RND = 1 gives U = A + B – A = B.

The program reads in M, the number of times that the uniformly distributed random variables **U**(A,B) are to be multiplied together, and it repeats the multiplication M times over in a simple loop (lines 320–350 in Figure 8.4), generating the product of M uniformly distributed random variables:

$$C = U_1 U_2 \ldots U_M$$

Prior to entering the multiplication loop, the program sets C = CONST, the constant multiplier of the product, which denotes the initial concentration c_o. Thus, C will be the product of a constant and M random variables and the initial concentration c_o. The entire process—generating the random variables **U**(A,B) and multiplying them together M times—is repeated NUM times over in a larger loop (lines 320–350). As each random product is created, it is printed on the screen of the computer, along with J, a counter (line 360). Each value, as it is produced, also is sorted into a histogram with interval width DELTA (lines 420–460). The histogram of the simulated concentration in beaker 4 (Figure 8.5) is for the case in which A = 0.12, B = 0.28, M = 4, NUM = 1000, and DELTA = 0.5. On the standard IBM-PC, this program required less than one minute to generate the full set of 1000 random products.

Development of Successive Random Dilutions (SRD) Theory

Examples such as the one above readily produce frequency distributions that appear lognormal to the eye, but the resulting frequency distribution of the concentrations in beaker 4 is not truly lognormal in the sense described in Chapter 9. Rather, it is asymptotically lognormal as the number of successive pourings from beaker to beaker increases toward infinity. To see this, it is necessary to develop a theoretical basis for this process and its resulting distribution. Fortunately, convergence to a lognormal distribution occurs quite rapidly as the number of successive dilutions increases beyond 10–12. Also, the theory is based on the Central Limit Theorem and applies whether the probability distributions of the dilution factors are normal, uniform, or of another type, or even if the dilution factors have different distributions, provided that the distributions have finite variance. It is customary to represent random variables by capital letters, so we shall change the notation and substitute D_i for α_i. For simplicity, we usually consider the number of dilutions m to be fixed.

Let D_i be an independent random variable between 0 and 1 denoting the dilution factor of the ith successive pouring such that $0 < D_i \le 1$. As with the above example, the random variable C_m denoting the final concentration after m successive pourings will be the product of the initial concentration c_o and m individual dilution factors:

$$C_m = c_o D_1 D_2 \ldots D_m = c_o \prod_{i=1}^{m} D_i$$

Our analysis proceeds by taking logarithms of both sides, giving the following result:

$$\log C_m = \log c_o + \log D_1 + \log D_2 + \ldots + \log D_n = \log c_o + \sum_{i=1}^{m} \log D_i$$

The random variable D_1 is assumed to be independent of the random variable D_2, and D_i is assumed to be independent of D_{i+1} for all i. The random variable

D_i may have the same probability distribution as D_{i+1}, or each D_i may have a different distribution, but the mean and variance must be finite because D_i lies between 0 and 1. Since $0 < D_i \leq 1$ for all i, then the logarithms of the dilution factors will be random variables ranging from slightly greater than negative infinity to 0; that is, $-\infty < \log D_i \leq 0$. Introducing the notation $R_i = \log D_i$, each R_i will be a new random variable independent of R_{i+1} for all i. Using this notation, the above equation can be simplified by writing it as the sum of m independent random variables and a constant, *const.*

$$\log C_m = const + R_1 + R_2 + \ldots + R_m = const + \sum_{i=1}^{m} R_i$$

The left-hand side of this equation is the logarithm of the final concentration after the mth successive dilution, and the right-hand side is the sum of a constant and m independent random variables. As discussed in Chapters 6 and 7, the Central Limit Theorem implies that the process of adding independent random variables gives rise asymptotically to a normal distribution; the conditions for convergence are discussed by Feller.[2,3] In the limit, as m approaches infinity, $\log C_m$ will equal a normally distributed random variable plus a constant. Because the logarithm of the concentration will be asymptotically normally distributed, the *concentration itself, in the limit, will be logarithmic-normally distributed*, or *lognormally distributed.* This completes the formal argument, and it shows why lognormal distributions arise so readily from dilution processes.

Some words about the constant are appropriate. For each i, $R_i = \log D_i$ is a negative number, because D_i ranges between 0 and 1. If the sum $S = R_1 + R_2 + \ldots + R_m$ denotes the stochastic part of the right-hand side of the above equation, then S will be asymptotically normally distributed with negative mean $\mu < 0$. The additive constant $const = \log c_o$ will translate the mean of the normal distribution to the right (positive) side of the origin if the initial concentration c_o is large enough such that $|\log c_o| > |\mu|$. Thus, the initial concentration c_o affects the location parameter of the normally distributed random variable $Y = \log C_m$.

Following the rules presented in Chapter 2, the mean and variance of $\log C_m$ are computed as follows:

$$E[\log C_m] = const + E[R_1] + E[R_2] + \ldots + E[R_m]$$

$$Var(\log C_m) = Var(\log R_1) + Var(\log R_2) + \ldots + Var(\log R_m)$$

The role of the constant as a location parameter is readily apparent in the above equation for the expected value $E[\log C_m]$.

Gamma Distribution

Although $\log C_m$ is asymptotically normally distributed, it will not converge immediately, and it is instructive to examine the properties of a particular case for small m before convergence occurs. In the beaker example above, each dilution factor was assumed uniformly distributed over the relatively narrow range between 0.12 and 0.28. Now consider a case of extreme error. Suppose that each individual dilution factor D_i is allowed to take on values over its entire feasible range $0 < \alpha_i \leq 1$ and that it is uniformly distributed with PDF $f_D(\alpha) = 1$. To simplify the analysis, let the initial concentration $c_o = 1$ ppm. For convenience, we use natural logarithms, and $\ln C_m$ will be written as the sum of m logarithms of uniformly distributed random variables, in which each D_i is assumed to be distributed as **U**(0,1) for all i:

$$Y = \ln C_m = \ln D_1 + \ln D_2 + \ldots + \ln D_i + \ldots + \ln D_m$$

Denoting the logarithm of the random variable as $R_i = \ln D_i$ in the same manner as before, this sum also can be written as follows:

$$Y = \ln C_m = \quad R_1 + R_2 + \ldots + R_i + \ldots + R_m$$

Here, each $R_i = \ln D_i$ will be an exponentially distributed random variable defined over the negative axis with the following PDF:

$$f_R(r) = e^r \qquad \text{where } -\infty < r \leq 0$$

Thus, Y will be the sum of a series of exponentially distributed random variables, and the distribution of Y is known exactly. As discussed by Johnson and Kotz,[4] the sum of a series of exponentially distributed random variables will have a *gamma distribution* (Table 8.1). As noted by Hahn and Shapiro,[5] adding the logarithms of unit uniformly distributed random variables is one of the common computer simulation techniques for generating gamma distributions.

The two parameters of the gamma distribution **G**(η,λ) are its shape parameter η and its scale parameter λ (Table 8.1). The PDF is right-skewed when defined over the positive axis $0 \leq X < \infty$, as is usually the case (Figure 8.7). The expected value of the distribution is given by η/λ and the variance is given by η/λ^2. The mode is given by $x_m = (\eta - 1)/\lambda$. For the case of $\eta = 2$ and $\lambda = 1$, the distribution has its mode at $x_m = 1$ and exhibits considerable right-skewness. As η increases, the mode shifts to the right, and the distribution soon resembles a symmetrical, bell-shaped curve, the normal distribution. Hahn and Shapiro[5] discuss use of the normal distribution as an approximation to the gamma distribution. The gamma distribution is the appropriate probability model for representing the time period required for a total of η independent events to take place if the events occur at a constant arrival rate λ. If the shape parameter η is a positive integer, the gamma distribution is called an *Erlang distribution.* For positive integers, the gamma function $\Gamma(\eta)$ becomes a simple factorial, $\Gamma(\eta) = (\eta - 1)(\eta - 2) \ldots 1 = (\eta - 1)!$

The distribution given in Table 8.1 is called the "incomplete gamma function," or sometimes the "incomplete gamma function ratio." Tables for plotting the CDF of this distribution are available in the literature.[6,7] A BASIC computer program (Figure 8.8) was written to generate the values plotted in Figure 8.7. This program can be used to plot similar curves of the incomplete gamma function PDF for any values of its parameters η and λ.

Table 8.1. The Gamma Distribution $\mathbf{G}(\eta,\lambda)$

Probability Density Function:

$$f_X(x) = \begin{cases} \dfrac{\lambda^\eta}{\Gamma(\eta)} x^{\eta-1} e^{-\lambda x}, & x \geq 0,\ \eta > 0,\ \lambda > 0 \\ 0 \text{ elsewhere} \end{cases}$$

$$\textit{where } \Gamma(\eta) = \int_0^\infty u^{\eta-1} e^{-u} du \quad \text{or, for integers, } \Gamma(\eta) = (\eta - 1)!$$

Expected Value:

$$E[X] = \frac{\eta}{\lambda}$$

Variance:

$$Var(X) = \frac{\eta}{\lambda^2}$$

Coefficient of Skewness:

$$\sqrt{\beta_1} = \frac{2}{\sqrt{\eta}}$$

Coefficient of Kurtosis:

$$\beta_2 = \frac{3(\eta + 2)}{\eta}$$

The more general form of the gamma probability density function contains three parameters η, λ, and γ, in which γ is a location parameter:

$$f_X(x) = \frac{(x-\gamma)^{\eta-1} e^{-(x-\gamma)/\lambda}}{\lambda^\eta \Gamma(\eta)}, \quad x > \gamma,\ \eta > 0,\ \lambda > 0$$

If we set $\gamma = 0$, $\eta = \nu/2$, and $\lambda = 1/2$, the following equation, which is the PDF for the gamma distribution $\mathbf{G}(\nu/2,1/2)$, will result:

$$f_X(x) = \frac{x^{(\nu/2 - 1)} e^{-x/2}}{2^{\nu/2} \Gamma(\nu/2)}, \quad x \geq 0,\ \nu = 1, 2, \ldots$$

This result is known as the *chi-square distribution* with ν degrees of freedom,[8] and it often is written with χ^2 substituted for X (Table 8.2).

The chi-square distribution is a special case of the gamma distribution. It is important in many statistical applications. If Y_1, Y_2, … , Y_ν are ν independent normally distributed random variables with distributions $\mathbf{N}(0,1)$, then the sum

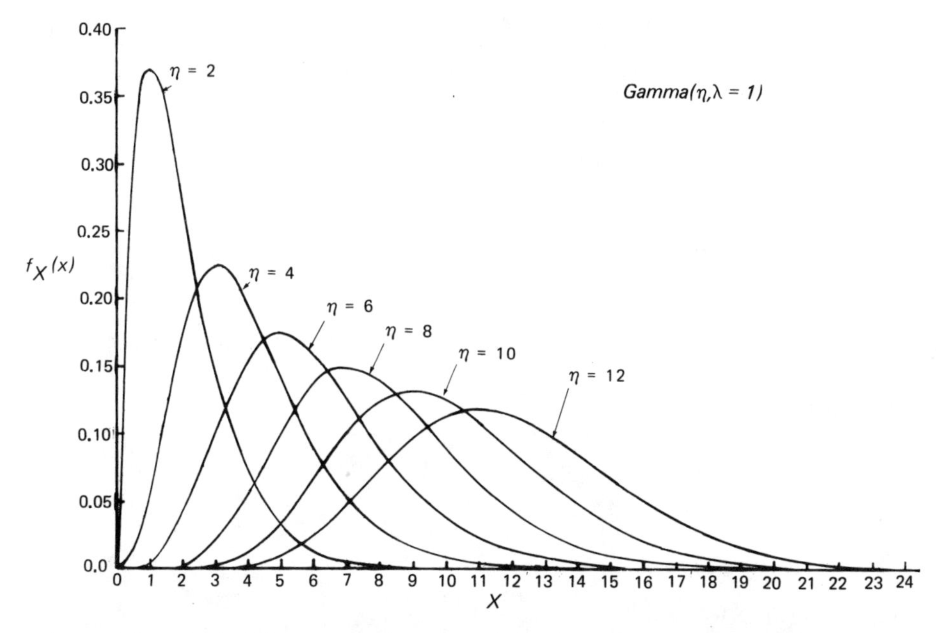

Figure 8.7. Probability density function of the gamma probability model $\mathbf{G}(\eta,\lambda)$ for $\lambda = 1$ and different values of η, showing tendency of the distribution to approach a symmetrical, bell-shaped curve for large η (Source: Ott[11]).

```
100 REM Program to compute the Gamma probability distribution.
105 PRINT
106 KEY OFF
110 PRINT "PROGRAM TO COMPUTE THE GAMMA PROBABILITY MODEL"
120 PRINT : INPUT "ENTER VALUE FOR AIDA"; AIDA: PRINT
130 PRINT : INPUT "ENTER VALUE FOR LAMBDA"; LAMBDA: PRINT
135 PRINT : INPUT "Incremental steps of variate for each computation"; DELTA
140 PRINT : INPUT "Upper Range of Variate to be Computed"; RANGE: PRINT : PRINT
150 REM Compute denominator.
160 DENOM = 1
170 VALUE = AIDA - 1
180 IF VALUE = 0 GOTO 220
190 DENOM = DENOM * VALUE
200 VALUE = VALUE - 1
210 GOTO 180
220 RATIO = (LAMBDA ^ AIDA) / DENOM
225 LPRINT "GAMMA PDF:  AIDA ="; AIDA; " AND LAMBDA ="; LAMBDA
230 CUM = 0
240 X = 0
250 I = 1
260 P = RATIO * EXP(-LAMBDA * X) * X ^ (AIDA - 1)
270 CUM = CUM + P
280 LPRINT I, X, P
290 I = I + 1
300 X = X + DELTA
310 IF X <= RANGE THEN GOTO 260
320 END
```

Figure 8.8. BASIC computer program for computing tables for plotting the gamma probability density function $\mathbf{G}(\eta,\lambda)$.

Table 8.2. The Chi-Square Distribution $\chi^2(\nu)$

Probability Density Function:

$$f_{\chi^2}(x) = \begin{cases} \dfrac{x^{(\nu/2 - 1)} e^{-x/2}}{2^{\nu/2}\,\Gamma(\nu/2)}, & x \geq 0,\ \nu = 1,2,\ldots \\ 0 \text{ elsewhere} \end{cases}$$

$$\text{where } \Gamma(\nu/2) = \int_0^\infty u^{(\nu/2 - 1)} e^{-u} du \text{ or, for even integers, } \Gamma(\nu/2) = (\nu/2 - 1)!$$

Expected Value:

$$E[\chi^2] = \nu$$

Variance:

$$Var(\chi^2) = 2\nu$$

Coefficient of Skewness:

$$\sqrt{\beta_1} = \frac{2\sqrt{2}}{\nu}$$

Coefficient of Kurtosis:

$$\beta_2 = 3 + 12/\nu$$

---

of their squares, $S = Y_1^2 + Y_2^2 + \ldots Y_\nu^2$, will have a chi-square distribution with ν degrees of freedom. If U_1, U_2, ... , U_ν are uniformly distributed random variables with distributions $\mathbf{U}(0,1)$, then one-half their sum of squares, $S = 1/2(U_1^2 + U_2^2 + \ldots U_\nu^2)$, will have a chi-square distribution with ν degrees of freedom. Tables of the cumulative distribution function of the chi-square distribution appear in most standard statistical reference texts.[9,10] Like other forms of the gamma distribution, the chi-square distribution approaches the normal distribution for large ν. For $\nu > 30$, the normal distribution is recommended as a quite accurate approximation to the chi-square distribution.[9]

Because each D_i ranges from 0 to 1, each exponentially distributed random variable $R_i = \ln D_i$ will be negative, and the sum Y of the m exponentially distributed random variables also will be negative, causing the resulting gamma distribution to be defined over the negative axis, $-\infty < Y \leq 0$ (Figure 8.9). For $m = 1$, the distribution is a negative exponential, as it must be, since $Y = R_1 = \ln D_1$, and D_1 is uniformly distributed as $\mathbf{U}(0,1)$. For $m = 2$, the distribution has a single mode at $y = -1$ and is left-skewed. For $m = 4$, the distribution has a single mode at $y = -3$ and considerable left-skewness. The gamma distribution for Y is obtained from Table 8.1 by setting $X = -Y$, $\lambda = 1$, and $\eta = m$, giving the following PDF:

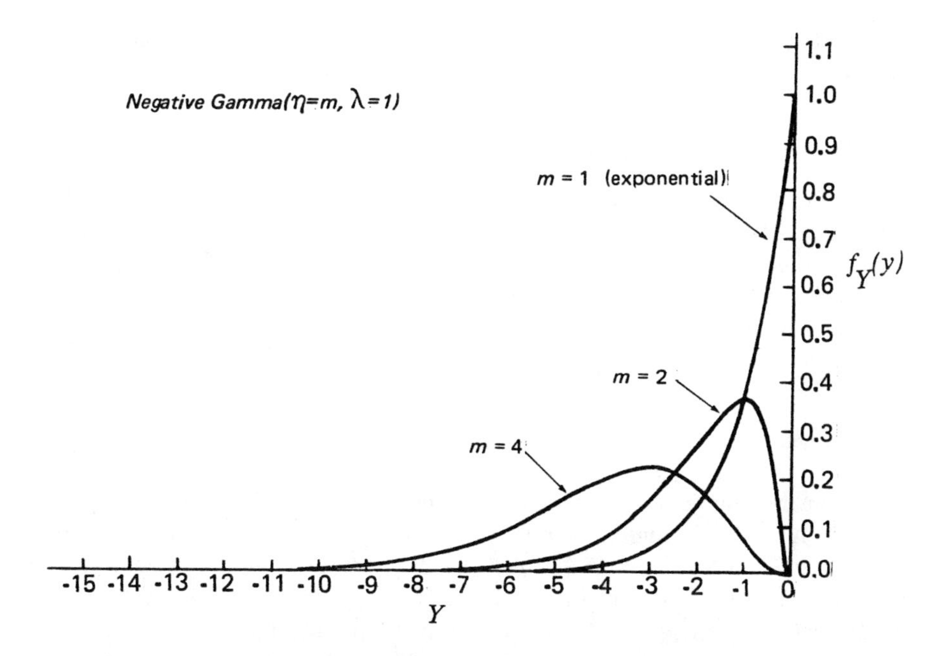

Figure 8.9. Gamma distribution of the logarithm of the product of m unit uniformly distributed random variables.

$$f_Y(y) = \frac{1}{\Gamma(m)}(-y)^{m-1}e^y \quad where - \infty < y < 0$$

Because m is a positive integer in this case, $\Gamma(m) = (m-1)!$ In practical situations, it is convenient to redefine Y as a positive random variable and to plot it on the positive axis.

In summary, we began by considering the following successive random dilution process in which $c_o = 1$, and each D_i is an independent random variable distributed as **U**(0,1):

$$C_m = c_o \prod_{i=1}^{m} D_i$$

After taking natural logarithms, it was seen that the sum $-Y = -\ln C_m = \ln D_1 + \ln D_2 + \ldots + \ln D_m$ has a gamma distribution **G**$(m,1)$ exactly. In the limit, as m becomes large, $-Y$ will approach a normal distribution, because the gamma distribution asymptotically approaches the normal distribution for large m. Since $Y = \ln C_m$ has a (negatively defined) gamma distribution, then C_m can be said to have a logarithmic-gamma distribution exactly. As m becomes large, C_m approaches a logarithmic-normal distribution and can be said to be asymptotically lognormally distributed.

Thus, for this extreme case of pouring error in which the dilution factors are uniformly distributed as **U**(0,1), C_m is asymptotically lognormally distributed.

Since $E[D_1] = E[D_2] = \ldots = E[D_m] = 1/2$ for this uniform distribution, the expected value of C_m will be given by:

$$E[C_m] = c_o E[D_1]E[D_2]\ldots E[D_m] = c_o(1/2)^m$$

If the dilution factors $D_1, D_2, \ldots, D_m$ have some other distribution in which the variance is less than the unit uniform distribution, or if the dilution factors all have different distributions with finite variance, then C_m also will be asymptotically lognormally distributed. This result illustrates the tendency for successive random dilution processes to give rise to lognormal distributions under rather general conditions.

Examples Based on Monte Carlo Simulation

To illustrate these concepts, we can examine the theory in light of computer simulation results. Using the random number generator in the BASIC computer program described earlier (Figure 8.4, page 198) to generate the product of m uniformly distributed random variables, we consider two cases: $m = 5$ and $m = 12$. Instead of $c_o = 1$ ppm, we choose $c_o = 1{,}000$ ppm for the case of $m = 5$, giving an expected value for C_5 of $E[C_5] = (1000)(1/2)^5 = 31.25$ ppm. The computer program was run $n = 100{,}000$ times, giving 100,000 values for C_m. The cumulative frequencies of C_m were plotted on logarithmic-probability paper as dots (Figure 8.10). To obtain the necessary detail and resolution, this three-cycle logarithmic-probability paper was constructed from Keuffel and Esser two-cycle papers.

To obtain the results predicted by theory, tables for the gamma CDF published by Wilk, Gnanadesikan, and Huyett[6] were used. According to the theory, the scaled concentration $C_5/1000$ will have a negative logarithmic-gamma distribution, or $Y = \ln(C_5/1000)$ will have a negative gamma distribution. Then $X = -\ln(C_5/1000)$ will have a positive gamma distribution **G**(5,1). Since $f_X(x)$ is the PDF of this gamma distribution, tables were used to find x for each appropriate cumulative frequency and then to compute $C_5 = 1000e^{-x}$. The resulting smooth theoretical curve (Figure 8.10) shows excellent agreement with the experimental (computer generated) results, since nearly all of the computer simulation data points lie on the curve. The curve exhibits considerable concave-downward curvature.

For the case of $m = 12$, the computer simulation program also was run $n = 100{,}000$ times. To make the expected value, $E[C_{12}] = 31.25$ ppm, the same, it was necessary to set $c_o = 128{,}000$ ppm. Again, the theoretical curve based on the gamma distribution **G**(12,1) showed excellent agreement with experimental results, as can be seen by the large number of data points that lie almost exactly on the theoretical line. The results for $m = 12$ showed much less curvature than the case for $m = 5$, illustrating the tendency for the distribution to approach a straight line for large m when plotted on logarithmic-probability paper. This straightness reflects the tendency of the logarithmic-gamma distribution to approach a lognormal distribution asymptotically.

These examples are for the rather extreme case of pouring error in which each dilution factor is allowed to vary over its full range $0 < D_i \leq 1$ for all i. In

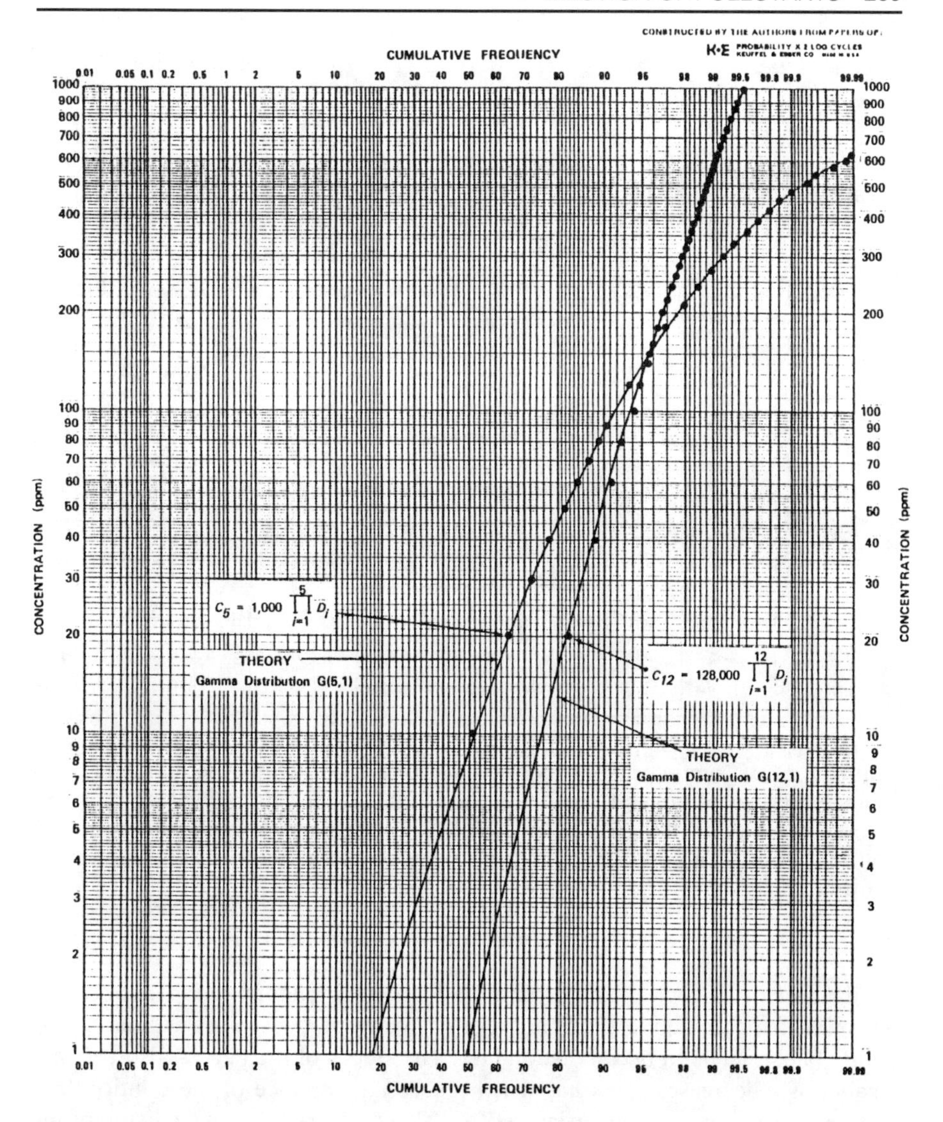

Figure 8.10. Computer simulation of successive random dilution experiments with n = 100,000 runs and dilution factors uniformly distributed as **U**(0,1), compared with distributions predicted by the gamma probability model. The curves both have the same means, 31.25 ppm, and are plotted on logarithmic probability paper for the cases of m = 5 and m = 12 dilutions.

real situations, the dilution factors are likely to be more restricted, with distributions other than the uniform. To see how the distribution of C_5 varies when the dilution factors are restricted to a more narrow range $r = b - a$, in which $r \leq 1$, the computer simulation was repeated for five uniformly distributed random variables **U**(a,b) multiplied together. Five cases of the uniform distribution were considered, **U**(0.2,0.8), **U**(0.25,0.75), **U**(0.3,0.7), **U**(0.35,0.65), and **U**(0.4,0.6), and the product was generated n = 10,000 times for each case (Figure 8.11) using the computer program mentioned earlier (Figure 8.4, page 198).

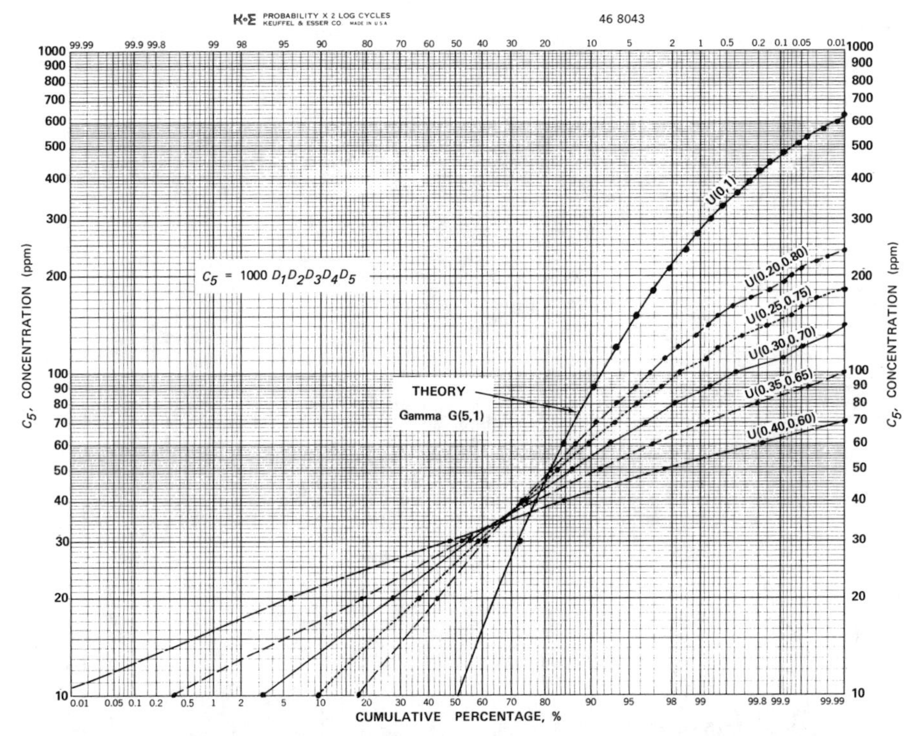

Figure 8.11. Computer simulation results for the product of five uniformly distributed random variables **U**(a,b) multiplied together for different values of a and b. For comparison, the results for five **U**(0,1) distributions multiplied together and the theoretical curve predicted by the gamma distribution **G**(5,1) also are shown (top curve).

As the range r of each of the uniformly distributed random variables becomes narrower, the cumulative distribution of C_5, when plotted on logarithmic-probability paper, tends to become straighter. The slopes of the curves also become smaller as r decreases. This happens because the variance of the dilution factors decreases, causing the variance of C_5 also to decrease. For the cases of restricted$r < 1$, the lines are drawn through the points by "eye" and are not theoretical curves.

If the number of successive random dilutions were increased beyond $m = 5$, the straightness of these curves would be more pronounced. Although the dilution factors are assumed here to be uniformly distributed, similar curves would result from a whole class of distributions other than the uniform. These examples illustrate the tendency of successive random dilution processes to generate distributions that appear "approximately lognormal" and resemble those commonly found in the environment.

Successive Random Dilutions: Single Beaker Case

What may be surprising is that the same theory applies to a single beaker, provided that the excess contents are removed on each step after the fresh water is poured into the beaker. In this situation, a single beaker acts as a *single mixing compartment.* Indeed, for many environmental problems, the single-compartment model is the most important case.

In a single-beaker experiment, the investigator pours fresh water into just one beaker, and the contents become mixed and diluted by the added water. As the beaker becomes filled, the excess mixture is removed. One way to achieve removal of the excess liquid is to allow the beaker to overflow; the mixture simply spills over the edges. Another way is to remove the excess liquid mixture manually: the investigator pours off the excess liquid until the original volume is restored. Each cycle of adding fresh water, allowing for mixing to occur, and removing the overflow constitutes one stage of the dilution process.

Consider a single beaker filled with v = 250 ml of water containing a pollutant at concentration c_o. A small quantity of fresh water diluent is added, allowed to mix, and the excess mixture is poured off. What will be the new concentration C_1 inside the beaker?

Before the fresh water is added, the total quantity of pollutant present in the system is

$$q_o = c_o v$$

Assume that a small volume W_1 ml of fresh water is added to the beaker (without overflowing). After the water has been added, the new volume of the contents (provided there is space for W_1 inside the beaker) will be the sum of the original volume v plus the new volume added W_1, giving a total volume $v + W_1$. By conservation of mass, the new concentration C_1 times this volume must equal the original quantity of pollutant in the system q_o:

$$C_1(v + W_1) = q_o = c_o v$$

Solving this equation for C_1, we obtain

$$C_1 = c_o \frac{v}{v + W_1}$$

This equation shows that the original concentration c_o in the beaker has been reduced by multiplying it by the dilution factor $D_1 = v/(v + W_1)$. Using this dilution factor, the above equation also can be written as

$$C_1 = c_o D_1$$

$$\text{where} \quad D_1 = \frac{v}{v + W_1}$$

At this point, the beaker is somewhat full, so we can manually pour off the excess water (liquid volume W_1) and bring the contents of the beaker back to its original volume of 250 ml.

Now suppose the entire process is repeated again; another incremental amount of fresh water W_2 is added to the 250 ml contents of the beaker. The result quickly mixes and the concentration present in the beaker is once again reduced slightly. The excess volume once again is poured off, leaving the 250 ml in the beaker. The new concentration C_2 in the beaker will become C_1 multiplied by $D_2 = v/(v + W_2)$. If this process is repeated over and over again, each time with a different incremental volume $W_1, W_2, \ldots, W_n$, then the final concentration inside the beaker will be given by the following result, which is also the general equation for successive random dilutions:

$$C_m = c_o D_1 D_2 \ldots D_m = \prod_{i=1}^{m} D_i$$

$$\text{where} \quad D_i = \frac{v}{v + W_i}$$

C_m = concentration in the beaker after m steps
c_o = initial concentration in the beaker
W_i = incremental liquid diluent added on step i
v = volume of the initial contents of the beaker
m = number of dilution steps in the single beaker

If the quantities $W_1, W_2, \ldots, W_n$ added on each mixing step are random variables, then the equation above for the single-beaker case yields the same result as for the multiple-compartment case described earlier [see pages 195–208 on the multiple beaker example] and the two cases are mathematically equivalent.

The development of the single-compartment model just discussed assumed that the small quantity of pollutant added to the beaker rapidly became fully mixed, and that the fully-mixed contents then were poured off. Thus, the discarded liquid contained the pollutant at concentration C_1. An alternative formulation of this model has been published[11] that assumes that the discarded quantity does not have time to mix. Therefore, as W_1 is added on the first step, the quantity W_1 leaves the system at the old concentration c_o. This might occur, for example, if the beaker were already full, and the new quantity W_1 was poured in on one side of the beaker, while an equal quantity W_1 spilled over the other side at the prior concentration c_o. We now develop this alternate version of the model to show that both forms are mathematically equivalent when the amount of the diluent added is small.

In the alternate formulation of the model, the amount of pollutant present prior to dilution is the same as for the above case: $q_o = c_o v$. After adding the small amount of fresh water diluent W_1, the total mass to be accounted for in the system (inside and outside the beaker) is $q_o = C_1 v + c_o W_1$. Using the concept of conservation of mass (sometimes called "mass balance"), we now set these before-and-after quantities equal:

$$c_o v = q_o = C_1 v + c_o W_1$$

If we solve this expression for the new concentration C_1, we obtain:

$$C_1 = \frac{c_o v - c_o W_1}{v} = c_o\left(\frac{v - W_1}{v}\right) = c_o\left(1 - \frac{W_1}{v}\right)$$

If this process is repeated in the same manner by adding another small quantity of fresh water W_2, then the quantity C_1 will be multiplied by $D_2 = (1 - W_2/v)$ to give C_2. If this process is repeated over and over, it will give a general expression similar to the previous general expression for the SRD, except that D_i is different:

$$C_m = c_o D_1 D_2 \ldots D_m = \prod_{i=1}^{m} c_o D_i$$

$$\text{where } D_i = \left(1 - \frac{W_i}{v}\right)$$

and C_m = pollutant concentration after the mth dilution
c_o = initial pollutant concentration present
n = number of dilution steps
v = mixing volume of beaker
W_i = volume of diluent added on the ith step

Once again, we obtain the same general expression for a SRD process. However, we need to explain the different equations for D_i obtained on the two alternative formulations:

$$D_i = \left(\frac{v}{v + W_i}\right) \text{ and } D_i = \left(1 - \frac{W_i}{v}\right)$$

Are these two formulations mathematically equivalent? If we divide the expression for D_i in the first formulation above by v to expand it as a series, we obtain:

$$\left(\frac{v}{v + W_i}\right) = \left(\frac{1}{1 + \frac{W_i}{v}}\right) = 1 - \left(\frac{W_i}{v}\right) + \left(\frac{W_i}{v}\right)^2 - \left(\frac{W_i}{v}\right)^3 + \ldots.$$

If the quantity W_i is very small compared with the original volume of v in the container—that is, $W_i \ll v$—then all terms of power 2 and above are extremely small; if these higher powered terms are ignored, then the two alternative formulations of the single-compartment model are mathematically equivalent.

This analysis shows the remarkable result that a single-compartment system can act as a complete SRD process. Thus, if we had a number of different beakers, each beginning with a pollutant concentration c_o, and we diluted each of them by 20 random pourings of a diluent, the frequency distribution of the concentrations from all the beakers would follow, to a good approximation, a lognormal distribution (Chapter 9). The single-compartment analysis is especially important because, as we shall see, the single-compartment model is ubiquitous in the environment. Furthermore, our theoretical prediction that the resulting concentration distribution is approximately lognormal is not obvious and does not follow automatically from common sense.

Continuous Mass Balance Model

The reason for first considering these two alternative discrete formulations is that dilution occurs essentially on a *continuous basis.* Thus, the spilloff of the beaker actually begins at concentration c_o at the instant that pouring begins, and changes from c_o to C_1 gradually as the diluent is added. A *continuous mass balance model* is needed to bridge the gap between the two alternative discrete formulations just presented. Because continuous pouring can be represented as a large number of steps that are infinitesimally small, the dilution process in the continuous model can be described by a differential equation.

To derive this differential equation, consider a slightly modified experiment. The beaker begins completely full with water containing pollutant concentration c_o. Instead of pouring off the excess liquid at each dilution step, the experimenter pours the diluent continuously into the beaker. As soon as the pouring begins, the beaker begins to overflow, and the concentration inside the beaker gradually declines. The continuous pouring case can be viewed as the continuous version of many small dilutions in the two discrete cases just discussed.

Derivation of the continuous model describing this system begins by considering two states in time: the original state (at $t = 0$) and a subsequent state after some time has elapsed (at $t = T$). Just as pouring is about to start, at $t = 0$, the total mass of pollutant in the system is contained entirely within the beaker as $q_{original} = c_o v$. Instead of adding very small quantities of diluent in successive increments, the investigator pours the diluent into the beaker continuously (although he may pour erratically) and the contents spill over the edges of the beaker. At the elapsed time $t = T$, if we take a "snapshot" of the system we will find that the total mass of the pollutant is distributed in two places: the contents present in the beaker at time T, $q_{inside}(T)$, and the amount that has spilled off after time T, $q_{spilled}(T)$.

If we consider the mass balance before and after the passage of time T, we set the original mass in the beaker equal to the sum of the quantity that is present at the two locations at time T. The following is the "mass balance equation" for this system:
The mass balance equation accounts for all mass everywhere in the system

$$\begin{matrix}\text{Original} \\ \text{Mass}\end{matrix} = \begin{matrix}\text{Mass Inside} \\ \text{at Time } T\end{matrix} + \begin{matrix}\text{Mass Spilled} \\ \text{at Time } T\end{matrix}$$

$$q_{original} = q_{inside}(T) + q_{spilled}(T)$$

at time T. The original mass of pollutant in the system prior to pouring was $q_{original} = c_o v$. Because mass cannot be created or destroyed in this system, the sum of the pollutant mass inside the compartment at time T and the pollutant mass that has spilled out must be equal to the original total.

We can represent the quantity that has spilled over the beaker's edges mathematically by integrating the product of the time varying concentration $c(t)$ and the flow rate $w(t)$ between time $t = 0$ and $t = T$:

$$q(T)_{spilled} = \int_0^T c(t)w(t)dt$$

where $q(T)_{spilled}$ = total quantity spilled over edge of single beaker after time T [M]
$c(t)$ = time varying concentration spilling over beaker's edges [typically M/L^3]
$w(t)$ = flow rate of liquid spilling over beaker's edges [L^3/T]

Assume that the flow rate is held constant in a particular experiment; that is, $w(t) = w$. If the beaker remains well-mixed as the diluent is added, then the quantity inside the compartment will be the product of the concentration at time T and the volume of the beaker, or $q_{inside}(T) = c(T)v$. If we substitute both this expression for $q_{inside}(T)$ and the above expression for $q_{spilled}(T)$ into the mass balance equation of the system with $q_{original} = c_o v$, we obtain the following specific mass balance equation:

$$c_o v = c(T)v + \int_0^T c(t)wdt$$

Differentiating this equation, we obtain

$$\frac{v}{w}\frac{dc(t)}{dt} + c(t) = 0$$

If we rearrange the terms, this equation can be written as

$$\frac{dc(t)}{dt} = -\frac{w}{v}c(t)$$

and, integrating this equation to obtain a solution we obtain:

$$\ln c(t) = -\frac{w}{v}t + B$$

Here, B is a constant of integration, and, substituting the constant $A = e^B$ in place of B and solving for $c(t)$, we obtain the following solution:

$$c(t) = e^{-\frac{w}{v}t}e^B = Ae^{-\frac{w}{v}t}$$

The initial conditions require that $c(0) = c_o$ at $t = 0$, and therefore $A = c_o$. The resulting solution shows that the concentration inside the beaker decays exponentially from its initial value of c_o approaching, but never reaching, zero as time passes:

$$c(t) = c_o e^{-\frac{w}{v}t} = c_o e^{-\phi t}$$

If we represent the ratio of the flow rate to the volume as a single parameter $\phi = w/v$, then ϕ will be the number of "beaker volumes" per unit time of fresh water that is added (which equals the quantity of water that spills out). Here ϕ is analogous to the "air exchange rate" used to characterize indoor air quality problems, and v is analogous to the mixing volume of a room (discussed in the following pages 227–235 on indoor air quality).

Stochastic Flow Rate

If the flow rate w is constant, then the exponential solution gives a smooth curve, and the result is a deterministic system. If, however, the flow rate W is a random variable, then this system will become a stochastic process. How will variability of pouring the diluent change the behavior of this system?

Consider an example in which the random variable W takes on discrete values, $w_1, w_2, \ldots, w_m$ over m time segments $(0, t_1), (t_1, t_2), \ldots, (t_{m-1}, t_m)$. For the first segment, the solution will be

$$c(t) = c_o e^{-\left(\frac{w_1}{v}\right)t} \quad \text{for } 0 \le t \le t_1$$

The concentration during the second segment will be an exponentially decaying function also; its initial condition c_1 will be given by the above expression for $t = t_1$, or

$$c(t) = c_1 e^{-\left(\frac{w_2}{v}\right)(t-t_1)} = c_o e^{-\left(\frac{w_1}{v}\right)t_1} e^{-\left(\frac{w_2}{v}\right)(t-t_1)} \quad \text{for } t_1 \le t \le t_2$$

Finally, if this process is repeated once more for a third interval, we obtain

$$c(t) = c_2 e^{-\left(\frac{w_2}{v}\right)(t-t_1)} = c_o e^{-\left(\frac{w_1}{v}\right)t_1} e^{-\left(\frac{w_2}{v}\right)(t_2-t_1)} e^{-\left(\frac{w_3}{v}\right)(t-t_2)} \quad \text{for } t_2 \le t \le t_3$$

Suppose, for simplicity, that each segment is of the same exact time duration $(t_m - t_{m-1}) = \delta$ and that the process continues over m discrete, equally spaced intervals. We can say that this is a continuous process whose exchange rate parameter is allowed to vary in time such that it is "piecewise-constant." The resulting concentration C_m will be the product of factors that are "piecewise exponential."

Then, an expression for the concentration at the end of the endpoint of interval m is written as

$$C_m = c_o e^{-\left(\frac{w_1}{v}\right)\delta} e^{-\left(\frac{w_2}{v}\right)\delta} e^{-\left(\frac{w_3}{v}\right)\delta} \ldots e^{-\left(\frac{w_m}{v}\right)\delta}$$

Notice that this equation is a product of m exponential terms which are themselves random variables since they depend on the pouring rate W. Thus, if we substitute the random variable D_i for each term i above, this equation also can be written as follows:

$$C_m = c_o D_1 D_2 \ldots D_m$$

$$\text{where } D_i = e^{-\frac{w_i}{v}}$$

This result is in the form of the general product law required by the theory of successive random dilutions. This result shows that the continuous process model of the beaker experiment with variable diluent flow rate, like the two alternate discrete forms described earlier, gives rise to concentrations that approach the lognormal distribution under certain circumstances. This model is illustrated in greater detail on pages 227–235 on indoor air quality.

Theory of Successive Random Dilutions

An idealized successive random dilution process, such as the beaker experiments, consists of the following events:

1. A source is released into a carrier medium.
    - Thereupon becomes diluted.
    - Completely mixes.

2. Result becomes diluted again.
    - Completely mixes.

3. Above result becomes diluted again.
    - Completely mixes.

•
•
•

m. Result becomes diluted for the mth time.
    - Completely mixes, giving concentration C_m.

If this entire process were repeated with the same degree of dilution at each stage, then exactly the same final concentration C_m would result. Real processes in the environment seldom give the same final concentrations when they are repeated, because many complex factors affect the dilutions, causing them to change. If the entire process is repeated again and again with independent dilution factors and the resulting final concentrations plotted on a histogram, the distribution will have a single mode and will be right-skewed. If the cumulative frequencies are plotted on logarithmic-probability paper, the result will exhibit a certain degree of straightness, appearing similar to the examples given in this chapter.

A 4-stage successive random dilution process can be viewed as a series of boxes connected by inputs and outputs (Figure 8.12). Each box represents an independent dilution factor. The initial concentration c_o is the input to the system. The output of each box is the input to the next box. The output of the last box C_4 is the concentration resulting from the four dilutions in this successive random dilution process.

As m increases, ln C_m approaches a normal distribution in accordance with the Central Limit Theorem, and the variance of ln C_m increases with m. Some readers may be troubled by the tendency of the variance to increase as m increases. As with sums governed by the Central Limit Theorem, the variance increases without limit as m approaches infinity. However, in real dilution processes, m is finite; otherwise C_m would be zero. For finite m, C_m is *approximately* lognormally distributed. Fortunately, m does not have to be very large for ln C_m to be very nearly normally distributed (say, $m > 5$), and the approximation of C_m to a lognormally distributed random variable usually is quite good. (The logarithmic-gamma distribution arises only for the special case in which the dilution factors are distributed as **U**(0,1); in this case, the distribution is approximately lognormal if $m > 12$.) In general, if an m-stage dilution process with finite m is repeated many thousand times, the frequency distribution plots as a nearly straight line—or a line with slight downward curvature—on logarithmic-probability paper.

In summary, a series of m independent successive random dilutions of an initial concentration creates a distribution which is approximately lognormal, regardless of the distributions of the dilution factors. This concept, the *Theory of Successive Random Dilutions* (SRD), is published in an earlier paper[11] and is described in greater detail in this book. Briefly, it can be stated as follows:

> A concentration undergoing a series of independent random dilutions tends to be lognormally distributed.

The language, "tends to be lognormally distributed," is meant to express a strong tendency toward lognormality, rather than pure lognormality in which all the asymptotic conditions are met. Thus, the Theory of Successive Random Dilutions includes the case in which m is small, and the lognormality is not pronounced. For larger m, the lognormality of the variate C_m rapidly becomes more evident, and the Theory applies reasonably well in most real situations. As indicated above, the Theory applies both to single compartments and to multiple compartments.

The Theory of Successive Random Dilutions applies best to physical or chemical substances that do not participate in biological processes or chemical reactions. A biological process may create new material, and a chemical reaction may consume the original ingredients. Both situations will alter the distrib-

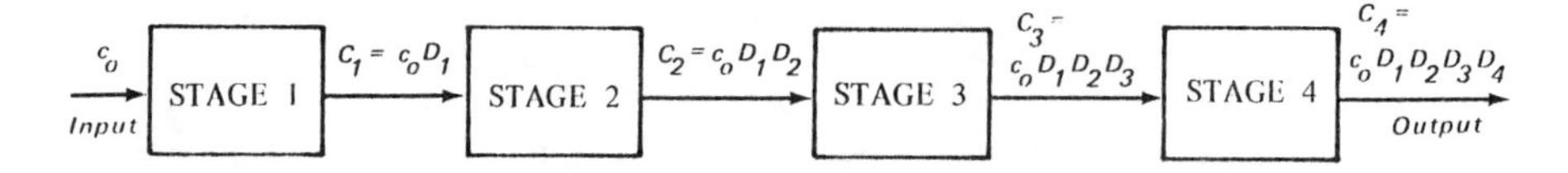

Figure 8.12. Components of a 4-stage successive random dilution process.

utional form. Hence, even in a chemical reaction, the Theory still will apply if the chemical products are a linear function of the original or diluted ingredients. The Theory of Successive Random Dilutions is especially appropriate for representing relatively inert substances initially released at very high concentrations into carrier media (air, water, soil), and undergoing considerable physical movement and agitation before they are measured.

There are two additional reasons why observed concentrations from real processes may deviate from a pure lognormal law: (1) the dilution factors may not be independent, and (2) error may be present in the measurement of the concentrations. The latter problem sometimes is overlooked. Most concentrations of environmental pollutants are measured by relatively complex chemical reactions or by instruments utilizing various principles of physics, and these techniques introduce error. Sometimes the error is normally distributed about the true value, causing the resulting concentration to be the sum of a normally distributed variate and a lognormally distributed variate. Sometimes the error distribution is not constant over the range of the concentration and depends on the value of the concentration measured. For example, measurement systems with "minimum detectable limits" respond very poorly, or not at all, at very low concentrations. Such measurement error can noticeably modify the form of the distribution, causing it to deviate from lognormality. Nehls and Akland[12] note that, although the lognormal is the distribution selected most often for representing air quality data, it seldom accurately describes the lower end of the observations near the minimum detectable limits.

With measurement error present in the observations, we would expect statistical "goodness-of-fit" tests for lognormality, which assume that the observations are independent and free of measurement error, to reject the lognormal hypothesis at an artificially high rate, even if the distribution really is lognormal. If the measurement error is not sufficiently great to alter the tendency toward lognormality, the distribution still will plot as a nearly straight line on logarithmic-probability paper, despite rejection by these tests.

The Theory of Successive Random Dilutions may be viewed as a special case of Kapteyn's *Law of Proportional Effect* (Chapter 9, pages 253–255). Aitchison and Brown[13] attribute the Law of Proportional Effect to a book published in 1903 by Kapteyn,[14] and they summarize the work of other investigators who have used this law over the years to explain a variety of phenomena. The Theory applies to any process in which the value of the variate in each state is a *random proportion* of the value of the variate in the previous state. Thus, the Theory applies to each process in this book that has been described as a random product ($\mathcal{RP}$-process). The $\mathcal{RP}$-process, which is a class of phenomena in its own right, is separate and distinct from the $\mathcal{RS}$-process described earlier (Chapters 6 and 7). The fact that the lognormal distribution is relatable to the normal distribution by a transformation, and therefore can be called a *derived* distribution, is irrelevant, since the genesis of pure normal and pure lognormal distributions is entirely different, resulting from different laws of nature.

Kahn[15] proposed Kapteyn's Law of Proportionate Effect (pages 253–255) in 1973 as a heuristic explanation of the common finding that air pollution concentration data appeared lognormally distributed. Similar propositions appeared in papers by Gifford,[16] Knox and Pollack,[17] and Pollack.[18] Kahn did not explain how the Law could be incorporated into the actual physical processes responsible for generating air pollutant concentrations. By representing diffusion processes by stochastic differential equations, Mage and Ott[19–21] sought to explain how the Law of Proportionate Effect might apply to the physics of diffusion under certain circumstances. My own investigation of a Brownian motion model[22] has suggested that diffusion processes could give rise to right-skewed, unimodal distributions in certain cases, but these distributions clearly were not lognormal. Benarie[23–25] has noted that wind speed distributions in any given direction may be approximated by a lognormal distribution, and has suggested that the lognormality of winds may explain the lognormality of air pollutant concentration data. Knox and Pollack[17] and Bencala and Seinfeld[26] have presented arguments similar to those of Benarie,[23–25] although the latter authors suggest other factors may be responsible as well: "The near lognormality of [air] pollutant concentration frequency distributions can be explained on the basis of the near lognormality of wind speed distributions, although this explanation does not establish that wind speed distributions are solely responsible for concentration distributions." Mage[27] has concluded that the lognormality of wind speed data does not sufficiently explain the lognormality of observed air quality data.

The reasons previously given in the literature to explain why air quality data tend to be lognormally distributed are not very satisfactory. Those arguments based on diffusion theory seem unnecessarily complex and without empirical proof. On the other hand, the arguments based on winds are straightforward but do not seem sufficiently general to extend to other fields. If the lognormality of air quality data can be attributed primarily to the lognormality of wind speeds, then why do so many other phenomena in the environment, which do not depend on winds, also have similar lognormal distributions? What physical properties, which might explain their lognormality, do these phenomena share in common? We need to identify a set of common traits present in these widely differing phenomena, which also can be embraced by a single, general law.

I believe these processes all share two common properties: (1) a substance initially is released into a carrier medium at a relatively high concentration and thereafter undergoes dilutions, and (2) considerable randomness takes place as the dilutions occur. These properties are necessary conditions for the Theory of Successive Random Dilutions, which, as discussed above, is a special application of Kapteyn's Law. These conditions are not very restrictive and occur in many processes commonly found in the world around us. To my knowledge, Kapteyn's Law has not previously been applied to dilution phenomena, and it is hoped that this book sheds some light on the reasons for the ubiquitous lognormal distribution.

Finally, consider the concentration time series generated by a successive random dilution process. The theoretical development presented above assumed that c_o represented a single release of pollutant, and c_o was subsequently diluted in m successive stages, resulting in concentration C_m. To examine the distribution of C_m, the entire process was repeated n times, beginning each time with a fresh system and the same value of c_o. Instead of a single release of c_o, suppose that the source operates *continuously*. For example, a water pollutant may be discharged at a constant rate (volume of liquid effluent per second) at some point upstream, and the concentrations resulting from this source monitored at

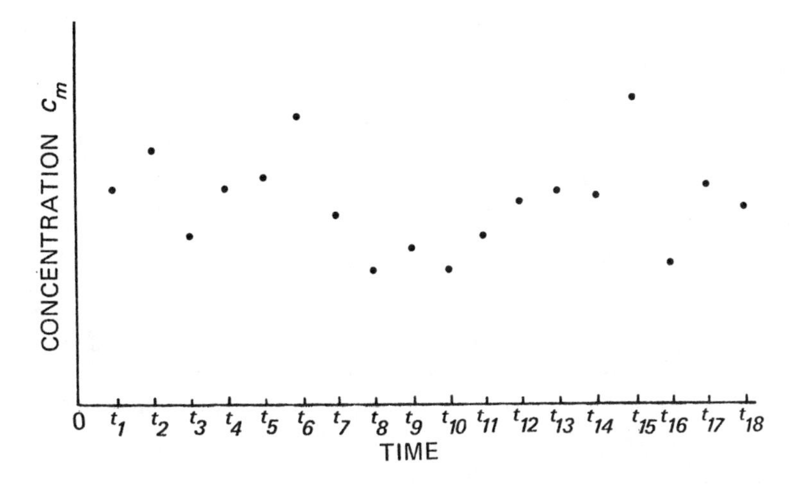

Figure 8.13. Typical concentration time series from an environmental process.

some location downstream at successive, equally spaced times $t_1, t_2, \ldots, t_n$ (Figure 8.13). When the effluent discharge initially begins, there will be a time delay until the pollutant reaches the downstream monitoring point. Thereafter, the pollutant concentration will rise from zero, and, after reaching steady state conditions (with a constant discharge rate), will vary about some mean.

Once steady-state conditions have been reached, the process will be stationary; its mean and variance will be constant. If the water pollutant is chemically inert, then any variation in concentration must be due to different dilutions experienced by the pollutant at different times as it travels from the source to the receptor. If the time interval between the observed concentrations is sufficiently large that the dilutions are independent of each other, then each observation is the result of a separate successive random dilution process, and the model described above will apply. The concentrations making up this time series will be independent random variables governed by the Theory of Successive Random Dilutions.

Often, governmental agencies choose to form averages from the raw data to store the observations more compactly. This practice is very common in the air pollution field, where continuous monitoring is customary for many pollutants. If the raw concentration time series consists of independent lognormally distributed random variables, and if arithmetic averages are formed, what will happen to the resulting distribution? In an m-stage successive random dilution process, the raw observations $C_{m1}, C_{m2}, \ldots, C_{mn}$ will comprise the concentration times series. Because m usually is unknown in real processes, we shall simplify the analysis by letting $X_1 = C_{m1}$, $X_2 = C_{m2}$, $\ldots$, $X_n = C_{mn}$. Then the concentration time series $X_1, X_2, \ldots, X_n$ will represent the independent, lognormally distributed random variables generated by this successive random dilution process. Suppose that the average concentration $\overline{C}$ is formed by averaging k raw observations in this time series:

$$\overline{C} = \frac{X_1 + X_2 + \ldots + X_k}{k}$$

If successive averages are formed, the averages $\overline{C}_1$, $\overline{C}_2$, and $\overline{C}_3$ will be given as follows:

$$\frac{X_1 + X_2 + \ldots + X_k}{k}, \frac{X_{k+1} + X_{k+2} + \ldots + X_{2k}}{k}, \frac{X_{2k+1} + X_{2k+2} + \ldots + X_{3k}}{k}$$

For example, the raw observations $X_1, X_2, \ldots, X_n$ might represent CO concentrations observed in a city's air at intervals 5 minutes apart, and the $\overline{C}_1, \overline{C}_2, \ldots, \overline{C}_j$ would represent the hourly average concentrations computed by setting $k = 12$. If the raw observations were independent, then the Central Limit Theorem would apply. If the process generating the raw observations were stationary with mean μ and variance σ^2, then resulting averages would be asymptotically normally distributed with mean μ and variance $Var(\overline{C}) = \sigma^2/k$.

Notice that the successive random dilution model is a $\mathcal{RP}$-process, while the action of forming averages is a $\mathcal{RS}$-process (Chapter 6). Because the resulting averages are a consequence of forming sums of products, it is a mixed process and can be called a random sum-product ($\mathcal{RSP}$-process). Under the conditions stated above, the averages will approach normality despite the lognormality of the raw observations as k increases. In the intermediate range, when k is not very large, the distribution will be right-skewed and will have a lognormal appearance. As k increases, the distribution will appear increasingly normal and symmetrical, accompanied by a decrease in its variance as $1/k$.

In air quality data, for example, the observations often are close together in time and exhibit considerable serial correlation. For example, adjacent pairs of hourly average air pollutant concentrations usually are highly correlated, and the autocorrelation decreases rapidly if the elapsed time between pairs exceeds 24 hours. Weather variables, such as wind and temperature, behave in a similar manner. Because of the autocorrelation of hourly outdoor temperature readings, this hour's temperature usually is a good predictor of the next hour's temperature, but it is a poorer predictor of tomorrow's or next week's temperature. Temperature and winds affect the dilution factors, which, in turn, affect the observed air pollutant concentrations and help explain their autocorrelation.

How does the successive random dilution model incorporate serial correlation? Suppose that the dilution factors change slowly relative to the time between successive observations such that the dilution factors are correlated. Let D_{11} denote the first dilution factor on the first realization of the process, and D_{12} denote the first dilution factor on the second realization of the process. Then D_{1n} denotes the first dilution factor in n realizations of the process generating concentrations $X_1, X_2, \ldots, X_n$, and the series of dilution factors $D_{11}, D_{12}, \ldots, D_{1n}$ is assumed to be correlated. Similarly, the second dilution factor gives the series $D_{21}, D_{22}, \ldots, D_{2n}$, and this series also is assumed to be correlated. If the series generated by the mth dilution factor $D_{m1}, D_{m2}, \ldots, D_{mn}$ also is correlated, then the resulting observations $X_1, X_2, \ldots, X_n$ will be serially correlated. This happens because the observed concentrations are a linear function of the dilution factors.

Because of the autocorrelation of $X_1, X_2, \ldots, X_n$, the resulting averages $\overline{C}_1, \overline{C}_2, \ldots, \overline{C}_j$ computed from this correlated series will not obey the Central Limit Theorem. Rather, the distribution will retain its right-skewness and will not converge to a normal distribution as rapidly as the Central Limit Theorem predicts. The variance $Var(\overline{C})$ usually will not decrease as rapidly as σ^2/k due to the autocorrelation. This autocorrelation explains why ambient air quality data often appear right-skewed and lognormal even after a great number of instantaneous readings have been averaged. It also explains why empirical averaging time models have been necessary to explain the properties of environmental quality data.[38–39, 42–43]

In the above development of the Theory of Successive Random Dilutions, the simplifying assumption was made that the source c_o either was a constant or was released at a constant rate. Pollutant sources released from an industrial plant usually tend to vary with changes in the firm's operations, productive volume, or other factors. Suppose that the source is not constant but is itself a random variable that is independent of the dilution factors in the carrier medium. Let the random variable C_o denote the source, in which $C_o > 0$. Then the final concentration C_m after m dilutions will be given by the following law:

$$C_m = \prod_{i=1}^{m} C_o D_i$$

Notice that C_o becomes just another of the many random variables being multiplied together to form the product. By the reasoning given earlier, the distribution of C_m will be approximately lognormal regardless of the distribution of the source concentration C_o. (Here, we assumed m was a constant to simplify the model.) Thus, the Theory of Successive Random Dilutions is robust to assumptions about the distributions of the sources or the dilution factors, and there is a tendency toward lognormality no matter what these distributions happen to be.

Thus, if the source experiences erratic changes, the distribution of the instantaneous observations will tend to be approximately lognormal, regardless of whether the observations are collected over a week, a month, a season, or a year. The mean and variance of the distribution for a season may differ from the mean and variance of the distribution for the entire year, but the distributional form for both cases will be the same: approximately lognormal. The Theory of Successive Random Dilutions offers a simple, yet surprisingly general, explanation of why approximately lognormal distributions arise so often in environmental phenomena.

APPLICATIONS TO ENVIRONMENTAL PHENOMENA

The Theory of Successive Random Dilutions can be applied to a large variety of common environmental phenomena. Several examples are presented to illustrate these concepts.

Air Quality

When an air pollutant is emitted from a source, it is ordinarily transported by mechanical air movements and by winds. Due to the many physical obstacles present in cities (houses, walls, buildings, and other edifices), the air movement will be quite erratic, and the air will experience many changes in direction. Due to these motions, a given parcel of air containing a pollutant is likely to experience an erratic trajectory as it travels from the source to some measuring point, or receptor. A typical path might consist of several straight-line segments (Figure 8.14). Beginning with the initial concentration c_o, the pollutant will be diluted as it travels along the first segment, reaching concentration C_1, such that $C_1 < c_o$. The resulting concentration can be expressed as a function of the initial concentration and the dilution factor $C_1 = D_1 c_o$. If the wind suddenly changes direction at C_1, the parcel will undergo additional dilution as it travels along a second segment, giving concentration $C_2 = D_2 C_1 = D_2 D_1 c_o$. As the parcel travels along other segments, it will experience additional successive dilutions. After m segments, the concentration $C_m = D_1 D_2 \ldots D_m c_o$ will result. If a concentration is observed some time later at the same monitoring location, it will be the result of a different trajectory, and a different set of dilution factors will apply. (The number of dilutions m has been assumed constant here only to simplify the analysis.) By the Theory of Successive Random Dilutions, the resulting distribution of concentrations observed at this monitoring location will be approximately lognormal.

A large number of technical papers have been published suggesting that empirical distributions of ambient air quality data are approximately lognormal. A literature review by Mage[28] summarizes the history of the application of probability models to air quality data, and he cites a paper by Harris and Tabor[29] in 1956 as one of the first to apply the lognormal distribution to aerometric data. In 1959, a paper by Zimmer, Tabor, and Stern[30] described concentrations of total suspended particulates measured in both urban and nonurban atmospheres as approximately lognormal. In 1961, Larsen[31] used logarithmic plots to examine the properties of hourly oxidant concentrations measured in Los Angeles, and subsequent papers by Larsen[32] and Zimmer and Jutze[33] extended this work. In 1965, a paper by Zimmer and Larsen[34] examined air quality data in seven U.S. cities and proposed an empirical model relating the maximum concentration to the averaging time. For all averaging times, the observations were assumed to obey a two-parameter lognormal distribution: "For a given averaging time ..., concentration versus frequency of occurrence plots as a straight line

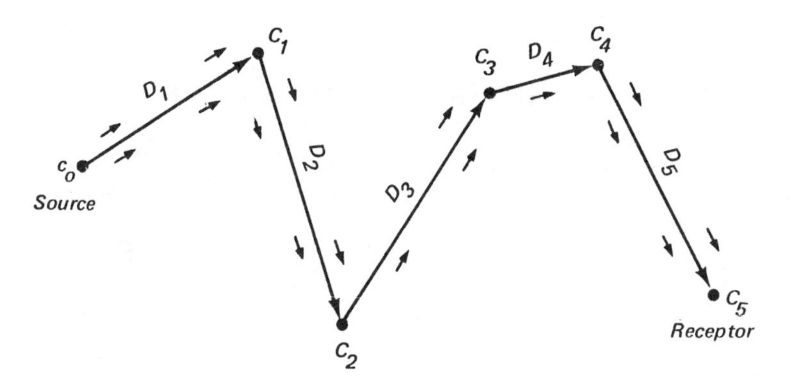

Figure 8.14. Air pollution trajectory in an atmospheric Successive Random Dilution process.

on logarithmic by normal (lognormal) probability paper."[34] Discussions and extensions of this work appeared in papers by Larsen[35,36] and his colleagues[37] in the late 1960s. After analyzing continuous air pollutant concentration data for 7 pollutants in 6 cities covering a 3-year period,[37] the model was improved, and Larsen[38] presented it as a formal model linking together the statistical parameters of air quality frequency distributions with averaging times. The primary assumptions of this model were as follows:

1. Air pollutant concentrations are approximately lognormally distributed for all pollutants in all cities for all averaging times.
2. The median concentration (50 percentile) is proportional to averaging time raised to an exponent (and thus plots as a straight line on logarithmic probability paper).

Application of the lognormal model to air quality data is described in considerable detail in an early technical report by Larsen.[39] In 1973, Patel[40] noted that, by not properly taking into account serial dependence of air quality observations, some of the model's statistical formulas gave incorrect estimates. Larsen[41] responded that the model was completely empirical, and, if used to calculate source reduction "design values," the model would not give very great errors. Applications of this model to make predictions are discussed on pages 283–84 of Chapter 9. The important point is that the development of this model was stimulated by the somewhat striking tendency of empirical air quality data, when plotted on logarithmic-probability paper, to appear as a straight line, and the ubiquitous nature of these straight lines.

As noted by Mage,[28] from 1972 to 1975 Larsen[42,43] continued to apply the two-parameter lognormal model to air quality data and continued to restate the basic assumptions contained in his model. Few investigators considered any other distributional forms for air quality data. An exception was Lynn,[44] who applied the normal, two- and three-parameter lognormal, gamma, and Pearson Types I and IV distributions to total suspended particulate data from Philadelphia. He concluded that "... the two-parameter lognormal does overall slightly better than the three-parameter and in fact does the best of all four distributions (LN2, LN3, gamma and four-parameter Pearson)." It was difficult to understand why the two-parameter lognormal distribution would perform better than the three-parameter lognormal distribution, because the former is just a special case of the latter. Mage[28] concluded that the poor fit of the three-parameter lognormal model resulted from failure to censor the model when the third parameter is negative, and in 1974 Mage[45] proposed the censored, three-parameter lognormal model (Chapter 9, pages 272–276). This model explained the downward curvature observed so often in ambient air quality data sets.[19–21, 45]

Bencala and Seinfeld[26] accepted the two-parameter lognormal distribution as a convenient practical model for representing ambient air quality frequency distributions, although they noted that many other common statistical models are available which fit the data as well as or better than the lognormal. One difficulty is that right-skewed distributions tend to resemble each other, and one has difficulty finding a single model which fits the data set best under all conditions. When considering these models, the authors usually are empirically fitting existing data sets, rather than attempting to understand the underlying stochastic process in order to make predictions about the future.

In subsequent years, more complicated models have been applied to ambient air quality data. Some investigators have taken into account serial correlation of observed data, and some modelers have applied goodness-of-fit criteria to their efforts to quantify how well these models fit observed air quality data sets. The more recent history of applications of these models to ambient air quality data is described in the next chapter.

The above discussion applies to ambient air quality. The Theory of Successive Random Dilutions also can be applied to pollutants released into small microenvironments, such as those pollutants emitted by motor vehicles traveling on highways. If we observe the smoke released by a moving automobile with a defective exhaust, or a vehicle that is burning oil, we will be struck by the complex eddies and intricate patterns of the smoke. For a single parcel of air, many changes in direction occur as it leaves the exhaust tailpipe and is diluted. These changes result from air movements caused by temperature gradients, mechanical mixing in traffic, and winds striking the vehicles, roadway structures, and surrounding buildings. By the time the pollutant reaches the edge of the roadway, many dilutions have occurred, and many more dilutions will take place before the pollutant reaches a nearby ambient air monitoring station.

The pollutants present in the exhaust gases of vehicular tailpipes ordinarily are emitted at very high concentrations, but concentrations measured nearby may be several thousand times lower due to dilution. For example, carbon monoxide (CO) leaving an automobile's tailpipe may constitute 1% of the exhaust gases, or 10,000 parts per million (ppm). However, CO concentrations measured at nearby monitoring stations, or several feet away from vehicles, usually lie in the range of 1–100 ppm, representing an overall dilution factor ranging from about 10^{-2} to 10^{-4}. As the pollutant moves from the source to the receptor, many independent, successive dilutions occur, and the Theory of Successive Random Dilutions will apply.

Between January 1980 and February 1981, a total of 93 drives, ranging from 35 minutes to 1 hour, were made on a 11.8-mile segment of an urban arterial highway, El Camino Real, in Palo Alto, CA.[46] During each drive, CO concentrations were measured inside the passenger compartment of the test vehicle. A special study determined that the test vehicle was free of contamination from its own exhaust, and, since ambient CO concentrations usually were low, the CO exposure inside the vehicle resulted primarily from the exhaust gases released by surrounding vehicles in traffic. A CO observation was taken every 12 seconds, and the instantaneous readings were averaged to form 1-minute averages. The frequency distribution of 3,403 1-minute averages from all trips, when plotted on logarithmic-probability paper, exhibits impressive straightness for all but the lowest 20% of the observations (Figure 8.15). This figure illustrates downward curvature occurring at the lower end of the distribution, and this phenomenon is described in greater detail in the next chapter. Despite its downward curvature, this distribution has a single mode, is right-skewed, and has a strong tendency toward lognormality. Downward curvature at the lower frequencies can be handled by the three-parameter lognormal model suggested by Mage[45] (Chapter 9, pages 272–276). Downward curvature may show up either at the lower end or the upper end of the distribution, or at both ends simul-

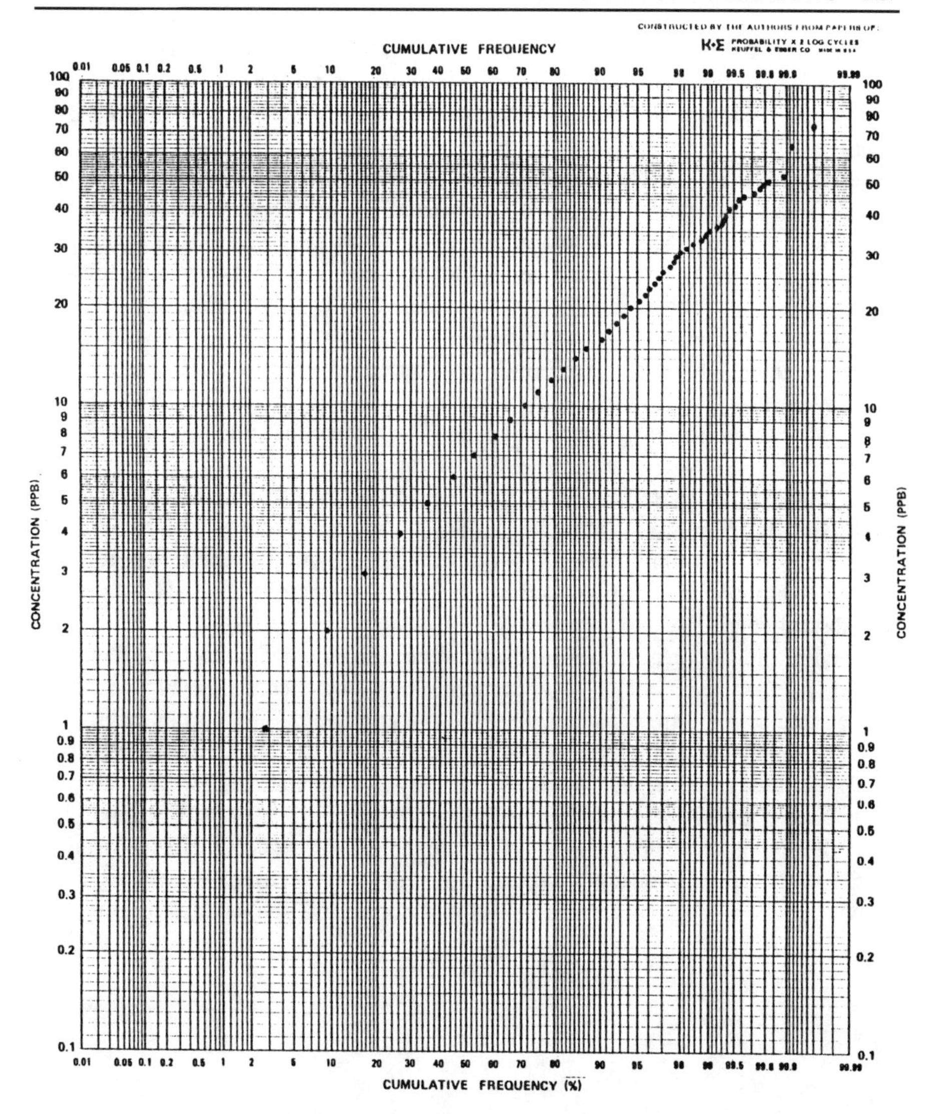

Figure 8.15. Cumulative frequencies of CO exposures (1-min averages; N = 3403) on El Camino Real, January 1980 to February 1981, plotted on logarithmic-probability paper.

taneously. (Figure 9.9 does not show the same downward curvature because it is based on trip averages rather than 1-minute averages, and the longer trip averages show less variability and a narrower range.)

Indoor Air Quality

Like the motor vehicle example above, the Theory of Successive Random Dilutions also can be applied to pollutants released into other small microenvironments, such as houses, buildings, and interior rooms. Many pollutants inadvertently are released at relatively high concentrations into the air of indoor dwellings and offices, and the ordinary motion of the air causes the pollutants to undergo successive dilutions as they travel through these mixing volumes, ultimately reaching low concentrations as they mix with fresh air from outdoors.

Some gas appliances, such as ovens and stoves, emit carbon monoxide (CO), particularly if their burners are poorly adjusted. In an experiment conducted by the author, a friend cooked a turkey dinner in the kitchen of her small one-bedroom home while a continuously recording instrument placed inside the kitchen measured indoor carbon monoxide levels. The turkey was cooked in the oven, and the potatoes, artichokes, and gravy were cooked on the top burners of the gas stove. The CO levels in the home were recorded from just before the cooking started at 4:10 pm until 2:00 pm the next day (Figure 8.16). The CO concentrations in the room began to rise when the oven was turned on at 5:00 pm, and the CO levels increased further as the top burners were turned on to cook the artichokes and potatoes. Finally, opening the oven door before removing the turkey caused a further jump to 16 ppm at 9:00 pm. Soon afterward, with all the burners off and the turkey removed, the concentration in the home began a gradual decline over a 14-hour period (from 9:00 pm that night to 11:00 am the next day) that appeared to be an exponential decay in shape.

Usually, indoor air pollutant concentrations are modeled as though the home (or a particular room) behaves as a large, well-mixed chamber (Figure 8.17). Here, w denotes the fresh air flowing into the chamber (for example, cubic meters of air per hour), and the air flow rate is directly analogous to the flow rate w of fresh water poured into the beaker in the experiment described earlier (see page 214 on the Continuous Mass Balance Model). The only difference between a room and the beaker is that the "fluid medium" indoors is air instead of water. Also, unlike the beaker experiment, real indoor settings may contain internal sources (for example, a cigarette in a room or a gas stove in a kitchen),

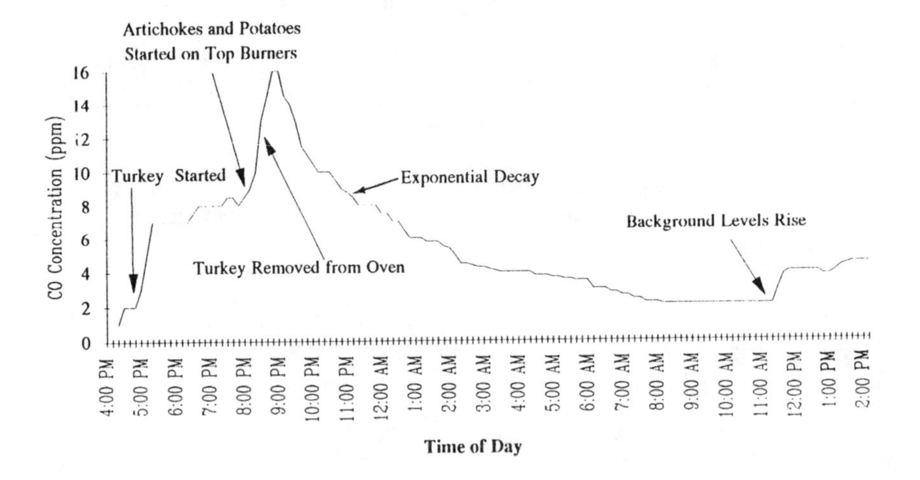

Figure 8.16. Observed indoor carbon monoxide (CO) concentration measured in a home with a gas stove during and after cooking a turkey dinner. Soon after the turkey was removed from the oven and the gas burners were turned off, the concentration began to decay in a curve that was approximately exponential from 9:00 pm to 11:00 am the next day.

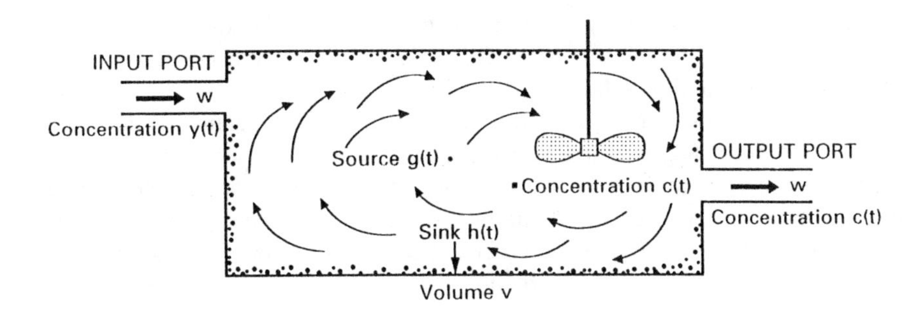

Figure 8.17. A well-mixed chamber representing the general form of the mass balance equation in an indoor setting (for example, a room or an automobile). Air flowing into the system at flow rate *w* contains the pollutant at concentration $y(t)$. The chamber contains an internal pollutant source $g(t)$ (single dot) and a removal sink $h(t)$ (dots on the inside surfaces of the chamber). Because there is no compression, air leaves the chamber at the same flow rate *w* as it entered the chamber. Because the chamber is well-mixed, the air leaving through the output port also has the same concentration $c(t)$ as the air inside the chamber.

and the walls or surfaces of rooms may act as "sinks" for pollutants (for example, the plating out of particles on walls and furniture).

The most general form of the model of indoor air quality takes all these factors into account and contains five terms:

$$\int_0^t y(u)w\,du + \int_0^t g(u)\,du = \int_0^t h(u)w\,du + \int_0^t c(u)w\,du + vc(t)$$

where

$y(u)$ = pollutant concentration in the input flow stream [M/L^3]
w = flow rate of air into system [L^3/T]
$g(u)$ = internal pollutant source [M/T]
$h(u)$ = internal pollutant sink [M/T]
$c(t)$ = pollutant concentration in output flow stream [M/L^3]
v = volume of chamber [L^3]

Normally concentration units can be expressed either on a parts-per-million (ppm) or mass-per-volume (for example, mg/m^3); for illustrative purposes, we assume that the units are mass per unit volume [M/L^3]. Since there is no compression of air inside the chamber, the flow rate of air into the system w equals the flow rate out of the system. If we examine each term in this equation, we can see that the two terms on the left side of the equal sign represent the sources of pollutant entering the chamber: (a) the pollution contributed by the input port, and (b) the source located inside the chamber. Similarly, the three terms on the right side of the equal sign reflect the destination of the pollutant among three components: (c) the amount absorbed by the internal sink, (d) the amount departed through the output port, and (e) the amount currently inside the chamber. The result is a general statement of the mass balance equation for indoor settings: the quantity of pollutant generated by the sources must equal the quantity of pollutant accounted for by the chamber (absorption plus contents) plus the quantity spilled out.

This model is analogous to the continuous beaker mass balance model (pages 214–217), except that we have added the possibility that the beaker has an internal source of pollution and we have added surfaces on which the pollutant can be absorbed. As we did for the beaker equation, we now differentiate the above equation to obtain a general differential equation for indoor settings, and divide by w and then rearrange terms:

$$\frac{v}{w}\frac{dc(t)}{dt} + c(t) = y(t) + \frac{g(t)}{w} - \frac{h(t)}{w}$$

Here $\phi = w/v$ ordinarily is defined as the "air exchange rate" of the system since it is the "number of chamber volumes of air per unit time" entering the system through the input port (and leaving through the exit port). In real indoor settings, the fresh air enters through a variety of openings (windows, doors, wall leakage) rather than a single input port, and ϕ reflects the sum of all these input flows. To simplify the mathematical analysis, the air infiltration from outside the chamber has been "lumped" into a single input port in Figure 8.17, but the results for the lumped and distributed forms are mathematically equivalent, since the interior air is assumed to be well-mixed in both cases.

If all the sources on the right side of the equation suddenly stop generating pollution at time $t = 0$ and the right side becomes zero, this equation has an exponential solution with c_o as the initial concentration at time $t = 0$:

$$c(t) = c_o e^{-\phi t} \quad \text{for } t \geq 0$$

where

$\phi = w/v$ = air exchange rate of system [l/T]
$c(t)$ = pollutant concentration as a function of time
c_o = initial concentration at time $t = 0$

Notice that this is the same solution that was obtained for the deterministic form of the continuous beaker example: if an initial concentration is present in the beaker and fresh water is added at a continuous, constant rate, then the concentration decays exponentially with its exponent determined by the product of time t and the fresh water rate w flowing into (or leaving) the system, divided by the volume v of the beaker.

This decaying exponential function is the same pattern that emerged in the turkey cooking experiment (Figure 8.16). There, the concentration curve began at 16 ppm just before the oven and burners were turned off. As the CO concentration in the kitchen decayed, it reached $16e^{-1} = 16/e = 16/2.7183 = 5.9$ ppm approximately 3.5 hours later. Thus, we say that the "residence time"—the time required to replace one full volume of air—was approximately 3.5 hours. The air exchange rate therefore was $\phi = 1/(3.5 \text{ hr}) = 0.29$ air changes per hour (ach), which is reasonable for a small, tightly sealed home. Although we have assumed that all sources of pollution are zero in the turkey cooking experiment once the cooking stopped, we can see that the concentration in the kitchen rises at about 11:00 am the next morning due to ambient levels $y(t)$ infiltrating indoors into the indoor mixing volume from outdoors.

Stochastic Air Exchange Rate

The air flowing into a room or a house is a function of the outdoor winds, the opening and closing of windows, heating and air conditioning systems that turn on and off, and other factors; therefore the air exchange rate ϕ is likely to vary. If we treat the air flow rate W as a random variable, then $\phi = W/v$ also is a random variable, and the results outlined above in the section on Stochastic Flow Rate (page 216) will hold, giving a SRD process.

An example using Monte Carlo simulation illustrates these concepts. To simplify the analysis, we assume that the air exchange rate change makes abrupt changes every hour (Figure 8.18) and that the value of each air exchange rate is an independent random variable. This variation in air exchange rates might occur if a ventilation system operated with completely different flow rates each hour that bore no relationship to the previous settings. To consider the behavior of a particular stochastic process, we further assume that the air exchange rates are uniformly distributed between 0 and 1.

A BASIC computer program was written that assumes the initial concentration is c_o and then takes samples for the air exchange rate ϕ from a unit uniform distribution $\mathbf{U}(0,1)$ (Problem 5). Figure 8.18 shows, for example, a single realization for a SRD process with $m = 5$ dilution stages. In this case, the random number generator gave an air exchange rate of 0.53304 ach for the first hour; it gave an air exchange rate of 0.19483 ach for the second hour, and so on. The resulting concentration with respect to time is a "piecewise exponential" function consisting of five exponential curve segments joined at their endpoints.

If the air exchange rate had been fixed at the mean value of the uniform dis-

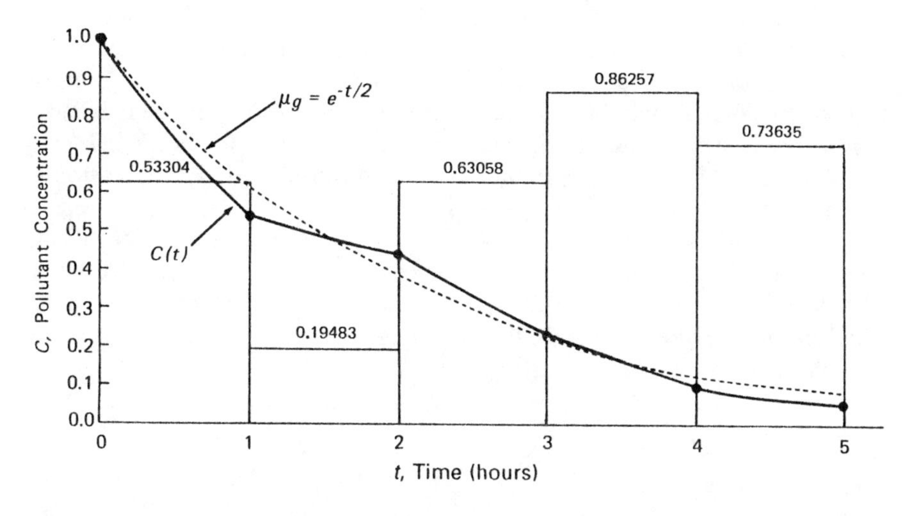

Figure 8.18. Concentration predicted from the mass balance equation for a well-mixed chamber with uniformly distributed air exchange rates $\mathbf{U}(0,1)$ and $m = 5$ dilution stages.

tribution, or $\bar{\phi} = 1/2$, then the concentration as a function of time would follow a smooth exponential curve $c(t) = e^{-t/2}$. At the end of the fifth hour, the concentration would be $e^{-2.5} = 0.082085$, which is the geometric mean of this stochastic process with $m = 5$. For the particular random numbers shown in Figure 8.18, the final concentration after five random dilutions is, on this realization, lower than the mean value. If the process were repeated over and over, the final concentration would exhibit a distribution with a geometric mean of 0.082085 that we can predict exactly.

Mathematically, if each ϕ_i has a unit uniform distribution, then the final concentration at the end of the fifth hour is given by

$$C_5 = c_o e^{-\phi_1} e^{-\phi_2} e^{-\phi_3} e^{-\phi_4} e^{-\phi_5}$$

Because c_o can be regarded as a scaling factor, we can simplify the analysis of this example by setting $c_o = 1$. Taking logarithms, the resulting expression can be written as

$$\ln C_5 = -\phi_1 - \phi_2 - \phi_3 - \phi_4 - \phi_5 = -\sum_{i=1}^{5} \phi_i$$

The left-hand side of this equation is the logarithm of the final concentration and the right-hand side is the negative sum of 5 random variables that are unit uniformly distributed. The distribution of the sum of m random variables that have uniform distributions **U**(0,1) is known exactly and consists of m polynomial arcs.[47] Therefore, it can be called the "arcpolynomial" distribution, although its properties are beyond the scope of this book. For $m = 2$, the PDF consists of two straight-line segments forming a triangle. For m greater than 2, the PDF consists of m curved arcs approaching the normal distribution. Since the right-hand side of the equation approaches the normal distribution for large m, the logarithm of the concentration approaches the normal distribution. Because the concentrations are always less than 1, the normal distribution is defined on the negative axis due to the minus sign. Because the logarithm of the concentration has an arcpolynomial distribution, the concentration C_m has a "log-arcpolynomial" distribution exactly. Because the log-arcpolynomial distribution approaches the two-parameter lognormal distribution for large m, we once again find that the concentration itself approaches the lognormal distribution in this process if m is large.

Because each air exchange rate ϕ_i for each i is assumed to be independent, we can apply Rule 3 in the section "Moments, Expected Value, and Central Tendency" in Chapter 3 (pages 38–41) to obtain the expected value. For sums of independent random variables, the expected value of the basic process (with $c_o = 1$) is given by

$$E[\ln C_5] = -E[\phi_1] - E[\phi_2] - E[\phi_3] - E[\phi_4] - E[\phi_5] = -5E[\phi]$$

Since $E[\phi] = 1/2$ for a unit uniform distribution, then $E[\ln C_5] = (-5)(1/2) = -2.5$. Similarly, the variance of the logarithm of the concentration is obtained as the sum of the variances:

$$Var(\ln C_5) = \sum_{i=1}^{m} Var(\phi_i) = 5Var(\phi)$$

From Table 3.2, the variance of a unit uniform distribution is computed as $Var(\phi) = (1 - 0)^2/12 = 1/12$, so the standard deviation of ln C_5 is $(5/12)^{0.5}$ = 0.645497. Because $m = 5$ is not large, we know that C_5 has a distribution that is the "log-arcpolynomial" exactly, which has similarities to the two-parameter lognormal distribution (Chapter 9). Even though the distribution for $m = 5$ is known to be a log-arcpolynomial exactly, it is reasonable to ask, "How close does the two-parameter lognormal probability model come to representing the concentration distribution from this stochastic indoor air model?"

This process has been repeated on the computer 10,000 times in a Monte Carlo simulation, and the frequency distribution, when plotted on logarithmic-probability paper, appears to be nearly a straight line over much of the range (Figure 8.19). Because $c_o = 1$, the resulting values plotted on the vertical axis in Figure 8.19 are normalized and also can be written as C_5/c_o. Thus, the results will apply to any initial concentration by multiplying the vertical axis by any concentration c_o.

The mean of the logarithm of the concentration was calculated above to be −2.5, and the standard deviation of the logarithm was calculated to be 0.645497. Using the notation of Chapter 9, we set the normal parameters

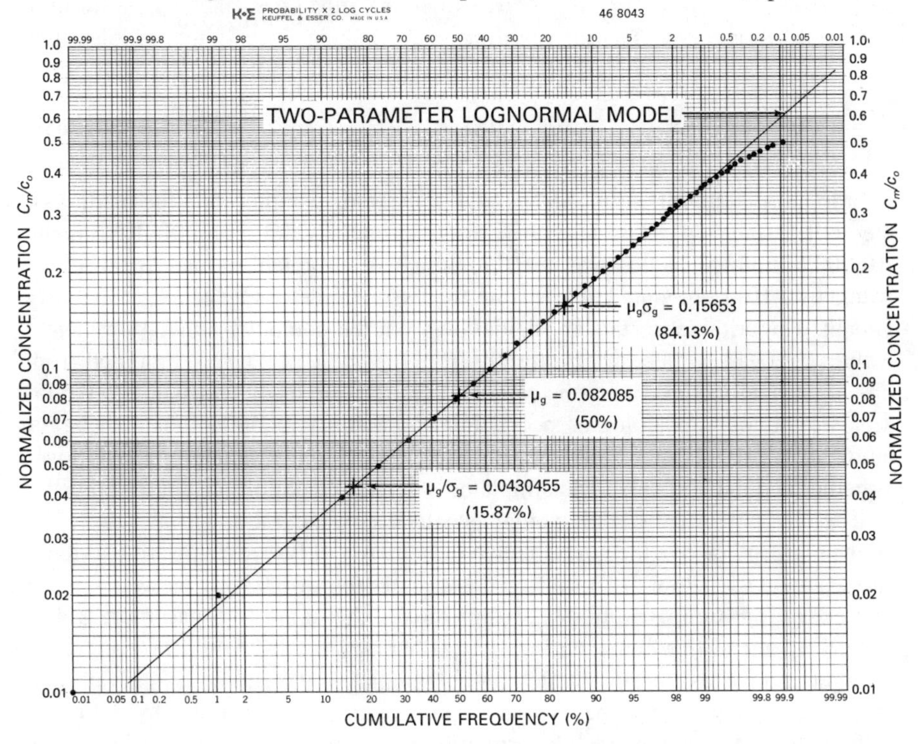

Figure 8.19. Computer simulation of final pollutant concentration in 10,000 experiments in an enclosed indoor chamber with an air exchange rate that is uniformly distributed as **U**(0,1), with $m = 5$ dilution stages, and an initial concentration of $c_o = 1$.

$\mu = -2.5$ and $\sigma = 0.645497$, and using the conversion formulas in Table 9.2 (page 262), the geometric mean μ_g and geometric standard deviation σ_g for the two-parameter lognormal distribution are calculated as follows:

$$\mu_g = e^{\mu} = e^{-2.5000} = 0.082085$$

$$\sigma_g = e^{\sigma} = e^{0.645497} = 1.90693$$

Using the graphical techniques described in Chapter 9, we plot the lognormal distribution on this paper by drawing a straight line through two points. Choosing the geometric mean μ_g and geometric standard deviation σ_g as the two plotting points, we plot $\mu_g = 0.082085$ at the frequency 0.50 and we plot $\mu_g\sigma_g = (0.082085)(1.90693) = 0.15646$ at the frequency 0.8413 (or 84.13%), which corresponds to one standardized normal deviate. When a straight line is drawn through these points, we see that it matches the frequency distribution of the simulated concentrations quite well (Figure 8.19).

Using the formulas for the arithmetic mean α and arithmetic standard deviation β in Table 9.2, we can calculate these parameters for the two-parameter lognormal probability model:

$$\alpha = \mu_g e^{\frac{1}{2}(\ln \sigma_g)^2} = 0.082085 e^{\frac{1}{2}(\ln[1.90693])^2} = 0.101098$$

$$\beta = \mu_g \sqrt{e^{2(\ln \sigma_g)^2} - e^{(\ln \sigma_g)^2}} = 0.082085\sqrt{e^{2(\ln[1.90694])^2} - e^{(\ln[1.90694])^2}} = 0.072685$$

In a lognormal distribution, the arithmetic mean α is higher than the median or geometric mean μ_g. The actual arithmetic mean value computed from the 10,000 simulated values was 0.1017, which is reasonably close to the mean value of 0.1011 computed above from the two-parameter lognormal probability model. The arithmetic standard deviation from the computer simulation gave a value of 0.0714, which differs slightly from the value of 0.0727 computed from the lognormal model. Using the calculated arithmetic mean and standard deviation for the more exact log-arcpolynomial probability model and using a larger number of simulations should bring the computed and simulated means closer together and the computed and simulated standard deviations closer together.

The data obtained from the computer simulation show a downward curvature at the high end (above the normalized concentration of 0.4) and upward curvature at the low end (below the normalized concentration of 0.03). If we were to use the exact log-arcpolynomial distribution for the same two parameters and $m = 5$, then the line representing the probability model would no longer be straight at the ends. Indeed, the log-arcpolynomial would follow the curvature at the extremes better than the two-parameter lognormal probability model does. Although the two-parameter lognormal distribution does not fit the data in this example well at the extreme end points, it provides a good fit over much of the data, even for the case of only $m = 5$ dilution stages. For larger values of m, the data at the end points will follow the straight line much more closely, because the process will satisfy the asymptotic conditions for a true lognormal distribution.

The computer simulation results given in Figure 8.19 might apply to the distribution of concentrations measured in 10,000 homes in which an indoor tracer pollutant is released at concentration c_o and $m = 5$ random dilutions take place in each home. Even though the air exchange rates are not lognormal, the distribution of concentrations in the resulting population of homes would be very nearly lognormal, as shown by Figure 8.19. Release of a tracer pollutant indoors followed by dilution and uniform mixing is an exact analog of the beaker experiment described earlier (see pages 216–217 on "Stochastic Flow Rate"). Thus, if a "puff" of a tracer gas were released in each home, our analysis based on the mass balance model with stochastic inputs shows that the concentration distributions across homes would be approximately lognormal. This result is somewhat surprising, because it implies that the lognormal distribution will arise naturally, even though the other variables in the process do not themselves necessarily have lognormal distributions. This analysis assumes that m is fixed and each home experiences 5 unit uniformly distributed air exchange rates. By the Central Limit Theorem, similar overall results are likely to be obtained when the air exchange rate is sampled from other distributions. In real homes, m is likely to vary from home to home. I have carried out computer simulations in which m is allowed to vary from 3 to 9. This approach yields distributions that resemble the lognormal very closely but the case of variable m needs additional research.

Although the lognormal hypothesis has not been tested on many indoor air pollutants, considerable data are available on indoor radon levels. Radon enters homes from water or natural gas, from the natural emissions of some building materials, or through cracks or openings in walls or floors in contact with the soil. The source of radon is radium in underlying soil and rocks. Alter and Oswald[48] report indoor air concentrations measured in Northern California, Eastern Pennsylvania, Maine, Canada, and other locations, and they conclude, "The data for these and all other surveys reported here have a lognormal distribution." The measurements are made by counting the damage tracks produced by alpha particles in a plastic film detector. These studies represent a total of about 30,000 indoor radon observations.

Water Quality

Pollutants often are released into water bodies at relatively high concentrations and undergo numerous stages of dilution. A simple example is an effluent discharged from a pipe into a pond (Figure 8.20). If the pipe discharges the pollutant at high concentration over a very short time (for example, a few minutes), and if fresh water is flowing into the pond, the pond will cause the pollutant to become well-mixed just like the beaker example discussed earlier or the indoor air chamber. The fresh water entering the pond from upstream is like the fresh water being poured continuously into the beaker by the experimenter using a masked cylinder to introduce variability. The water leaving the pond is like the well-mixed contents leaving the beaker. Immediately after the pollutant has been released, the concentration in the stream will decline exponentially, following the exponential solution of the mass balance equation. If the process were repeated many times over and the fresh water added were allowed to vary, then the pollutant concentration in the stream after a sequence of these dilutions would approach very nearly a two-parameter lognormal distribution.

Not only does a single mixing volume, such as the pond in Figure 8.20 help bring about natural lognormality, but the flow of water leaving the mixing volume usually travels into another body of water: a larger pond or a river. During the flow, some of the water containing the pollutant is lost to other tributaries and some seeps into the ground. Considerable mixing and dilution occurs, and one can imagine the pollutant concentration after many points downstream to be some fraction of the initial concentration. Each of these dilution stages along the path can be viewed as a stage of a SRD process (Figures 8.12 and 8.21). For example, if the source has released an effluent with pollutant concentration c_o at Point A, then the concentration at Point B will be diluted and can be expressed as $C_1 = D_1c_o$ where $0 < D_1 < 1$. Similarly, the concentration at Point C will be $C_2 = D_1C_1 = D_1D_2c_o$, and so on, such that the final concentration C_m after m dilutions is given by the usual result for a SRD process:

$$C_m = c_o \prod_{i=1}^{m} D_i$$

The SRD process results from the cascading effect of these dilution stages as well as the effect of the stochastic mass balance action whenever water containing the pollutant encounters fresh water and a mixing volume. In summary, one can expect the resulting downstream concentrations measured after pollutant discharges on different dates to be approximately lognormally distributed.

The above analysis assumed that the flow rates were constant on a particular

Figure 8.20. An effluent discharged into a pond becomes mixed by the stream entering and leaving.

Figure 8.21. Network of tributaries and streams illustrating a successive random dilution process. Pollutant released at point A is diluted in successive stages as it travels downstream to point F.

date. If a sample were collected at another time of the year with the same source c_o, then the stream flow rates would be different due to differences in rainfall and other factors. If the network of tributaries is sufficiently large, the flow rate at each junction will be affected primarily by local phenomena. If each tributary joins an extremely large stream, then the dilution factor will be almost entirely a function of the freshwater flow rates. If the flow rates are assumed to behave as independent random variables, then the dilution factors $D_1, D_2, \ldots, D_m$ will be independent, and the process will be governed by the Theory of Successive Random Dilutions. Thus, the concentrations of the tracer pollutant measured on different dates at a downstream location (Point F in Figure 8.21) will have a frequency distribution that is approximately lognormal. This model is analogous to the beaker pouring experiment described earlier in the chapter, except that continuous fluid flow occurs in the stream model in place of individual beaker pourings.

Unlike the air pollution field, in which several major air pollutants are monitored continuously, water quality monitoring usually is conducted sporadically at widely scattered locations on streams. For example, the Colorado State Health Department operates 127 sampling stations on 18,000 miles of stream, with samples at each location collected 6 to 24 times per year. Thus, water quality monitoring activities seldom generate the large data bases on a single pollutant at a single site found in the air monitoring field. As a result, only a few intensive studies have been undertaken of the statistical properties of water quality variables.

Sherwani and Moreau[49] analyzed water quality data from monitoring stations on two streams to determine the best fit of three different probability models—the normal, lognormal, and gamma—to the observed frequency distributions. The parameters for each model were estimated by the "method of moments" (Chapter 3, page 60), and the chi-square goodness-of-fit test was applied to each distribution. They found that a majority of the observations did not follow a normal distribution. Only pH (which already is logarithmic), temperature, and, occasionally, dissolved oxygen were found to follow normal distributions. All the other variables had right-skewed distributions. Water quality variables highly correlated with flow rate (turbidity, suspended solids, and sometimes conductivity) were best represented by lognormal distributions. These authors concluded that "... total coliforms, nitrate-nitrite nitrogen, and total phosphorus for all stations follow a lognormal distribution." For one station, the gamma distribution provided a somewhat better fit than the others for total coliform and nitrate-nitrite nitrogen, and the gamma model provided a reasonably good fit for several other cases. These findings support the prediction from the Theory of Successive Random Dilutions that water quality variables reflecting concentrations of constituents will be approximately lognormally distributed. Water quality variables such as temperature, which do not reflect physical or chemical concentrations, show little tendency toward lognormality and often yield symmetrical distributions.

The lognormal distribution also appears in other water quality investigations. Reckhow[50] studied phosphorus concentrations from the Environmental Protection Agency's National Eutrophication Survey, which collected field data on 800 lakes throughout the U.S. in the early 1970s. Although he felt that more study is needed before a definitive statement can be made about the distributions of phosphorus in lakes, he hypothesized that "... as a lake becomes enriched from oligotrophic to mesotrophic to eutrophic, its phosphorus concentration distribution goes from normal to lognormal to more skewed still (perhaps a shifted gamma distribution)." Reckhow[51] believes that the lognormal may be the best model for representing the distributions of variables such as calcium, aluminum, alkalinity, and phosphorus in lakes, but he feels more studies are needed of large water quality data bases.

Davis[52] examined the frequency distribution of dissolved solids in groundwaters from a variety of geologic terrains. He selected dissolved solids because he felt it reflects the influence of all dissolved constituents and is an important descriptor of water quality for most uses. His data came from three areas: the Potomac River Basin (e.g., marine sediments of the Coastal Plain, crystalline rocks of the Piedmont-Blue Ridge, carbonate rocks of the Great Valley), the San Joaquin Basin (e.g., crystalline igneous and metamorphic rocks of the Sierra Nevada), and the Salt Lake Basin (e.g., alluvial deposits of the Wasatch Range). He analyzed 1,205 ground water samples from 11 different terrain units and concluded, "The frequency distribution of dissolved solids content of ground waters in geologically and climatologically homogeneous terrain units commonly approximates a lognormal distribution."

Concentrations in Soils, Plants, and Animals

Pollutants released into the soil from point sources are apt to experience successive random dilutions, giving rise to concentration distributions that are right-skewed and appear lognormal. Little and Whicker[53] studied the distributions of plutonium concentrations in the soil of Rocky Flats, a nuclear facility about 12 miles northeast of Denver, CO. The contamination occurred after a fire at the site and was attributed to leakage from drums containing plutonium-laden cutting oil that had been stored outdoors. They believed that the mechanism responsible for transport of the pollutant was movement from the surface soil by wind, followed by weathering, microdispersal, and penetration into the soil. They cite several other investigators who have reported lognormal radionuclide distributions in soils: "Generally, radionuclide contamination of the environment results in lognormal distributions." Their own data did not result in truly lognormal distributions, as judged by goodness-of-fit tests, but the logarithmic transformation reduced the skewness for observations taken from 7 different depths at various plots tested and resulted in nearly symmetrical distributions.

As the concentrations of trace contaminants in soils are taken up by plants, they undergo additional dilutions, becoming dispersed throughout the plant fibre. Cattle and other animals may graze on these plants, absorbing a small proportion of the original substance in their tissue. As the pollutant moves through the food chain, it undergoes additional successive dilutions, and the concentration distribution in plants and animals should show a tendency toward lognormality. However, for some substances, bioaccumulation may occur, reducing the skewness of the distribution and its tendency toward lognormality.

Pinder and Smith[54] examined the concentrations of radiocesium (primarily ^{137}Cs with a small fraction of ^{134}Cs) in 33 data sets of soils, plants, and animals from contaminated environments of the Savannah River Nuclear Power Plant near Aiken, SC. The data sets contained from 20 to 235 observations from creeks and ponds, with concentrations measured in soils, plants, arthropods, fishes, amphibians, birds, and mammals. The observed concentration distributions were highly right-skewed in plants, although they were less so in animals. Pinder and Smith[54] evaluated the fit of four probability models—the normal, lognormal, exponential, and Weibull—to these observations. Of the four models tested, the lognormal gave the best fit in a majority of the cases (18 out of 33). The Weibull model was next (12 out of 33 cases), and the normal and exponential models did very poorly. The normal distribution gave the best fit of the four models in only 3 of the 33 cases, and the exponential distribution never gave the best fit.

Although the lognormal model fit the observations best, it was rejected by goodness-of-fit criteria at the 99.9% confidence level in 24.2% of the data sets. The rejection rate for the Weibull distribution was 33.3%, and the normal and exponential distributions were rejected in 66.7% and 69.7% of the cases, respectively. Real-world conditions make it difficult to meet the stringent requirements of these goodness-of-fit tests. Despite such rejections, the match of the model to the empirical distribution can be surprisingly good.

Concentrations in Foods and Human Tissue

Randecker[55] has reported unimodal, right-skewed frequency distributions for the concentrations of nitrosamines and other food additives found in meats surveyed by the U.S. Department of Agriculture. Although these findings are not published, it was possible to obtain a portion of the observed data, and a plot of the cumulative frequencies of 2,476 measurements of nitrosoprolidine in bacon (Figure 8.22) shows the characteristic "straightness with mild downward curvature" on logarithmic-probability paper. It should be noted that the processes

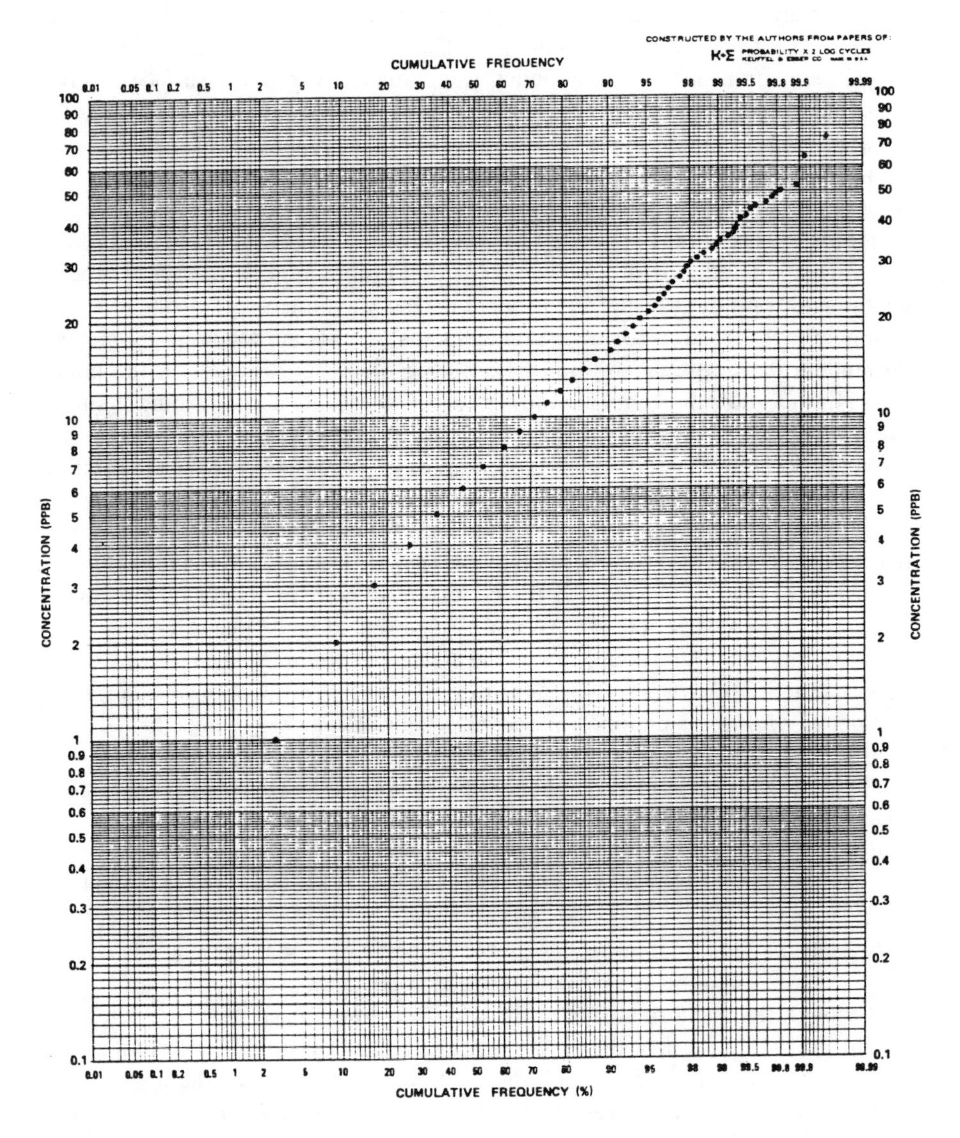

Figure 8.22. Frequency distribution of 2,476 measurements of nitrosoprolidine concentrations in bacon from the U.S. Deparment of Agriculture's nitrosamine monitoring program (Source: Randecker[55]).

leading to the resulting levels of nitrosamines in bacon are more complex than the simple dilution experiments considered above. Suppliers of the food add nitrite as a preservative. The nitrite is diluted in the bacon many times from its initial concentration as it is absorbed. In the presence of heat, the nitrite combines with amines to form nitrosamines such as nitrosoprolidine. Although these chemical reactions are intrinsic to the process and may affect the distribution, the dilution components of the process appear to be dominant, as suggested by the characteristic form of the distribution that results.

Rustagi[56] has proposed the lognormal model for representing the frequency distribution of the concentrations of trace metals found in human tissue, blood, and feces. He examines data collected by other investigators and concludes that the concentrations of trace substances—lead in food, lead in feces, cadmium in blood, creatinine in urine—all follow a lognormal law. He notes that two lognormally distributed random variables, when multiplied together, yield a lognormally distributed product. He assumes that both the quantity of food consumed and the concentration of trace materials in the food are lognormally distributed: "The total amount of a trace substance in intake is essentially a product of two random variables: the amount of substance taken in, multiplied by the per unit concentration. Hence, if the total amount consumed is lognormal and the concentration is lognormal, the total amount of the trace substance will also be lognormal."

Schubert, Brodsky, and Tyler[57] review the data and findings from studies throughout the world on the concentrations of trace elements (e.g., strontium-90, radium, cesium) found in human tissue (e.g., bones, skeletal ash, stillborn fetuses and samples of lung, kidney, and spleen) and they conclude that the observations follow a lognormal rather than a normal law:

> Data on the concentrations of strontium-90, radium-226, cesium-137, and some of the stable trace elements in human tissue as well as environmental samples have been found, without exception, to follow lognormal distributions rather than normal distributions. Also, while the median of the strontium-90 concentration in humans shifts for different time periods, areas of the world, and population age distributions, the geometric standard deviation has seemed recently to remain within the relatively narrow range of 1.55–1.67 for large population groups, so that about the same proportion of each population contains concentrations in excess of a given multiple of the median (or the average) concentration. The lognormal distribution fits the data better than the normal distribution, particularly in the "tail" of higher concentrations. This is the region most critical for the evaluation of the proportions of the population exceeding recommended limits of individual exposure. The repeated suitability of the lognormal permits the recommended limits of environmental contamination to be established with more confidence in predicting the proportions of the population exceeding a given level.[57]

Ore Deposits

Many of the heavier metals found in the earth's surface once were present in the central core of the earth in a molten state. As the planet cooled, these materials were transported to the earth's surface by volcanic action. Geological movements caused them to be scattered and dispersed in the earth's crust. Over the long history of the earth's development, these materials underwent many dilution stages, changing from highly concentrated forms deep inside the earth to relatively low concentrations at the surface.

The near-surface deposits of precious metals to which man has access are the result of a multitude of geological actions (volcanic activity, earthquakes, uplift and fracture, glacial movement, floods, and erosion). Gold deposits, for example, commonly are found in quartz veins, because the molten quartz-gold mixture intruded into the rock and hardened as it cooled. Earth movements cause these veins to fracture; erosion breaks the quartz and gold into smaller pieces; streams carry the materials along and disperse them widely in stream beds. In these successive steps, the concentration of gold is reduced, because the gold becomes more widely dispersed in three dimensions. Thus, by the Theory of Successive Random Dilutions, one would expect to find the concentrations of rare metals found in different mines, or in different sites within the same mine, to be approximately lognormally distributed.

As indicated by Krige,[58] statistical methods of ore valuation employed in South African gold mines, except for quality control sampling, all have been based on the lognormal distribution, which was first applied to gold values by Sichel[59] in 1947. Krige analyzed a large number of gold, uranium, and pyrite distributions from mines in South Africa. The frequency distributions generally were lognormal in appearance, but some departure from perfect lognormality also was noted. The characteristic distribution had a larger frequency of occurrence of very low values than does the two-parameter lognormal distribution. As a result, the curves exhibited gentle downward curvature when plotted on logarithmic-probability paper, particularly at the lower concentrations, and the empirical distributions were found to be better represented by a three-parameter lognormal distribution (Chapter 9).

SUMMARY AND CONCLUSIONS

This chapter was intended as a transition from the discussion of (symmetrical) normal processes (Chapters 6 and 7) to the discussion of lognormal (right-skewed) processes (Chapter 9). It has suggested a general mechanism for generating the right-skewed concentration distributions commonly observed in nature and the environment. This mechanism, called the Theory of Successive Random Dilutions (SRD), applies to a pollutant that is released at initially high concentrations into a carrier medium and then undergoes dilution in successive, independent stages. If a mixing stage occurs between each independent dilution stage, the final concentration will be the product of the initial concentration and a series of independent dilution factors. As the number of successive random dilutions becomes large, the distribution of the final concentration will be approximately lognormal. The Theory of Successive Random Dilutions is a special application of the Law of Proportionate Effect originally proposed by

Kapteyn in 1903. It is especially appropriate for modeling substances released into environmental carrier media (e.g., air, water, soil) experiencing considerable physical movement and agitation.

Real-world conditions are likely to meet the requirements for the Theory of Successive Random Dilutions only approximately. Nevertheless, distributions with a characteristic lognormal appearance have been observed in a variety of fields. Examples discussed in this chapter included outdoor air pollutants; indoor air pollutants; water pollutants measured in streams, lakes, and ground waters; radionuclides in soils, plants, fishes, birds, and mammals; additives in foods; radionuclides and trace metals in human tissue, blood, and feces; and precious metals in the earth.

One can imagine many other cases in which the Theory of SRD might apply, but the data have not yet been collected or analyzed to verify the predicted tendency toward lognormality. The following list includes examples given in this chapter, as well as other variables for which few data were available or whose distributions have not been studied intensively. It is predicted that distributions similar to those discussed above will emerge when the data are studied. The list is not intended to be exhaustive, and the reader no doubt can think of additional examples to include.

Air

- Concentrations of gases and particles measured in the ambient outdoor air.
- Cigarette smoke components (for example, CO, particles, and organic concentrations) observed in a room.
- Radon concentrations measured indoors.
- A chemical solvent released inside a house or building and allowed to evaporate.
- Vehicle exhaust gases measured near traffic.
- Smoke concentrations released from a fire.

Water

- Concentrations of physical and chemical pollutants observed in streams.
- Phosphorus and metals observed in lakes.
- Dissolved solids and organics in groundwater.
- Pesticide concentrations in agricultural runoff.
- Concentrations of pollutants in the ocean caused by wastes dumped at sea.
- Metals naturally found in seawater.
- Sulfate concentrations in rainwater.

Earth

- Toxic chemical concentrations measured in landfills.
- Toxic compounds sprayed or dumped near roadways.

- Chemicals contributed to the soil from leaking drums.
- Concentrations of trace metals in soils.
- Pollutant concentrations in sediments.
- Lead concentrations in roadway dust.
- Trace metals and pesticides in cultivated farmland.
- Trace metals in house dust.

Food and Drinking Water

- Chemical additive concentrations in foods.
- Concentrations of organic compounds in beverages.
- Pesticides and trace metals in foods.
- Asbestos concentrations in drinking water.
- Ethylene dibromide concentrations in wheat products.

Ecology

- Metals and pesticides in plants.
- DDT in the eggshells of birds.
- Concentrations of pollutants measured in animal tissue.
- Mercury and other pollutants in fish.
- Radionuclide concentrations in plants and animals near nuclear power facilities.

Physiology

- Contaminant concentrations in the bloodstream.
- Pollutant concentrations in samples from tissue banks.
- Radionuclides and trace metals in bones.
- Concentrations of metals in human and animal feces.
- Pollutant concentrations in urine.

PROBLEMS

1. By differentiating the PDF of the gamma probability model and setting the result equal to zero, show that the mode is given by the equation $x_m = (\eta - 1)/\lambda$.
2. Plot the gamma distribution for the case of $\eta = 4$ and $\lambda = 0.5$, and show that the mode occurs at $x_m = 6$.
3. Using the BASIC computer program (Figure 8.4) for generating the product of m uniformly distributed random variables $\mathbf{U}(a,b)$ multiplied together, generate histograms of $C_5 = c_o D_1 D_2 D_3 D_4 D_5$ for the case of $c_o = 1{,}000$ ppm and each dilution factor distributed as $\mathbf{U}(0,1)$ for each of the following number of trials: (a) $n = 10$, (b) $n = 50$, (c) $n = 100$, (d) $n = 1{,}000$. [Hint: Set the interval width at 10 ppm and use 50 intervals.]
4. Plot the PDF of the gamma distribution for the case of $\eta = 5$ and $\lambda = 1$.

5. Modify the BASIC computer program (Figure 8.4, page 198) so it takes the logarithm of the resulting concentration, $-\ln C_m$, and sorts the result into intervals of the histogram. Generate a histogram for $n = 1{,}000$ random concentration logarithms for $c_o = 1$ and $m = 5$, and compare the result with the PDF plotted in Problem 4. [Hint: Insert line #355, U = –1*LOG(U) in the program, and use 40 intervals with an interval width of 0.5.] Compare the observed and theoretical mean and variance.
6. Repeat Problem 5 for the case of $m = 12$. Discuss the symmetry of $-\ln C_{12}$ and plot the observed cumulative frequencies on normal probability paper.
7. The end of the chapter contains a partial list of examples of environmental processes likely to obey the Theory of Successive Random Dilutions and give rise to distributions that are approximately lognormal. Give 10 additional examples of environmental variables that could be added to the list.
8. Obtain real data from a process likely to be governed by the Theory of Successive Random Dilutions. Plot the data on logarithmic-probability paper and compare it with some of the distributions in this chapter.

REFERENCES

1. Ott, Wayne R., David T. Mage, and Victor W. Randecker, "Testing the Validity of the Lognormal Probability Model: Computer Analysis of Carbon Monoxide Data from U.S. Cities," U.S. Environmental Protection Agency, Washington, DC, EPA-600/4-79-040 (June 1979).
2. Feller, William, *An Introduction to Probability Theory and Its Applications*, Volume I (New York: John Wiley and Sons, 1968).
3. Feller, William, *An Introduction to Probability Theory and Its Applications*, Volume II (New York: John Wiley and Sons, 1970).
4. Johnson, Normal L., and Samuel Kotz, *Continuous Univariate Distributions–1* (Boston: Houghton Mifflin Company, 1970).
5. Hahn, Gerald J., and Samuel S. Shapiro, *Statistical Models in Engineering* (New York: John Wiley and Sons, 1967).
6. Wilk, M.B., R. Gnanadesikan, and M.J. Huyett, "Probability Plots for the Gamma Distribution," *Technometrics*, 4(1):1–20 (February 1962).
7. Harter, H. Leon, "More Tables of the Incomplete Gamma-Function Ratio and of Percentage Points of the Chi-Square Distribution," U.S. Air Force, Office of Aerospace Research, NTIS AD 607403 (August 1964).
8. Hastings, N.A.J., and J.B. Peacock, *Statistical Distributions* (New York: John Wiley and Sons, Halsted Press, 1975).
9. Beyer, William H., ed., *Handbook of Tables for Probability and Statistics*, 2nd ed., (Boca Raton, FL: CRC Press, Inc., 1968).
10. Abramowitz, Milton, and Irene A. Stegun, "Handbook of Mathematical Functions with Formulas, Graphs, and Mathematical Tables," National Bureau of Standards Applied Mathematics Series No. 55 (Washington, DC: U.S. Government Printing Office, December 1972).

11. Ott, Wayne R., "A Physical Explanation of the Lognormality of Pollutant Concentrations," *J. Air & Waste Manage. Assoc.* 40(10):1378–1383 (October 1990).
12. Nehls, Gerald J., and Gerald G. Akland, "Procedures for Handling Aerometric Data," *J. Air Poll. Control Assoc.* 23(3):180–184 (March 1973).
13. Aitchison, J., and J.A.C. Brown, *The Lognormal Distribution* (London: Cambridge University Press, 1973).
14. Kapteyn, J.C., *Skew Frequency Curves in Biology and Statistics* (Groningen, Noordhoff: Astronomical Laboratory, 1903).
15. Kahn, Henry D., "Note on the Distribution of Air Pollutants," *J. Air Poll. Control Assoc.* 23(11):973 (November 1973).
16. Gifford, Frank A., Jr., "The Form of the Frequency Distribution of Air Pollution Concentrations," *Proceedings of the Symposium on Statistical Aspects of Air Quality Data*, ed., Lawrence D. Kornreich, U.S. Environmental Protection Agency, Research Triangle Park, NC, EPA-650/4-74-038, October 1974, pp. 3–1 to 3–7.
17. Knox, Joseph B., and Richard I. Pollack, "An Investigation of the Frequency Distributions of Surface Air-Pollutant Concentrations," *Proceedings of the Symposium on Statistical Aspects of Air Quality Data*, ed., Lawrence D. Kornreich, U.S. Environmental Protection Agency, Research Triangle Park, NC, EPA-650/4-74-038, October 1974, pp. 9–1 to 9–17.
18. Pollack, Richard I., "Studies of Pollutant Concentration Frequency Distributions," U.S. Environmental Protection Agency, Research Triangle Park, NC, EPA-650/4-75-004 (January 1975).
19. Mage, David T., and Wayne R. Ott, "An Improved Statistical Model for Analyzing Air Pollution Concentration Data," Paper No. 75-51.4 presented at the 68th Annual Meeting of the Air Pollution Control Association, Boston, MA (June 1975).
20. Mage, David T., and Wayne R. Ott, "An Improved Model for Analysis of Air and Water Pollution Data," *International Conference on Environmental Sensing and Assessment*, Vol. 2, IEEE No. #75-CH 1004-1 ICESA, University of Las Vegas, NV, p. 20–5 (September 1975).
21. Ott, Wayne R., and David T. Mage, "A General Purpose Univariate Probability Model for Environmental Data Analysis," *Comput. and Ops. Res.*, 3:209–216 (1976).
22. Ott, Wayne R., "A Brownian Motion Model of Pollutant Concentration Distributions," SIMS Technical Report No. 46, Department of Statistics, Stanford University, Stanford, CA (April 1981).
23. Benarie, M., "Le Calcul de la Dose et de la Nuisance du Polluant Emis par Une Source Ponctuelle," *Atmos. Environ.* 3:467 (1969).
24. Benarie, M., "Sur la Validite de la Distribution Logarithmico-Normale des Concentrations de Pollutant," *Proceedings of the 2nd International Clear Air Congress*, 1970, Washington, DC, pp. 68–70 (New York: Academic Press, 1971).

25. Benarie, M., "The Use of the Relationship Between Wind Velocity and Ambient Pollutant Concentration Distributions for the Estimation of Average Concentrations from Gross Meteorological Data," *Proceedings of the Symposium on Statistical Aspects of Air Quality Data*, ed., Lawrence D. Kornreich, U.S. Environmental Protection Agency, Research Triangle Park, NC, EPA-650/4-74-038, October 1974, pp. 5–1 to 5–17.
26. Bencala, Kenneth E., and John H. Seinfeld, "On Frequency Distributions of Air Pollutant Concentrations," *Atmos. Environ.* 10:941–950 (1976).
27. Mage, David T., personal communication, Washington, DC, 1983.
28. Mage, David T., "A Review of the Application of Probability Models for Describing Aerometric Data," *Environmetrics 81: Selected Papers*, Selections from a Conference in Alexandria, VA, Society for Industrial and Applied Mathematics, Philadelphia, PA, April 8–10, 1981.
29. Harris, E.D., and E.C. Tabor, "Statistical Considerations Related to the Planning and Operation of a National Air Sampling Network," Proceedings of the 49th Annual Meeting of the Air Pollution Control Association, Buffalo, NY (1956).
30. Zimmer, Charles E., Elbert C. Tabor, and Arthur C. Stern, "Particulate Pollutants in the Air of the United States," *J. Air Poll. Control Assoc.*, 9:136 (1959).
31. Larsen, Ralph I., "A Method for Determining Source Reduction Required to Meet Air Quality Standards," *J. Air Poll. Control Assoc.*, 11:71–76 (February 1961).
32. Larsen, Ralph I., "United States Air Quality," *Arch. Environ. Health*, 8:325–333 (February 1964).
33. Zimmer, Charles E., and George A. Jutze, "An Evaluation of Continuous Air Quality Data," *J. Air Poll. Control Assoc.*, 14:262–266 (July 1964).
34. Zimmer, Charles E., and Ralph I. Larsen, "Calculating Air Quality and Its Control," *J. Air Poll. Control Assoc.*, 15:565–572 (December 1965).
35. Larsen, Ralph I., "Determining Source Reduction Needed to Meet Air Quality Standards," *International Clear Air Congress Proceedings*, Part 1, pp. 60–64, National Society for Clean Air, London E. C. 4, England (October 1966).
36. Larsen, Ralph I., "Determining Reduced-Emission Goals Needed to Achieve Air Quality Goals—A Hypothetical Case," *J. Air Poll. Control Assoc.*, 17:823–829 (December 1967).
37. Larsen, R.I., C.E. Zimmer, D.A. Lynn, and K.G. Blemel, "Analyzing Air Pollutant Concentration and Dosage Data," *J. Air Poll. Control Assoc.*, 17:85–93 (February 1967).
38. Larsen, R.I., "A New Mathematical Model of Air Pollutant Concentration Averaging Time and Frequency," *J. Air Poll. Control Assoc.*, 19:24–30 (January 1969).
39. Larsen, Ralph I., "A Mathematical Model for Relating Air Quality Measurements to Air Quality Standards," U.S. Environmental Protection Agency, Research Triangle Park, NC, Publication No. AP-89 (November 1971).
40. Patel, Nitin R., "Comment on a New Mathematical Model of Air Pollution Concentration," *J. Air Poll. Control Assoc.*, 23:291 (April 1973).

41. Larsen, Ralph I., "Response," *J. Air Poll. Control Assoc.*, 23:292 (April 1973).
42. Larsen, Ralph I., "An Air Quality Data Analysis System for Interrelating Effects, Standards, and Needed Source Reductions," *J. Air Poll. Control Assoc.*, 23:933–940 (November 1973).
43. Larsen, Ralph I., "An Air Quality Data Analysis System for Interrelating Effects, Standards, and Needed Source Reductions—Part 2," *J. Air Poll. Control Assoc.*, 24:551–558 (June 1974).
44. Lynn, David A., "Fitting Curves to Urban Suspended Particulate Data," *Proceedings of the Symposium on Statistical Aspects of Air Quality Data*, ed., Lawrence D. Kornreich, U.S. Environmental Protection Agency, Research Triangle Park, NC, EPA-650/4-74-038, October 1974, pp. 13–1 to 13–28.
45. Mage, David T., "On the Lognormal Distribution of Air Pollutants,"*Proceedings of the Fifth Meeting of the Expert Panel on Air Pollution Modeling*, NATO/CCMS N. 35, Roskilde, Denmark, June 4–6, 1974.
46. Ott, Wayne, Paul Switzer, and Neil Willits, "Carbon Monoxide Exposures Inside an Automobile Traveling on an Urban Arterial Highway," *J. Air & Waste Manag. Assoc.*, 44(8):1011–1018 (August 1994).
47. Johnson, Norman L., and Samuel Kotz, *Continuous Univariate Distributions–2*, Chapter 25, "Uniform or Rectangular Distribution," p. 64 (Boston: Houghton Mifflin Company, 1970).
48. Alter, H. Ward, and Richard A. Oswald, "Results of Indoor Radon Measurements Using the Track Etch® Method," *Health Physics*, 45(2):425–428 (August 1983).
49. Sherwani, Jabbar K., and David H. Moreau, "Strategies for Water Quality Monitoring," Water Resources Research Institute of the University of North Carolina, North Carolina State University, Raleigh, NC, Report No. 107, June 1975.
50. Reckhow, Kenneth H., "Lake Analysis and Phosphorus Variability," paper presented at the North American Lake Management Conference, Michigan State University, East Lansing, MI, April 16–18, 1979.
51. Reckhow, Kenneth, personal communication, University of North Carolina, Chapel Hill, NC, September 6, 1984.
52. Davis, G.H., "Frequency Distribution of Dissolved Solids in Ground Water," *Ground Water*, 4(4):5–12 (1966).
53. Little, C.A., and F.W. Whicker, "Plutonium Distribution in Rocky Flats Soil," *Health Physics*, 34:451–457 (1978).
54. Pinder, John E., and Michael H. Smith, "Frequency Distributions of Radiocesium Concentrations in Soil and Biota," *Proceedings of a Symposium on Mineral Cycling in Southeastern Ecosystems*, eds., Fred G. Howell, John B. Gentry, and Michael H. Smith, U.S. Energy Research and Development Administration, CONF-740513, 1975, pp. 107–125.
55. Randecker, Victor, U.S. Department of Agriculture, Washington, DC, letter to Wayne Ott, January 23, 1981.
56. Rustagi, J.S., "Stochastic Behavior of Trace Substances," *Arch. Environ. Health*, 8:76–84 (January 1964).

57. Schubert, Jack, Allen Brodsky, and Sylvanus Tyler, "The Log-Normal Function as a Stochastic Model of the Distribution of Strontium-90 and Other Fission Products in Humans," *Health Physics*, 13:1187–1204 (1967).
58. Krige, D.G., "On the Departure of Ore Value Distributions from the Lognormal Model in South African Gold Mines," *Journal of the South African Institute of Mining and Metallurgy*, 231–244 (November 1960).
59. Sichel, H.S., "New Methods in the Statistical Evaluation of Mine Sampling Data," *Transactions of the Institute of Mining and Metallurgy*, London (February 1947).

9 Lognormal Processes

Chapters 6 and 7 considered independent random variables that were added together to form a sum, and the sum tended naturally toward the normal distribution, which is symmetrical and bell-shaped. These normal processes are rather common in everyday phenomena, as illustrated by the many examples discussed earlier. We introduced the phrase *random sum process*, or $\mathcal{RS}$-process, to describe these summing processes.

In the previous chapter, an important counterpart to the $\mathcal{RS}$-process was introduced. Instead of adding the independent random variables together, they were multiplied to form a product. Unlike the symmetrical distributions generated by the $\mathcal{RS}$-processes, the resulting *product* of independent random variables tended toward an asymmetrical distribution with a single mode and a long tail to the right. We introduced the phrase *random product process*, or $\mathcal{RP}$-process, to describe this phenomenon. If the logarithms of the product then are taken, the result once again tends toward the symmetrical, bell-shaped normal distribution, indicating that the variate itself is *lognormally* distributed. Like the $\mathcal{RS}$-processes, the $\mathcal{RP}$-processes are common in our daily lives, the world around us, and the natural environment.

In this chapter, we further develop the concept of the $\mathcal{RP}$-process introduced in Chapter 8, and we discuss the distributions to which they give rise in greater detail. Examples of common phenomena that may be classified as $\mathcal{RP}$-processes are numerous and varied:

1. Interest on a savings account is compounded daily and varies according to interest rates nationally, which behave in a random manner. If the amounts of interest earned (dollars) are plotted on a histogram, what distribution should they have approximately?
2. Consider a large tree with many leaves. Assume that each leaf on the tree grows as follows during the year: due to weekly variations in sunlight, water, and nutrients, the leaf's length increases by some random proportion of the length attained in the previous week. For a given leaf, assume that the random proportions are independent. At the end of the year, the lengths of many of the leaves are measured. What probability model will the distribution of leaf lengths resemble?
3. Forty years ago, there were a number of countries with similar economies that had approximately the same population. Since then, the yearly population growth rates of each country varied in response to annual changes in economic, health, and social conditions. Assume that the annual socioeconomic changes within a country from year to year are independent, causing the annual population growth rates

within a country to be independent. Assume, also, that the growth rates in different countries are independent of each other. At present, the populations of these countries no longer are equal. If the current populations are plotted on a histogram, what probability model will describe their distribution?

4. An indoor air pollutant tracer is released on the same date in a number of different homes. The air exchange rate in each home varies from hour to hour in a random fashion. When the experimenters return to measure the concentration of the tracer gas in each home 30 hours later, they find that the distribution of the concentrations for all the homes has a single mode and is right-skewed. What probability model does this distribution resemble?
5. A fixed amount of a tracer pollutant is discharged into a pond each week, and the resulting concentration in the pond is measured three days later. Two years of data are collected in this manner, giving 2(52) = 104 observations. Fresh water flows continuously into the pond from a number of streams, and their flow rates vary considerably from hour to hour. At any instant of time, the water in the pond is relatively well-mixed due to agitation and natural movement of the water into and out of the pond. If a histogram is plotted of the observed concentrations in the pond, what does it look like? When the cumulative frequencies of the concentrations are plotted, what type of probability paper is a good choice?

In the examples above, the quantity present in each state can be expressed as a *random proportion* of the quantity present in the immediately prior state. If each successive proportion is independent of the one before, and if many states occur between the initial state and the final state, then the final result can be expressed as a product of random variables. If the independence condition is met, each example is an $\mathcal{RP}$-process that will give rise naturally to a lognormally distributed random variable.

CONDITIONS FOR LOGNORMAL PROCESS

The above examples can be shown to exhibit the properties discussed in Chapter 8:

1. The variable of interest can be expressed as a *linear proportion* of the value it attains in each previous state.
2. Each linear proportion is assumed to be independent of all successive linear proportions.
3. Many successive states have occurred between the initial state and the point in time in which the variable is observed.

In the products generated by this process, the values of the linear proportions are bounded. If any linear proportion is zero, then the variable of interest will be zero in all successive states, and the process will stop. If any linear proportion is infinite, then the variable of interest will be infinite, an impossible condition for real physical systems. In general, the linear proportions are assumed to be nonzero, positive, and finite.

In general, the important property of a lognormal process is for there to be a product of independent random variables. That is,

> A lognormal process is one in which the random variable of interest results from the product of many independent random variables multiplied together.

In processes observed in the environment, the number of independent random variables multiplied together usually does not have to be very great before characteristic lognormal properties emerge. Because environmental concentrations usually depend on the number of molecules of a pollutant present per unit volume, they ordinarily are positive random variables.

DEVELOPMENT OF MODEL

Chapter 8 presented a newer mathematical development of the lognormal model: by taking logarithms of both sides of an equation involving a product, the left side became the logarithm of the pollutant concentration while the right side became a sum of the logarithms of random variables and an initial condition. Because the logarithms of independent random variables are themselves random variables, the sum on the right side of the equation approached a normal distribution by the Central Limit Theorem. Thus, the logarithm of the concentration (left side of equation) tended to be *normally distributed*, with the consequence that the concentration itself was *lognormally distributed.*

In this chapter, we present an alternate derivation to the one described in Chapter 8; it was originally proposed by Kapteyn,[1] is described in detail by Aitchison and Brown,[2] and was brought to the attention of the air pollution community by Kahn.[3] This approach considers a positive random variable that is the outcome of a discrete stochastic process. Most investigators conceive of this process as taking place at equally-spaced, successive points in time, as is typical for biological growth — for example, the length of a fish, the size of an organism, or the height of a tree — observed on successive days or weeks. However, Aitchison and Brown[2] emphasize that ordering of the sequence in time (and equal spacing of the time periods) is not a necessary feature of this process. Nevertheless, it is convenient to present its development as a time sequence, while keeping in mind that the successive events are important in themselves, regardless of their relationship to time.

Consider a random variable that is X_o initially. After the jth step in the process, it has the value X_j, reaching its final value X_n after n steps. At the jth step, the change in the random variable, $X_j - X_{j-1}$, is assumed to be a random proportion of some function $f(X_{j-1})$ of the value X_{j-1} already obtained:

$$X_j - X_{j-1} = \epsilon_j f(X_{j-1}) \quad \text{for} \quad j = 1, 2, \ldots, n$$

where the elements of the set $\{\in_j\}$ are independent of the elements of the set $\{X_j\}$. We call this expression the "general form" of the Kapteyn model.

Consider the important case in which $f(X_{j-1}) = X_{j-1}$. Here, the change in the variate X is a random proportion of the previous value of the random variable. This is the Law of Proportional Effect (LPE),[1] which was discussed in the previous chapter, and which Aitchison and Brown[2] describe as follows:

> A variate subject to a process of change is said to obey the law of proportionate effect (LPE) if the change in the variate at any step in the process is a random proportion of the previous value of the variate.

For this case, the general LPE equation above reduces to the following:

$$X_j - X_{j-1} = \in_j X_{j-1}$$

Solving this equation for $\in_j$, we obtain

$$\frac{X_j - X_{j-1}}{X_{j-1}} = \in_j$$

If we perform a summation of both sides of this equation, we obtain

$$\sum_{j=1}^{n} \frac{X_j - X_{j-1}}{X_{j-1}} = \sum_{j=1}^{n} \in_j$$

Now if we assume that the change $\Delta X_j = X_j - X_{j-1}$ is extremely small, then the left side summation of the above equation can be written as an integral, approximately:

$$\sum_{j=1}^{n} \frac{X_j - X_{j-1}}{X_{j-1}} \approx \int_{X_o}^{X_n} \frac{dX}{X} = \ln X_n - \ln X_o$$

Substituting this approximation into the equation above and integrating, we obtain

$$\ln X_n - \ln X_o = \sum_{j=1}^{n} \in_j$$

or,

$$\ln X_n = \ln X_o + \in_1 + \in_2 + \ldots + \in_n$$

By the Central Limit Theorem, $\ln X_n$ is asymptotically normally distributed, and X_n is asymptotically *lognormally distributed.* The development outlined here is simply an alternative way of arriving at a lognormal distribution from an $\mathcal{RP}$-process, which was discussed in greater detail in Chapter 8. As with the cases described there, the variance of $\ln X_n$ increases with n, so the resulting lognormality holds for a fixed value of n.

By "lognormally distributed," we refer here to the two-parameter lognormal distribution. Suppose the difference in the random variable at successive states is equal to the previous value of the random variable minus some fixed quantity, say a. Then $f(X_{j-1})$ in the general form of the Kapteyn LPE equation above becomes $f(X_{j-1}) = (X_{j-1} - a)$. If the same steps above are repeated, the final equation above will be written as

$$\ln(X_n - a) = \ln(X_o - a) + \epsilon_1 + \epsilon_2 + \ldots + \epsilon_n$$

If this result is expanded in the manner presented above, the result is a three-parameter lognormal distribution (see Problem 1 at the end of this chapter). Unless otherwise stated, "lognormally distributed" in this chapter refers to the two-parameter lognormal distribution.

LOGNORMAL PROBABILITY MODEL

The lognormal probability model describes a distribution in which the logarithm of the random variable of interest is normally distributed. In Chapter 7 ("Development of Model," pages 164–167), we considered the "standardized" random variable Z, which was normally distributed with mean 0 and standard deviation of 1. We also considered the transformation $Z = (Y - \mu)/\sigma$, which allowed us to consider a normally distributed random variable Y that was displaced from the origin, or the "displaced" normal distribution.

We now consider the logarithmic transformation $Y = \ln X$. Substituting this transformation for Y into the transformation for the displaced normal distribution, we obtain $Z = (\ln X - \mu)/\sigma$. In making this substitution, we note that the probability density function (PDF) is really a derivative of probability P; that is, $f_Z(z) = dP/dz$ or $dP = f_Z(z)dz$ for the standardized normally distributed random variable Z:

$$dP = f_Z(z)dz = \frac{1}{\sqrt{2\pi}} e^{-\frac{1}{2}z^2} dz$$

Differentiating the logarithmic transformation $z = (\ln x - \mu)/\sigma$ with respect to x, we obtain $dz = dx/\sigma x$. Substituting the transformation for z and $dz = dx/\sigma x$ into the above expression, we obtain the following expression for the lognormally distributed random variable X:

$$dP = \frac{1}{x\sigma\sqrt{2\pi}} e^{-\frac{1}{2}\left(\frac{\ln x - \mu}{\sigma}\right)^2} dx$$

Notice that the PDF above is just like the PDF for the normal distribution (Chapter 7, page 166), except that $\ln x$ substituted for y, and x now also appears in the denominator (Table 9.1). Here, μ is the mean and σ is the standard deviation of Y, while Y is the normal distribution describing the natural logarithm of X. Therefore, we refer to μ and σ as the "normal parameters" of this normal distribution. In summary, the normal parameters are the two parameters of the normal distribution describing the logarithm of the concentration X.

Table 9.1. The 2-Parameter Lognormal Distribution

Probability Density Function:

$$f_X(x) = \begin{cases} \frac{1}{x\sigma\sqrt{2\pi}} e^{-\frac{1}{2}\left(\frac{\ln x - \mu}{\sigma}\right)^2} & -\infty < \mu < \infty \\ 0 \text{ elsewhere} \end{cases}$$

Expected Value:

$$E[X] = e^{\mu + \frac{\sigma^2}{2}}$$

Variance:

$$Var(X) = e^{2\mu + \sigma^2}(e^{\sigma^2} - 1)$$

Coefficient of Skewness:

$$\sqrt{\beta_1} = (e^{\sigma^2} + 2)\sqrt{e^{\sigma^2} - 1}$$

Coefficient of Kurtosis:

$$\beta_2 = e^{4\sigma^2} + 2e^{3\sigma^2} + 3e^{2\sigma^2} - 3$$

The cumulative distribution function (CDF) of the lognormal distribution is the same as the CDF of the standardized normally distributed random variable Z described in Chapter 7, except that the quantity $(\ln x - \mu)/\sigma$ is substituted for z:

$$F_X(x) = F_Y(\ln x) = F_Z\left(\frac{\ln x - \mu}{\sigma}\right)$$

Thus, to determine the cumulative probability associated with any value of X, we can use standard normal probability tables developed for the normal distribution or the calculator approximations for the normal CDF given in the last section of Chapter 7 (pages 185–188). The analyst simply computes the value of the quantity $z = (\ln x - \mu)/\sigma$ and then looks up the corresponding value of the CDF of the normal distribution using tables, a computer program, or a hand calculator (see pages 185–188).

For example, suppose $\mu = 2.2$ and $\sigma = 0.41$, and we want to find the cumulative frequency associated with $x = 18$. In this case, $z = [\ln(18) - 2.2]/0.41 = 1.68$. The CDF of the normal distribution gives $F_Z(1.68) = 0.9535$, so the value $x = 18$ corresponds to a frequency of 95.35% if X is lognormally distributed (see Problems 4, 5, and 6 for additional examples).

Parameters of the Lognormal Distribution

Substituting particular values into the CDF allows us to derive useful equations for other parameters of the lognormal distribution. For example, consider the special case of the median x_{50}, which is the point at which the value of the CDF is $F_X(x_{50}) = 0.50$. To derive an equation for the median, we first set the CDF of the random variable equal to 0.5:

$$F_X(x_{50}) = F_Y(\ln x_{50}) = F_Z\left(\frac{\ln x_{50} - \mu}{\sigma}\right) = 0.5$$

At the median of the standardized normally distributed random variable Z, half the probability mass is included in the CDF, so the median occurs at the point in which $F_Z(0) = 0.5$; therefore, we set the expression for the argument in parentheses above equal to zero:

$$\left(\frac{\ln x_{50} - \mu}{\sigma}\right) = 0$$

Solving this equation for x_{50}, we obtain an equation for the median:

$$Median = x_{50} = e^{\mu}$$

Another parameter of importance is the geometric mean, or x_g. The geometric mean of a random variable is the antilogarithm of the arithmetic mean of the logarithmic transform of that random variable. Here, Z is the logarithmic transform of the random variable X, and its mean is μ. Thus, the geometric mean is given by the same expression as above for the median:

$$Geometric\ Mean = \mu_g = e^{\mu}$$

Because Z is symmetrical about $z = 0$, its median and mean coincide. Thus, for a lognormal distribution, the median and geometric mean are given by the same equation:

$$x_{50} = \mu_g = e^{\mu}$$

Solving this expression for μ as a function of μ_g, we obtain $\mu = \ln \mu_g$. Substituting this expression for μ into the PDF for the lognormal distribution given in Table 9.1, we obtain a slightly different version of the PDF of the lognormal distribution:

$$f_X(x) = \frac{1}{x\sigma\sqrt{2\pi}} e^{-\frac{1}{2}\left(\frac{\ln x - \ln \mu_g}{\sigma}\right)^2} = \frac{1}{x\sigma\sqrt{2\pi}} e^{-\frac{1}{2}\left(\frac{\ln\left[\frac{x}{\mu_g}\right]}{\sigma}\right)^2}$$

Because this form of the lognormal PDF contains the normal parameter σ and the geometric parameter μ_g, it can be called the "mixed-parameter" version of the lognormal PDF. The CDF of the mixed-parameter version of the lognormal probability model is written as

$$F_Z(z) = F_Z\left(\frac{\ln\left[\frac{x}{\mu_g}\right]}{\sigma}\right)$$

Now consider the case in which the argument of the CDF of the standardized normally distributed random variable Z is exactly +1; from Chapter 7 (Figure 7.1), this case (one standard deviation above the origin) gives the CDF value of $F_Z(+1) = 0.8413$. If we set the argument of the CDF $F_Z(z)$ above equal to 1 at the point corresponding to $X = x_{84.13}$ (denoting the 84.13% percentile value), we obtain the following:

$$\frac{\ln\left(\frac{x_{84.13}}{\mu_g}\right)}{\sigma} = +1$$

Solving this equation for the ratio $\{x_{84.13}/\mu_g\}$, we obtain an equation that defines a new and important parameter of this model, the *standard geometric deviation* (SGD) of the lognormal probability model represented by σ_g:

$$\frac{x_{84.13}}{\mu_g} = e^{\sigma} = \sigma_g$$

In a lognormally distributed random variable's frequency distribution, the SGD will be the *ratio of the value of the concentration observed at a cumulative frequency of 84.13% to the value observed at 50% (i.e., the median or geometric mean).*

For the case above, it was shown that $F_Z(+1) = F_X(x_{84.13}) = 0.8413$. An analogous case occurs in which the argument of the standardized normally distributed random variable Z is exactly –1; from Chapter 7 (Figure 7.1, page 165), this case (one standard deviation below the origin) gives the CDF value of $F_Z(-1) = 0.1587$. Setting the argument of the CDF $F_Z(z)$ above equal to –1 at the point corresponding to $X = x_{0.1587}$ (denoting the 15.87% percentile value), we obtain the following result:

$$\frac{\ln\left(\frac{x_{15.87}}{\mu_g}\right)}{\sigma} = -1$$

Solving the equation for the ratio $\{\mu_g/x_{15.87}\}$, we obtain a slightly different result for the SGD:

$$\frac{\mu_g}{x_{15.87}} = e^{\sigma} = \sigma_g$$

Here, we see that the SGD is also *the ratio of the value of the concentration observed at 50% (i.e., the median or geometric mean) to the concentration ob-*

served at a cumulative frequency of 15.87%. Putting these two different equations for the SGD together, we obtain:

$$\text{SGD} = \sigma_g = \frac{\mu_g}{x_{15.87}} = \frac{x_{84.13}}{\mu_g}$$

The importance of the SGD σ_g is seen easily when the two-parameter lognormal distribution is plotted on logarithmic probability paper (Figure 9.1, page 264). Random variables that are lognormally distributed plot nearly as straight lines on logarithmic-probability paper. The slope of the resulting straight line is directly related to the value of the SGD.

If the geometric parameters μ_g and σ_g are substituted into the PDF of the lognormal distribution, the "geometric" form of PDF of the distribution can be written as

$$f_X(x) = \frac{1}{x\sigma\sqrt{2\pi}} e^{-\frac{1}{2}\left(\frac{\ln\frac{x}{\mu_g}}{\ln\sigma_g}\right)^2}$$

The three forms of the PDF — the standard form (Table 9.1), the mixed form, and the geometric form — are the same basic equation written with different combinations of parameters.

To summarize the parameters introduced thus far, the lognormal distribution can be defined by specifying any of the following pairs of parameters: the normal parameters (μ,σ), the geometric parameters (μ_g,σ_g), or various combinations of the normal and geometric parameters: (μ,σ_g) and (μ_g,σ). The geometric parameters can be written as a function of the normal parameters as follows:

$$\mu_g = e^{\mu}$$

$$\sigma_g = e^{\sigma}$$

The expected value of the lognormally distributed random variable X is obtained by integrating the product of x and the PDF $f_X(x)$ listed in Table 9.1:

$$E[X] = \int_0^{\infty} x f_X(x) dx = \int_0^{\infty} \frac{x}{x\sigma\sqrt{2\pi}} e^{-\frac{1}{2}\left(\frac{\ln x-\mu}{\sigma}\right)^2} dx$$

To perform this integration, we need to substitute expressions involving z and dz for those involving x and dx. We begin with the transformation $z = (\ln x - \mu)/\sigma$ and $dz = dx/x\sigma$. Solving this expression for x and dx gives $x = e^{(\sigma z + \mu)}$ and $dx = \sigma x dz = \sigma e^{(\sigma z + \mu)} dz$. Substituting this result for z and dx into the above expression and changing the limits of integration:

$$E[X] = \int_{-\infty}^{\infty} \frac{1}{\sigma\sqrt{2\pi}} e^{-\frac{1}{2}z^2} \sigma e^{(\sigma z+\mu)} dz = \int_{-\infty}^{\infty} \frac{1}{\sqrt{2\pi}} e^{-\frac{1}{2}[z^2-2\sigma z]+\mu} dz$$

It is possible to form the perfect square for the quantity in brackets in the exponent, $[z^2 - 2\sigma z]$, if we can convert it to $[z^2 - 2\sigma z + \sigma^2]$ by adding σ^2, thus giving $[z - \sigma]^2$. Because one-half appears in front of the term in brackets, the perfect square can be formed only by subtracting the term $\sigma^2/2$ and then adding it again, giving:

$$E[X] = \int_{-\infty}^{\infty} \frac{1}{\sqrt{2\pi}} e^{-\frac{1}{2}[z-\sigma]^2 + \mu + \frac{1}{2}\sigma^2} dz = \left[e^{\mu + \frac{1}{2}\sigma^2} \right] \int_{-\infty}^{\infty} e^{-\frac{1}{2}[z-\sigma]^2} dz$$

The integrand on the right side of the equation above is just the PDF of a normally distributed random variable with location parameter σ. Integrating over the limits from $-\infty$ to $+\infty$ gives 1. Thus, we have derived a general equation for the expected value of the lognormal distribution:

$$E[X] = e^{\mu + \frac{1}{2}\sigma^2}$$

If we seek to determine the jth moment about the origin for the lognormal distribution, an algebraic approach similar to that employed above yields the following result for any moment j:

$$E[X^j] = e^{j\mu + \frac{j^2\sigma^2}{2}}$$

Thus, the second moment about the origin is obtained by substituting $j = 2$ into the above expression, giving

$$E[X^2] = e^{2\mu + 2\sigma^2}$$

The variance of the distribution, or second moment about its mean, is computed as follows:

$$Var(X) = E[X^2] - (E[X])^2 = e^{2\mu + 2\sigma^2} - \left[e^{\mu + \frac{1}{2}\sigma^2} \right]^2 = e^{2\mu + \sigma^2} (e^{\sigma^2} - 1)$$

We have now derived equations for the expected value $E[X]$ and the variance $Var(X)$ of the lognormally distributed random variable X. To simplify the notation, we introduce the Greek letters $\alpha = E[X]$ and $\beta = \{Var(X)\}^{1/2}$ to represent the arithmetic mean α and arithmetic standard deviation β of the lognormally distributed random variable X, and we will call these the "arithmetic" parameters of the lognormal distribution:

$$\beta = \sqrt{e^{2\mu + \sigma^2} (e^{\sigma^2} - 1)}$$

$$\alpha = e^{\mu + \frac{1}{2}\sigma^2}$$

Together with the pair of arithmetic parameters α and β, the pair of geometric parameters μ_g and σ_g, and the pair of normal parameters μ and σ, there are six different commonly used parameters for describing a lognormal distribution. Thus, in addition to the geometric and normal parameters and their combinations, the lognormal distribution can be specified by the pair of arithmetic parameters, (α,β), and by various combinations of these arithmetic parameters: (α,σ), (α,σ_g), (μ,β), and (μ_g,β).

A seventh useful parameter is the coefficient of variation $CV(X) = \nu$ of the lognormal distribution:

$$\nu = \frac{\sqrt{Var(X)}}{E[X]} = \frac{\beta}{\alpha} = \sqrt{e^{\sigma^2} - 1}$$

The coefficient of variation ν of the two-parameter lognormal distribution is a function of only the standard deviation of the normal parameter σ and does not depend on any other parameter. The coefficient of variation ν really is an arithmetic parameter, because it is defined as the ratio of β to α.

Often it is necessary to convert from one set of parameters to another. For example, one may know the values of the pair of arithmetic parameters α and β but may want to find the corresponding values of the pair of normal parameters μ and σ. Likewise, the analyst may be given the value of the arithmetic mean α and the standard geometric deviation σ_g and may wish to find the corresponding values of the normal parameters σ and μ. To facilitate the conversion from one set of parameters to another, equations have been derived in this book for all combinations of the most frequently used parameters (Table 9.2).

In addition to the seven parameters discussed above, the mode x_m of the lognormal distribution is of interest, since the mode does not coincide with the median or the arithmetic mean. An equation for the mode x_m is obtained by differentiating the PDF of the lognormal distribution and setting its derivative equal to zero: $f_X'(x) = 0$, giving the following result (see Problem 2):

$$x_m = e^{(\mu - \sigma^2)}$$

Plotting the Lognormal Distribution

To plot the lognormal probability model on logarithmic-probability paper, one needs values of the quantiles at two points, since two points will determine a straight line. A typical pair of points is the median and one (normal) standard deviation above the median, or the value of X at a cumulative frequency of 84.13%. In some cases, it may be necessary to obtain points that are farther from the median to plot the straight line with greater validity. A convenient way to obtain the values of these points for plotting the line is to consider multiples of the standard deviation of the standardized normally distributed random variable Z (that is, $z = \pm 2, \pm 3$, etc.). The results for one, two, and three standard deviations are summarized in Table 9.3, where "Cum. Freq." denotes the cumulative frequency given by the CDF.

Suppose we wish to use the lognormal model to represent a data set of carbon monoxide (CO) concentrations with a geometric mean of $\mu_g = 3$ parts-per-million (ppm). Suppose we wish to plot the lognormal probability model for

Table 9.2. Relationship Among Parameters of the Lognormal Probability Model

GEOMETRIC (μ_g, σ_g)

- *From Normal:*

$$\mu_g = e^{\mu}$$

$$\sigma_g = e^{\sigma}$$

- *From Arithmetic:*

$$\mu_g = \frac{\alpha^2}{\sqrt{\alpha^2 + \beta^2}} = \frac{\alpha}{\sqrt{1+v^2}}$$

$$\sigma_g = e^{\sqrt{\ln(1 + \beta^2/\alpha^2)}} = e^{\sqrt{\ln(1+v^2)}}$$

NORMAL (μ, σ)

- *From Geometric:*

$$\mu = \ln \mu_g$$

$$\sigma = \ln \sigma_g$$

- *From Arithmetic:*

$$\mu = \ln\alpha - \frac{1}{2}\sigma^2 = \ln\left(\frac{\alpha^2}{\sqrt{\alpha^2+\beta^2}}\right) = \ln\left(\frac{\alpha^2}{\sqrt{1+v^2}}\right)$$

$$\sigma = \sqrt{\ln(1+\beta^2/\alpha^2)} = \sqrt{1+v^2}$$

ARITHMETIC (α, β)

- *From Normal:*

$$\alpha = e^{\left(\mu + \frac{1}{2}\sigma^2\right)}$$

$$\beta = \sqrt{e^{[2\mu+\sigma^2]}(e^{\sigma^2} - 1)} = \alpha\sqrt{e^{\sigma^2} - 1}$$

- *From Geometric:*

$$\alpha = e^{[\ln \mu_g + \frac{1}{2}(\ln \sigma_g)^2]} = \mu_g e^{\frac{1}{2}(\ln \sigma_g)^2} = \mu_g e^{\frac{1}{2}\sigma^2}$$

$$\beta = \mu_g e^{\frac{1}{2}\sigma^2}\sqrt{e^{\sigma^2} - 1} = \mu_g\sqrt{e^{2(\ln \sigma_g)^2} - e^{(\ln \sigma_g)^2}}$$

μ_g = Geometric Mean (geometric; equivalent to the median x_{50})
σ_g = Standard Geometric Deviation (geometric)
μ = Mean of Logarithms (normal)
σ = Standard Deviation of Logarithms (normal)
α = Mean of Distribution (arithmetic)
β = Standard Deviation of Distribution (arithmetic)
v = Coefficient of Variation, β/α (arithmetic)

Table 9.3. Cumulative Frequencies at Various Plotting Positions for the Two-Parameter Lognormal Probability Model

z	−3	−2	−1	0	+1	+2	+3
x	x_{50}/σ_g^3	x_{50}/σ_g^2	x_{50}/σ_g	x_{50}	$x_{50}\sigma_g$	$x_{50}\sigma_g^2$	$x_{50}\sigma_g^3$
Cum. Freq.	0.135%	2.275%	15.87%	50%	84.13%	97.725%	99.865%

two different standard geometric deviations, $\sigma_g = 1.5$, and $\sigma_g = 2$ (Figure 9.1). We first locate the intercept for 3 ppm at the median value of 50%, since the median of the model is the same as its geometric mean. Next, we multiply the geometric mean by the SGD to obtain the first intercept at the cumulative frequency of 84.13%. That is, for the case of $\sigma_g = 1.5$, we obtain $x_{84.13} = (1.5)(3$ ppm$) = (4.5$ ppm$)$, and we plot this point on the graph. Drawing a straight line with a ruler through the two points 3 ppm and 4.5 ppm is sufficient to plot the model. When one actually tries to draw the straight line on logarithmic-probability paper with a ruler and a pen, often it is helpful to calculate additional points to position the line more precisely. Another point above 4.5 ppm at the cumulative frequency of 97.725% (Table 9.3) is obtained by multiplying the 4.5-ppm intercept by σ_g once again, or $x_{97.725} = \sigma_g x_{97.725} = \sigma_g^2 \mu_g = (1.5)(4.5$ ppm$) = 6.75$ ppm. Additional points along the line in Figure 9.1 are obtained by multiplying or dividing μ_g by σ_g according to Table 9.3.

The result of plotting the lognormal probability model for a *constant geometric mean* but for different standard geometric deviations will be a series of straight lines that all pass through the geometric mean but have different slopes. Thus, both lines plotted in Figure 9.1 pass through the median value at $\mu_g = 3$ ppm, but the line for $\sigma_g = 2$ is steeper than the line for $\sigma_g = 1.5$. If more cases were plotted for different standard geometric deviations but with the same geometric mean, the lines would appear to "rotate" counter-clockwise around the common median of 3 ppm as the value of σ_g increases. The lowest possible value of the standard geometric deviation is $\sigma_g = 1$, which would give a straight line that is horizontal and a distribution with zero variance.

Suppose, instead of a constant geometric mean, we wish to plot the lognormal distribution for a *constant arithmetic mean* of $\alpha = 3.0$ ppm, but for four different cases of the arithmetic standard deviation: $\beta = 0.5$ (Case 1), 1.0 (Case 2), 1.5 (Case 3), and 2.0 (Case 4) ppm (Table 9.4). For Case 1, $\nu = \alpha/\beta = 0.5/3 = 0.16667$.

Referring to the equations in Table 9.2 to calculate the geometric parameters, we obtain

$$\mu_g = x_{50} = \frac{\alpha}{\sqrt{1+\nu^2}} = \frac{3}{\sqrt{1+(0.16667)^2}} = 2.959$$

$$\sigma_g = e^{\sqrt{\ln(1+\nu^2)}} = e^{\sqrt{\ln(1+[0.16667]^2)}} = 1.1800$$

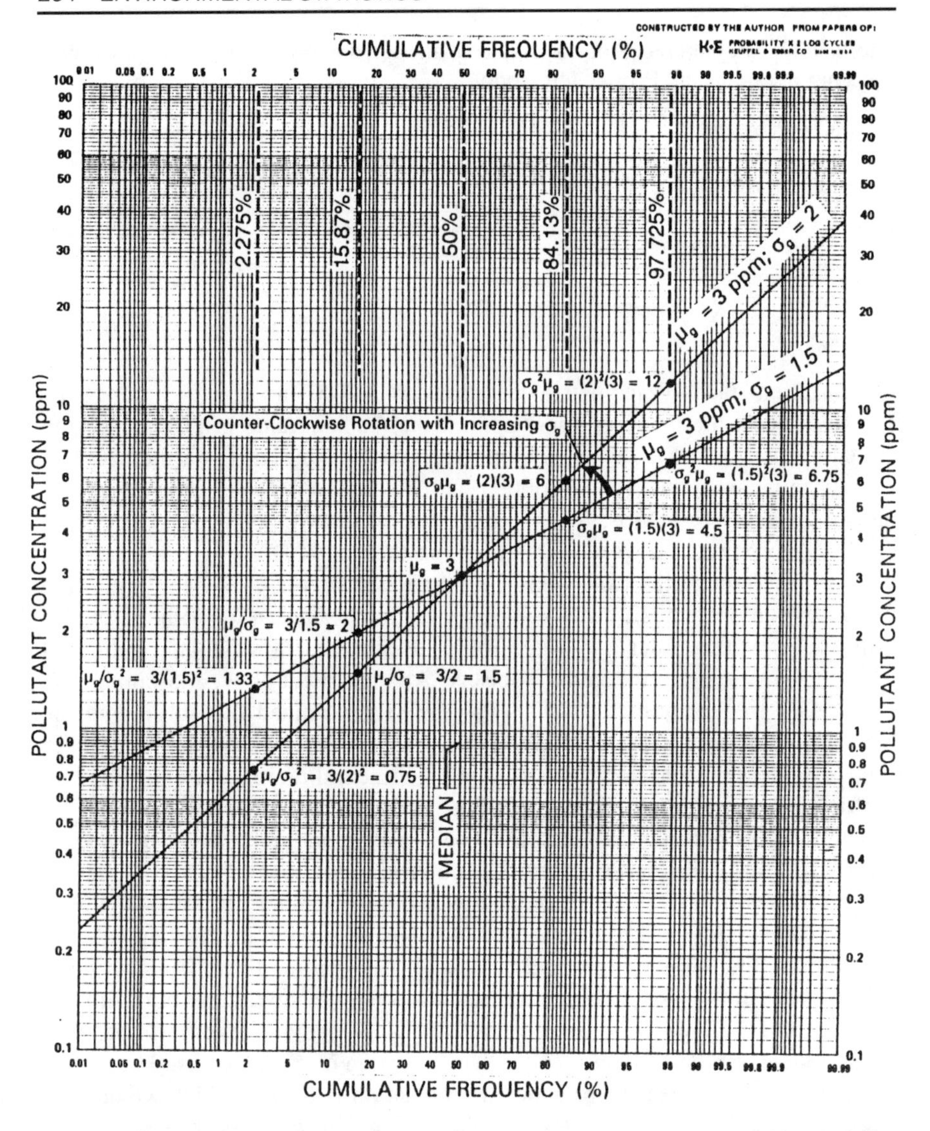

Figure 9.1. Plot on logarithmic-probability paper of the lognormal probability model for two different standard geometric deviations, $\sigma_g = 1.5$ and $\sigma_g = 2$, and the same geometric mean, $\mu_g = 3$ ppm.

The normal parameters μ and σ also are computed using the equations in Table 9.2:

$$\mu = \ln \mu_g = \ln(2.959) = 1.0849$$

$$\sigma = \ln(1.1800) = 0.1655$$

Finally, calculating the mode from the equation on page 261 at the end of the previous section, we obtain:

Table 9.4. Example of Parameter Values for Four Cases of the Lognormal Probability Model

Arith. Mean α (ppm)	Standard Deviation β (ppm)	Coeff. of Variation ν	σ	μ	Mode x_m (ppm)	Median x_{50}, μ_g (ppm)	SGD σ_g
3.0	0.5	0.16667	0.1655	1.0849	2.8792	2.959	1.1800
3.0	1.0	0.33333	0.3246	1.0459	2.5614	2.846	1.3835
3.0	1.5	0.50000	0.4724	0.9870	2.1466	2.683	1.6038
3.0	2.0	0.66667	0.6064	0.9148	1.7281	2.496	1.8338

$$x_m = e^{[\mu - \sigma^2]} = e^{[1.0849 - (0.1655)^2]} = 2.8792$$

Notice that the arithmetic mean (α = 3.000) is greater than the median (x_{50} = 2.959), which is greater than the mode (x_m = 2.8792). That is, $\alpha > x_{50} > x_m$, a general property of the lognormal distribution.

The PDF for Case 1 is plotted by substituting μ = 1.0849 and σ = 0.1655 into the equation for the PDF given in Table 9.1 (page 256) and then graphing the result on ordinary linear graph paper (Figure 9.2):

$$f_X(x) = \frac{1}{x(0.1655)\sqrt{2\pi}} e^{-\frac{1}{2}\left(\frac{\ln x - 1.0849}{\sigma}\right)^2}$$

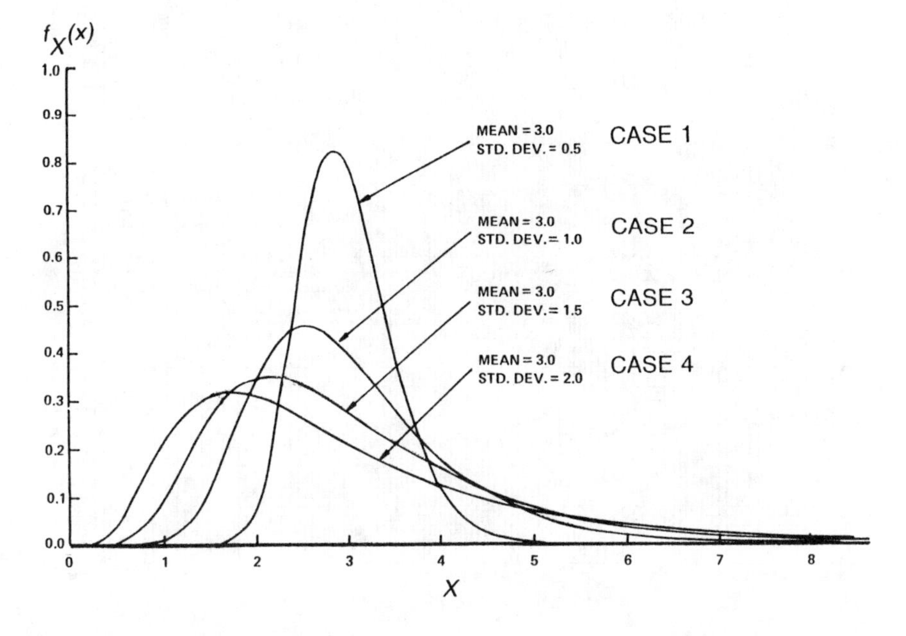

Figure 9.2. PDF of the lognormal probability model showing differences in shape for a constant arithmetic mean (α = 3) and four different arithmetic standard deviations ranging from β = 0.5 to β = 2.0.

Surprisingly, the resulting PDF for Case 1 resembles the normal distribution, although the slight difference between the mean, median, and mode shows that some right-skewness is evident. For values of ν less than about one-sixth (0.16667), the PDF of the lognormal probability model looks symmetrical and resembles the normal distribution. Cases 2, 3, and 4 show that the right skewness of the lognormal distribution increases as ν becomes larger, revealing an increasingly long tail to the right as the coefficient of variation goes from ν = 1/3 (Case 2) to ν = 1/2 (Case 3) to ν = 2/3 (Case 4).

When the CDF's for these four cases are plotted on logarithmic-probability

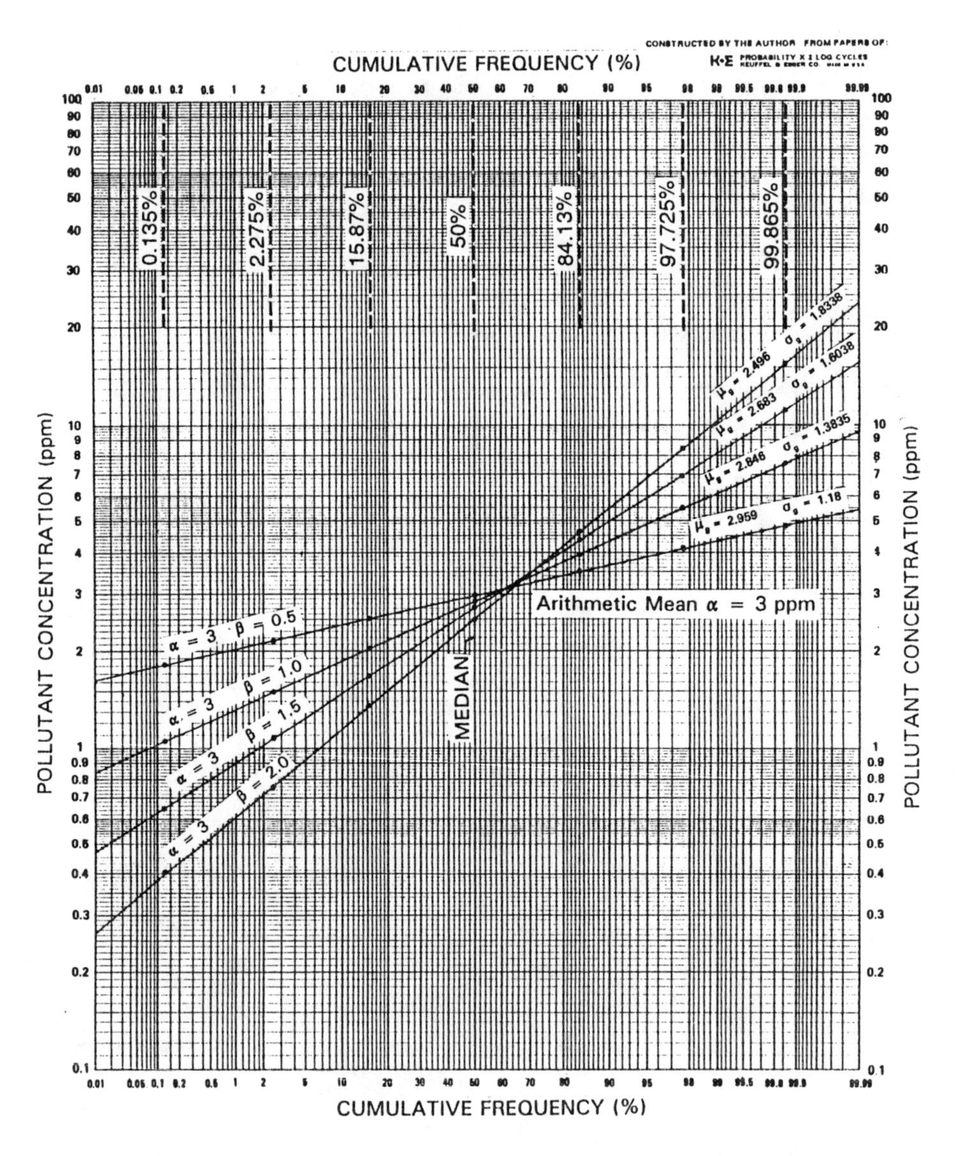

Figure 9.3. Example of four cases of the lognormal probability model plotted on logarithmic-probability paper for a constant arithmetic mean of α = 3 ppm and different standard deviations.

paper (Figure 9.3), we obtain straight lines similar to those in Figure 9.1. Because these models all have the same arithmetic mean but different medians, however, they no longer pass through the same point at the median. Pairs of lines intersect each other in the range of 60–70%, but the lines do not have a common intersection point. Although an observer can easily read the median value of each case plotted on the graph, it is difficult to determine visually where the arithmetic mean lies.

ESTIMATING PARAMETERS OF THE LOGNORMAL DISTRIBUTION FROM DATA

To "fit" the lognormal probability model to a set of environmental measurements, one must determine the values of the two parameters of this model.

Visual Estimation

Using the graphical techniques described above, one can obtain the values of the lognormal parameters visually from a straight line plotted on logarithmic-probability paper. It is necessary only to read the median value x_{50} and the intercept $x_{84.13\%}$ from the straight line, and then to use the above formulas to calculate the parameters of interest. The first step in fitting a distribution is to plot the data on logarithmic probability paper using the techniques described earlier in this book (Chapter 3: "Analysis of Observed Data," page 46) to visually determine if the cumulative frequencies lie approximately on a straight line. Obviously, the degree of "straightness" of the line is a matter of judgment. Consistent curvature of the data points may suggest that the two-parameter lognormal probability model may not be appropriate, and other models, such as the three-parameter lognormal probability model should be tried. Figure 9.9 (last section on "Field Study Example," pages 290–293) gives an example of visual estimation of the straight line drawn through the data points for the observations of air pollutant exposures measured in an automobile traveling on a highway.

Method of Moments

A straightforward approach for determining the parameters of a lognormal distribution from the data is by setting the moments calculated from the data equal to the moments of the model, or the "method of moments." To apply this approach, the analyst computes the arithmetic mean and arithmetic standard deviation from the set of observations $x_1, x_2, \ldots, x_n$:

$$\bar{x} = \sum_{i=1}^{n} x_i$$

$$s_2 = \frac{1}{n} \sum_{i=1}^{n} (x_i - \bar{x})^2$$

Usually, the biased estimate of the sample standard deviation is used instead of an unbiased estimate, although the two estimates are extremely close for large sample sizes (see Chapter 3: "Analysis of Observed Data: Computing Statistics from Data," pages 46–53; see also Table 3.4). To apply the method of moments, the arithmetic mean of the lognormal model is set equal to the arithmetic mean of the data:

$$\alpha = e^{\mu+\frac{\sigma^2}{2}} = \bar{x}$$

$$\beta = \sqrt{e^{2\mu+\sigma^2}(e^{\sigma^2}-1)} = s$$

Then the formulas given in Table 9.2 can be used to calculate μ, σ, or any of the other variables of interest.

For example, suppose $\bar{x}$ = 141 μg/m^3, while s = 80 μg/m^3. To apply the method of moments, we set $\alpha = 141$ and $\beta = 80$, and, using the equations in Table 9.2 for calculating the normal parameters (μ,σ) from the arithmetic parameters (α,β), we obtain

$$\mu = \ln\left(\frac{\alpha^2}{\sqrt{\alpha^2+\beta^2}}\right) = \ln\left(\frac{141^2}{\sqrt{141^2+80^2}}\right) = 4.809$$

$$\sigma = \sqrt{\ln\left(1+\frac{\beta^2}{\alpha^2}\right)} = \sqrt{\ln\left(1+\frac{80^2}{141^2}\right)} = 0.5283$$

Then, using the equations in Table 9.2 for μ_g and σ_g, we obtain

$$\mu_g = e^{\mu} = 122.6$$

$$\sigma_g = e^{\sigma} = 1.696$$

Method of Quantiles

An alternative to visual estimation from logarithmic-probability paper or the method of moments is to use the values corresponding to certain cumulative frequencies of the observations. From the frequency distribution of the observations, suppose that the value x_a corresponds to a cumulative frequency f_a, and the value x_b corresponds to a cumulative frequency of f_b. The value of the variate at a specified cumulative frequency is called a "fractile" or, more commonly, a "quantile." For example, if $f_b = 0.75$, then at least 75% of the observations are *below* the quantile x_b, and 25% of the observations are *above* the quantile x_b. The values of the two parameters of the lognormal model can be obtained from any two quantiles of a lognormal frequency distribution, although the quantiles selected should be as "representative" as possible of the range of interest of the random variable.

Fitting the lognormal distribution to two quantiles is called the "method of quantiles," and we now derive an expression for this approach. For the method of quantiles, we want to match the CDF values of the model, $F_X(x_a)$ and $F_X(x_b)$, to the two quantiles:

$$F_X(x_a) = f_a$$

$$F_X(x_b) = f_b$$

$$\text{where} \quad f_b > f_a$$

Using the CDF of the lognormal probability model appearing on page 256, these two equations are written as

$$F_Z\left(\frac{\ln x_a - \mu}{\sigma}\right) = f_a$$

$$F_Z\left(\frac{\ln x_b - \mu}{\sigma}\right) = f_b$$

We can solve each of these expressions for the quantity in parentheses by using $F_Z^{-1}(f_a)$ to denote the inverse CDF of the standardized normal distribution for the frequency f_a and $F_Z^{-1}(f_b)$ to denote the inverse CDF of the standardized normal distribution for the frequency f_b:

$$\ln x_a - \mu = \sigma F_Z^{-1}(f_a)$$

$$\ln x_b - \mu = \sigma F_Z^{-1}(f_b)$$

Subtracting these two quantile equations, we obtain

$$\ln x_a - \ln x_b = \sigma\left[F_X^{-1}(f_b) - F_Z^{-1}(f_a)\right]$$

Solving this expression for σ, we obtain a *general expression for σ using the method of quantiles*:

$$\sigma = \frac{\ln x_b - \ln x_a}{F_Z^{-1}(f_b) - F_Z^{-1}(f_a)} = \frac{\ln\left(\frac{x_b}{x_a}\right)}{F_Z^{-1}(f_b) - F_Z^{-1}(f_a)}$$

If we consider the pair of quantile equations before they were subtracted, we can solve each of them for μ to obtain a pair of *general expressions for μ using the method of quantiles:*

$$\mu = \ln x_a - \sigma F_Z^{-1}(f_a)$$

$$\mu = \ln x_b - \sigma F_Z^{-1}(f_b)$$

Either of these equations for μ and the one above for σ provide useful expressions for calculating the two parameters μ and σ for the lognormal probability model by the method of quantiles.

To apply the method of quantiles, the analyst chooses two frequencies f_a and f_b and obtains the corresponding quantiles x_a and x_b from the observed data using the plotting techniques described in Chapter 3 (see "Histograms and Frequency Plots," pages 53–59). Then the inverse CDF values $F_Z^{-1}(f_a)$ and $F_Z^{-1}(f_b)$ are obtained from the standardized normal distribution using the techniques described in Chapter 7 (pages 186–188). Finally, the results are substituted into the general expression above to calculate σ. Then σ and x_a [or x_b] are substituted into the general expression above to calculate μ.

Larsen[4–7] has recommended use of the two frequencies $f_a = 0.70$ and $f_b = 0.999$ for analyzing air quality data. Aitchison and Brown[2] show that the maximum efficiency for estimating μ occurs at $f_a = 0.27$ and $f_b = 0.73$, while the maximum efficiency for estimating σ occurs at $f_a = 0.07$ and $f_b = 0.93$ when symmetrical quantiles are chosen. Efficiency refers to the statistical properties of the estimators: in practice, an estimator with maximum efficiency, when calculated from lognormally distributed data, will have the smallest variability possible.

Maximum Likelihood Estimation (MLE)

If one is sampling from a true lognormal distribution, then it can be readily shown that the "minimum variance" estimate of the parameters is obtained by Maximum Likelihood Estimation (MLE), which was discussed briefly in Chapter 3 in connection with the "Tail Exponential Method" (see pages 75–76 on "Fitting Probability Models to Environmental Data"). By a minimum variance estimate, we mean that the difference between the true values of the parameters and those computed from the data will have the lowest variance that is theoretically possible. Thus, if many data sets were obtained from this same lognormal process and the parameters estimated by MLE, these estimates would tend to lie closer to the true values than any other estimation procedure. An example of the MLE approach with the lognormal distribution applied to environmental problems is published in considerable detail elsewhere.[8] For the lognormal distribution, the MLE procedure reduces to a fairly simple computational approach, and its derivation follows.

Assume that a data set of n observations is obtained from a true lognormal distribution whose two parameters are unknown. The MLE approach seeks to answer the question, "What are the *most probable* values of the two parameters, given this particular set of observations?" In other words, "What are the optimal values of μ and σ that maximize the probability $P\{A\}$ of the event $A = \{x_1, x_2, \ldots, x_n\}$, in which $x_1, x_2, \ldots, x_n$ are the values of a particular set of n observations?"

To derive the MLE estimators, we treat each observation as a probabilistic outcome obtained from the unknown lognormal PDF $f_X(x)$. Therefore, each observation x_i can be regarded as an "event" occurring with likelihood $f_X(x_i)$, and we introduce the "likelihood function," denoted by L, as the product of the n terms, one for each observation:

$$L = \prod_{i=1}^{n} f_X(x_i) = \prod_{i=1}^{n} \frac{1}{x_i \sigma \sqrt{2\pi}} e^{-\frac{1}{2}\left(\frac{\ln x_i - \mu}{\sigma}\right)}$$

$$= \left(\frac{1}{\sigma\sqrt{2\pi}}\right)^n \left[\prod_{i=1}^{n} \frac{1}{x_i}\right] e^{-\frac{1}{2\sigma^2}\sum_{i=1}^{n}(\ln x_i - \mu)^2}$$

The product L usually is a very small number. For computational purposes, therefore, it generally is more convenient to work with the logarithm of L. Thus, we form the "log-likelihood" function L^* by taking the logarithm of L:

$$L^* = n \ln\left(\frac{1}{\sigma\sqrt{2\pi}}\right) - \sum_{i=1}^{n} \ln x_i - \frac{1}{2\sigma^2}\sum_{i=1}^{n}(\ln x_i - \mu)^2$$

Maximizing the log-likelihood function L^* is equivalent to maximizing the likelihood function L. To find the values of μ and σ that maximize L^*, we take first partial derivatives of L^* with respect to μ and σ, and we then set the resulting functions equal to zero. Differentiating the above equation for L^* with respect to μ and setting the result equal to zero

$$\frac{\partial L^*}{\partial \mu} = -\frac{1}{2\sigma^2}\sum_{i=1}^{n} 2(\ln x_i - \mu)(-1) = 0$$

Solving this expression for μ, we obtain

$$\mu = \frac{1}{n}\sum_{i=1}^{n} \ln x_i$$

Similarly, differentiating L^* with respect to σ,

$$\frac{\partial L^*}{\partial \sigma} = \frac{-n}{\sigma} + \frac{1}{\sigma^3}\sum_{i=1}^{n}(\ln x_i - \mu)^2 = 0$$

Solving this expression for σ^2, we obtain

$$\sigma^2 = \frac{1}{n}\sum_{i=1}^{n}(\ln x_i - \mu)^2$$

These results show that the MLE estimators for the lognormal distribution are obtained by first taking logarithms of each observation. Then the mean of the logarithms is calculated to give the value for μ, and the (biased) estimate of the variance of the logarithms is calculated to give σ^2. The resulting values of μ and σ can be used in the equations in Table 9.2 to calculate the other parameter values of interest.

Because the MLE approach requires the logarithm of each observation, difficulties arise if any observation is zero. Often, a pollutant concentration reported as "zero" is not really zero but is simply below the minimum detectable limit (MDL) of the measuring method. Also, pollutants sometimes are rounded to zero. For example, an ambient air carbon monoxide concentration of "0 ppm" might be reported when the true value was 0.224 ppm. The stored value of zero either has been rounded down or the true value was too low to be detected by the measuring method. Sometimes observations below the MDL are stored in a special category as nondetectable. Nehls and Akland[9] propose that the midpoint between zero and the MDL of the measurement method be substituted for all observations at or below the MDL.

One approach for dealing with zero values is to sort the data into intervals, with observations of zero included in the first interval. For example, the first interval for x might be $0 \le x \le 0.5$ ppm, the second interval $0.5 < x \le 1.5$ ppm, the third interval $1.5 < x \le 2.5$ ppm, and so on. However, the MLE approach developed here is not appropriate for grouped data. An approximation to the MLE approach has been applied to a large number of grouped carbon monoxide data sets from throughout the U.S.[10] That study showed that the MLE approach usually gave the best fit of the lognormal model to the data.

THREE-PARAMETER LOGNORMAL MODEL

Mage[11] first noted that environmental quality data sets, such as those analyzed by Larsen[7], show consistent downward curvature when plotted on logarithmic-probability paper. In the basic logarithmic transformation described above, suppose a constant a is subtracted prior to taking the logarithm. Then, the following transform z will be a standardized normally distributed random variable:

$$z = \frac{\ln(x-a)-\mu}{\sigma} \qquad \text{where } x \ge a;\ -\infty < a < +\infty;\ -\infty < \mu < +\infty;\ \sigma > 0$$

Although σ is greater than zero, a is a location parameter that can be either positive or negative, and x must be equal to or greater than a. The PDF of the three-parameter lognormal distribution (LN3) is written as

$$f_X(x) = \frac{1}{\sigma(x-a)\sqrt{2\pi}} e^{-\frac{1}{2}\left(\frac{\ln(x-a)\ -\ \mu}{\sigma}\right)^2}$$

Here x represents a real physical variable, pollutant concentration, that cannot be negative. Thus, if a is negative, there will be an undefined region between a and the origin, and the PDF will be "cut off" or "censored" in this region. To make up for the missing area of the PDF, a single probability mass function is plotted at the origin, and the result is called the censored, three-parameter lognormal distribution (LN3C).[12,13]

Once the parameters of the LN3 or LN3C are known, the CDF is obtained from the standardized normally distributed random variable as follows:

$$F_X(x) = F_Z\left(\frac{\ln(x-a)-\mu}{\sigma}\right)$$

As with the two-parameter lognormal distribution, the analyst calculates the CDF of the three-parameter lognormal distribution by calculating the value of the argument above and then referring to tables of the standardized normal distribution, or to hand calculators or computers that compute the normal CDF.

Figure 9.4 shows an example of the LN3C distribution for the case of $a = -1.5$, $\mu = 0.987$, and $\sigma = 0.472$. Because x has a LN3C distribution, the quantity $x' = x - a = x - (-1.5) = (x + 1.5)$ has a two-parameter lognormal distribution, censored at 1.5. Because the two-parameter lognormal distribution is defined over the range $x' \geq 0$ (Table 9.1, page 256), a random variable with a LN3C distribution is defined over the range $(x + 1.5) \geq 0$, or $x \geq -1.5$. For this example, we plot LN3C PDF as a two-parameter lognormal distribution that has been shifted to the left of the origin by 1.5 units. However, since x represents pollutant concentration that is zero or greater, we cannot allow x to be negative, and we must "cut off," or censor, the portion of the PDF that lies to the left side of the origin. To make up for the lost probability in Figure 9.4, we plot a probability mass (0.109 in this example) at the origin.

A graphical approach using logarithmic-probability paper has been used for determining the value of the third parameter a of the LN3 (and LN3C) for any data set.[12,13] The analyst first plots the frequency distribution of the original data on logarithmic-probability paper. If the resulting distribution shows consistent curvature (Figure 9.5), then the three-parameter lognormal probability model may be a good choice. To find the value of the negative third parameter a if the curve is downward-concave, the analyst adds $-a$ to each data point, trying to find the value of a that removes the curvature and makes the resulting

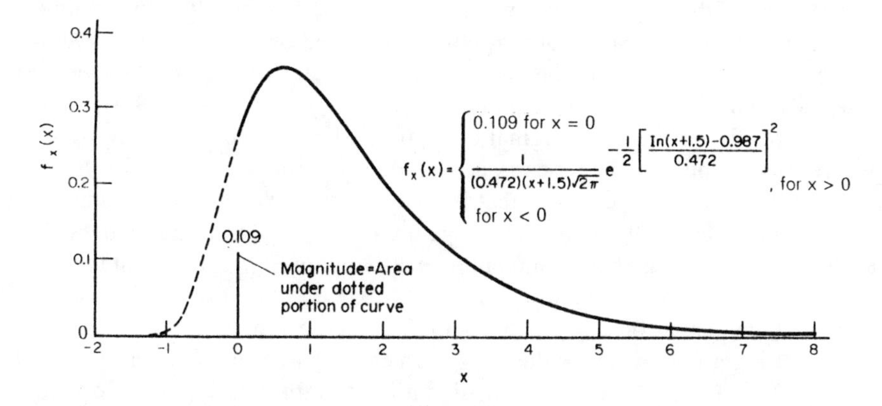

Figure 9.4. PDF of the censored 3-parameter lognormal probability model (LN3C) for $a = -1.5$, $\mu = 0.987$, and $\sigma = 0.472$.

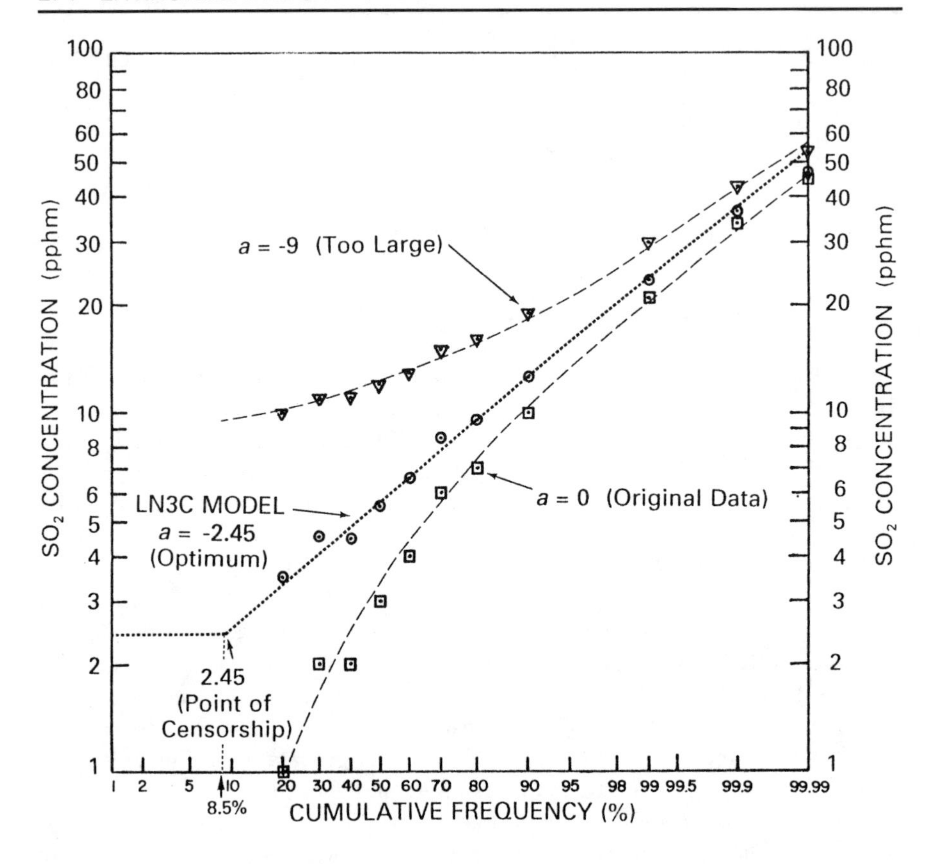

Figure 9.5. Logarithmic-probability plot of atmospheric sulfur dioxide concentrations observed in Washington, DC, showing original data (bottom), data transformed as $x' = x + 9$ (top; too large), and data transformed as $x' = x + 2.45$ (satisfactory) (Source: Reference 12).

plot a straight line. By plotting the distribution for different values of *a*, a good estimate of *a* can be obtained by trial and error.

To illustrate the approach, we examine data on 43,000 hourly average atmospheric sulfur dioxide (SO_2) concentrations measured over a number of years in Washington, DC, from a report by Larsen.[7] The data first are grouped into 11 intervals and plotted on logarithmic-probability paper (Figure 9.5). A line drawn visually through the original data points shows concave-downward curvature. If we arbitrarily add 9 parts-per-hundred-million (pphm; $a = -9$) to each observation, a new curve results that is concave-upward (top curve), indicating that the increment of 9 pphm is too large a correction to give straightness. The desired value for *a* lies between 0 pphm and –9 pphm. By repeating the process several times with different values of *a*, we examine the results and choose a "satisfactory" value of $a = -2.45$ pphm. The curve replotted with this value gives a straight line, which is the two-parameter logarithmic probability plot of $x' = x + 2.45$. Using the visual estimation approach, the median of the distribution can be read from the straight line as $x_{50}' = 5.5$ pphm, and the standard geometric deviation is obtained from the line as $\sigma_g = 1.80$. From these values, we calculate $\mu = \ln(5.5) = 1.7$ and $\sigma = \ln(1.80) = 0.588$. At the value of $x = 0$ pphm, the minimum physically realizable value of SO_2 concentration, $x' = 2.45$, and the CDF is given by

$$F_Z(z) = F_Z\left(\frac{\ln(0-a) - \mu}{\sigma}\right) = F_Z\left(\frac{\ln(2.45)-(1.7)}{0.588}\right) = F_Z(-1.37) = 0.085$$

This result shows that the censorship point corresponds to a frequency of 8.5%, indicating that the censored, three-parameter lognormal distribution predicts that 8.7% of the SO_2 concentrations are 0 pphm (see Figure 9.5).

The LN3C probability model also has been applied to water quality data.[12] The frequency distribution of total coliform (counts per hundred ml of water) measured in the Ohio River at Cincinnati shows slight concave-downward curvature when plotted on logarithmic-probability paper (Figure 9.6). When the

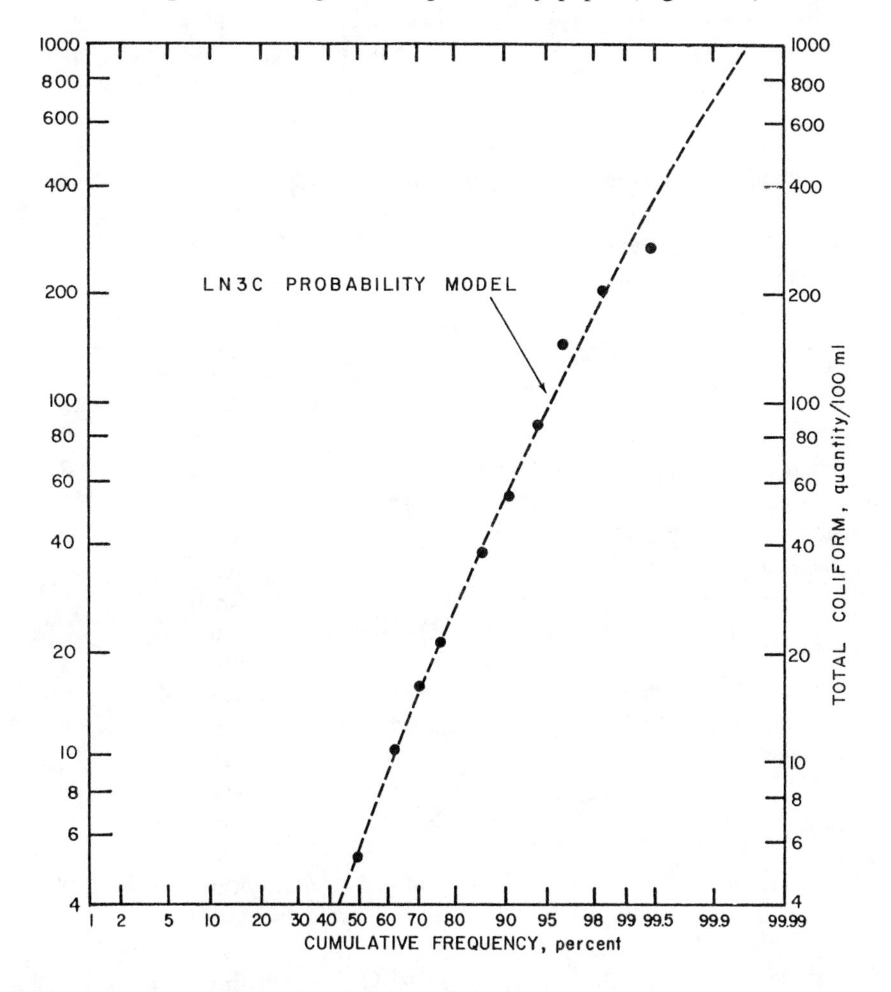

Figure 9.6. Logarithmic-probability plot of total coliform in the Ohio River at Cincinnati, showing both the raw data and the slight curvature of the LN3C probability model with $a = -2.8/100$ ml (Source: Reference 12).

LN3C probability model is applied to these data, the third parameter is a = –2.8/100 ml.

If an environmental concentration distribution shows consistent curvature, it may be difficult to fit a straight line through the data, and the three-parameter lognormal distribution often solves this problem. As we shall see in the next section, it makes little difference whether the distributions are straight or curved when analyzing the effect of source controls on pollutant concentrations.

STATISTICAL THEORY OF ROLLBACK (STR)

This section develops a theory for making predictions about the distribution of the concentrations of a pollutant after its source has been controlled by multiplying the source strength by a fixed proportion. To develop this theory, we first consider the general properties of two independent random variables multiplied together, examining how a multiplicative change in one random variable affects the overall product. Next we show that most source-receptor problems in the environment can be expressed as a product of two random variables, one representing the source and the other representing the dilution-diffusion processes. Finally, we show how linearly "rolling back" one of these random variables, the source term, affects the resulting statistics and predicted frequency distribution of the concentrations in the environment.

Suppose that the independent random variables Q and D are multiplied together yielding the product X:

$$X = QD$$

(Here X is general and represents any continuous random variable, but later in the chapter X represents pollutant concentration.) Because the random variables are independent, their expected value $E[X]$ is given by the product of the expected value of the two random variables using Rule 3 of the general rules for expected values (Chapter 3: "Moments, Expected Value, and Central Tendency," page 38):

$$E[X] = E[Q]E[D]$$

Similarly, the variance $Var(X)$ is given by the product of the variances of the two random variables using Rule 3 of the general rules for variances (Chapter 3: "Variance, Kurtosis, and Skewness," page 41):

$$Var(X) = Var(Q)Var(D)$$

The coefficient of variation $CV(X)$ is obtained by dividing the square root of the variance (standard deviation of X) by the expected value of X:

$$CV(X) = \frac{\sqrt{Var(X)}}{E[X]}$$

Substituting the equations for $E[X]$ and $Var(X)$ above into this expression for

$CV(X)$, we see that the coefficient of variation of X can be written as the product of the two coefficients of variation:

$$CV(X) = \frac{\sqrt{Var(Q)Var(D)}}{E[Q]E[D]} = \frac{\sqrt{Var(Q)}}{E[Q]}\frac{\sqrt{Var(D)}}{E[D]} = CV(Q)CV(D)$$

Suppose that the random variable Q undergoes some change, but its coefficient of variation $CV(Q)$ does not change. If the coefficient of variation $CV(D)$ also does not change, then the coefficient of variation $CV(X)$ will be unchanged. This result can be stated as a theorem:

Theorem 1. If $X = QD$, and Q and D are independent random variables that are modified in such a manner that $CV(Q)$ does not change and $CV(D)$ does not change, then $CV(X)$ also does not change.

Predicting Concentrations After Source Control

An important step in environmental pollution control is to estimate the effect on pollutant concentrations of controlling a particular source or group of sources. Often we are able to measure the pollutant concentrations in some detail *before* a source is controlled ("pre-control" state), and we would like to predict how the observed concentrations will change *after* the source has been controlled ("post-control" state). Ideally, we would like to be able to predict the entire distribution of concentrations in the post-control state from our knowledge of how much control is to be applied and our observations of concentrations in the pre-control state.

As we shall see later in this chapter, the transport of pollutants in the environment generally can be modeled as a product of a "source factor" Q and a "dilution-diffusion factor" D. Variability of the source Q usually results from mechanical factors; for example, emissions from a smokestack vary with the fuel combustion rate within a factory. By contrast, variability of the dilution-diffusion term D results from processes in nature (e.g., meteorological conditions, stream flow rates); thus, the source term and the dilution-diffusion term usually can be treated as independent. Although development of the theory assumes initially that the two random variables are independent, independence is not a necessary assumption, and we shall see that the theory applies even if the two random variables are correlated.

Suppose we multiply the random variable Q by a dimensionless "rollback factor" r to represent the proportion of the emissions remaining *after control is applied.* (This use of r should not be confused with the correlation coefficient, sometimes also denoted as r.) Here, r applies to a particular source, and we shall examine the change in the concentrations contributed by that source when that source alone is changed. We denote the new random variable as Q_{post} and the prior random variable as Q_{pre}:

$$Q_{post} = rQ_{pre}$$

Here we assume that all values of the random variable Q are scaled linearly by multiplying them by r. Thus, if the source emissions are reduced by 70%, then $r = 0.30$ reflects the proportion of the emissions remaining after control, or 30%.

By Rule 2 of the general rules for expected values (Chapter 3: "Moments, Expected Values, and Central Tendency," page 39), the new expected value $E[Q]_{post}$ will be the product of r and $E[Q]_{pre}$:

$$E[Q]_{post} = E[rQ_{pre}] = rE[Q]_{pre}$$

By Rule 2 of the general rules for variances (Chapter 3: "Variances, Kurtosis, and Skewness," page 42), the new variance $Var(Q)_{pre}$ will be the product of r^2 and $Var(Q)_{post}$:

$$Var(Q)_{post} = Var(rQ_{pre}) = r^2 Var(Q)_{pre}$$

If we now form the coefficient of variation of Q in the post-control state, we see that

$$CV(Q)_{post} = \frac{\sqrt{r^2 Var(Q)_{pre}}}{rE[Q]_{pre}} = \frac{\sqrt{Var(Q)_{pre}}}{E[Q]_{pre}} = CV(Q)_{pre}$$

Here, the rollback factor r appears in both the numerator and the denominator, so r divides out, and the coefficient of variation does not change; that is, $CV(Q)_{post} = CV(Q)_{pre}$. If the coefficient of variation due to natural geophysical processes (for example, long-term weather conditions) also does not change, then $CV(D)_{pre} = CV(D)_{post}$, and Theorem 1 applies. Consequently, the coefficient of variation of $CV(X)$ for the pre- and post-control states does not change:

$$CV(X)_{post} = CV(X)_{pre}$$

Thus, even though the mean value of X is multiplied by r, the coefficient of variation of X is unchanged. In summary, if one of the random variables forming a product is multiplied by a scaling factor r, then the coefficient of variation of the product does not change.

As we shall see later in this chapter (page 284 on "Environmental Transport Models in Air and Water"), equations for calculating the concentrations contributed by a particular source in the environment usually can be represented as the product of a source factor and dilution-diffusion factor. The dilution-diffusion factor results from conditions in nature and is assumed to be independent of the source factor. We also assume that no chemical reactions take place that would make these processes nonlinear. Finally, we are interested only in the share of the concentrations contributed by that particular source.

If we compare the pre-control and post-control states, and if the conditions of nature (represented by the dilution-diffusion factor D) are essentially the same in both states, the following theorems can be stated:

Theorem 2. If the random variable representing the source is multiplied by the rollback factor r, then the expected value of the concentrations contributed by that source in the post-control state will be the expected value of the concentrations in the pre-control state multiplied by r.

Theorem 3. If the random variable representing the source is multiplied by the rollback factor r, then the coefficient of variation in the post-control state will be the same as the coefficient of variation in the pre-control state.

From Theorem 3, we can state the following corollary for the variance:

Corollary 1. If the random variable representing the source is multiplied by the rollback factor r, then the variance in the post-control state will be multiplied by r^2, and the post-control standard deviation will be multiplied by r.

For an air pollutant, Theorems 2 and 3 imply that if all values of the source Q are multiplied by r, then all air pollutant concentrations represented by the random variable X also are multiplied by r. Similarly, for the water pollutant described in Chapter 8 (pages 235–239), Theorems 2 and 3 imply that if the source concentration c_o is multiplied by r, then the resulting concentrations also are multiplied by r. These results are "nonparametric" in that they apply to any distribution.

Multiplying all concentrations by the linear scaling factor r also has implications for the full distribution of X. Consider the CDF of air or water pollutant concentrations in the pre-control state:

$$P\{X \leq x\} = F_X(x) \qquad (\textit{pre-control state})$$

In the post-control state, the random variable X is multiplied by r, so the post-control CDF is written as

$$P\{X \leq rx\} = F_X(rx) \qquad (\textit{post-control state})$$

This result implies "geometric scaling": the value of the random variable at every frequency is multiplied by r. Thus, on logarithmic-probability paper (or any other probability paper with a logarithmic scale), the cumulative frequency distributions of the pre- and post-control state are expected to have exactly the same shapes, since all values along the curve will be multiplied by r. For example, all post-control values at a cumulative frequency of 10% are expected to be r times the values at the 10% frequency in the pre-control state. Geometric scaling will occur for all probability distributions, regardless of whether they are lognormal or not. This discussion gives rise to an additional theorem:

Theorem 4. If the random variable representing the source is multiplied by the rollback factor r, then the random variable representing the pollutant concentrations also will be multiplied by r, and the quantiles of the concentration distribution will be scaled geometrically.

Finally, multiplying any random variable by a coefficient r implies that the expected value will be multiplied by r and the standard deviation will be multiplied by the same coefficient r. Because the coefficient of variation is the ratio of the standard deviation to the expected value, the coefficient of variation will be unchanged. This result gives rise to another corollary:

Corollary 2. If any random variable is multiplied by a factor r, then its expected value and standard deviation also will be multiplied by r, and its coefficient of variation will be unchanged.

It is possible to imagine a slightly less restrictive set of assumptions about the source: instead of multiplying the random variable representing the source by the rollback factor r, all we know is that the *expected value* is multiplied by r. Theorems 1, 2, and 3 above still will apply, providing that the coefficient of variation of the source is the same in the pre- and post-control states. However, Theorem 4, which applies to the entire distribution, will not necessarily be valid. Because the linear scaling applies only to the expected value and the variance, it is useful to introduce two additional theorems:

Theorem 5. If the expected value of the random variable reflecting the source is multiplied by the rollback factor r, then the expected value of the concentrations observed in the post-control state will be multiplied by r.

Theorem 6. If the expected value of the random variable representing the source is multiplied by the rollback factor r, and if the coefficient of variation of the source remains the same, then the coefficient of variation of the final concentrations in the post-control state will be the same as the coefficient of variation in the pre-control state.

Notice that these theorems are completely general in that they do not require any of the distributions to be known. Indeed, these theorems apply to any distribution. Although the above theory applies to a single source and assumes that ambient background concentrations are zero, this theory also can be applied to multiple sources. In the multiple-source case, each source is treated individually and the results are added to give the final concentration. The process of adding individual concentrations sometimes is called "superposition," and is valid because molecules of the *same pollutant* do not interact chemically and are simply additive by the laws of physics. Superposition assumes that the physical and chemical processes involving the pollutant are linear with respect to the concentration.

Correlation

Thus far, development of the theory has assumed that the random variables representing the source Q and the dilution-diffusion phenomena D are independent and uncorrelated. Many sources, such as smoke emissions from a home burning wood for heating, are higher in the winter season when unfavorable meteorological conditions, such as frequent ground-based inversions, cause poor dilution and diffusion. The result is that both Q and D are higher in winter and lower in summer and therefore are correlated with each other. What effect does this correlation have?

Consider the case for the pre-control and post-control states for the product of the random variables $X = QD$ if Q and D happen to be correlated. The result is that the covariance in each case is not zero:

$$X_{pre} = QD, \quad Covariance\{Q, D\} \neq 0$$

$$X_{post} = rQD, \quad Covariance\{rQ, D\} \neq 0$$

If the random variable D representing the dilution-diffusion processes is unaffected by the rollback factor r, then we can substitute $QD = X_{pre}$ from the top equation into the bottom equation, and the bottom equation becomes:

$$X_{post} = rX_{pre}$$

This equation is the product of a constant and a random variable, and, by the laws of variances (see Chapter 3, "Variance, Kurtosis, and Skewness," page 41), its expected value and variance are given as follows:

$$E[X_{post}] = rE[X_{pre}]$$

$$Var(X_{post}) = r^2 Var(X_{pre})$$

Then the coefficient of variation is given by

$$CV(X_{post}) = \frac{\sqrt{r^2 Var(X_{pre})}}{rE[X_{pre}]} = \frac{\sqrt{Var(X_{pre})}}{E[X_{pre}]} = CV(X_{pre})$$

Thus, the coefficient of variation is the same for the pre- and post-control states, even though Q and D are correlated.

By the rules of covariances, the covariance in the pre-control state will be multiplied by r:

$$Covariance\{Q, D\}_{post} = Covariance\{rQ, D\}_{post} = rCovariance\{Q, D\}_{pre}$$

The correlation coefficient is the ratio of the covariance of two random variables to their respective standard deviations (i.e., the square root of their respective variances). Thus, for the post-control state, the correlation coefficient is given as

$$Correlation\{Q, D\} = \frac{rCovariance\{Q, D\}}{\sqrt{rVar(Q)}\sqrt{rVar(D)}} = \frac{Covariance\{Q, D\}}{\sqrt{Var(Q)}\sqrt{Var(D)}}$$

The right-hand side of this equation no longer includes r, and thus the correlation in the pre-control state is the same as the correlation in the post-control state:

$$Correlation\{Q, D\} = Correlation\{rQ, D\}$$

Thus, the correlation does not change, which gives rise to the following theorem:

> *Theorem 7.* If the random variable representing the source is correlated with the random variable representing dilution-diffusion phenomena, and if the source variable is multiplied by the rollback factor r, then the assumed correlation between the source and the dilution-diffusion phenomena in the post-control state will be the same as in the pre-control state.

A similar analysis can be undertaken to show that, if the pre-control concentration time series is correlated in time (that is, if successive concentrations are autocorrelated), then the post-control concentration time series, after multiplication by r, will have the same autocorrelation structure as the pre-control time series.

Overall, the changes that occur when the source strength is multiplied by the rollback factor r for correlated random variables are:

- The expected value of the controlled concentrations is multiplied by r.
- The variance of the controlled concentrations is multiplied by r^2.
- The coefficient of variation of the controlled concentrations is the same as the coefficient of variation of the uncontrolled concentrations.
- The covariance between the source and dilution-diffusion variable is multiplied by r.
- The correlation coefficient between the source and dilution-diffusion variables is the same for both the controlled and uncontrolled concentrations.

It may seem surprising that the above theory initially was developed for the case in which Q and D are uncorrelated random variables, and then they are allowed to be correlated. However, geometric scaling — multiplying one of the random variables comprising the product by the coefficient r — is a special type of transformation. An embedded assumption is that multiplying the source term by r does not change the random variable D. If r were actually a random variable, a necessary assumption would be that r and D are uncorrelated: $Covariance\{r, D\} = 0$. However, r is actually a fixed parameter of the pre- and post-control states. In the pre-control state, $r = 1$ for all values of Q and D; in the post-control state, r likewise is fixed (for example, $r = 0.30$) for all values of Q and D.

For air and water pollutants, the usual case is that the post-control concentration is linearly related to the source strength, and the values taken on by D are not affected by the values of r. For an outdoor air pollutant, D reflects the effect of meteorological conditions, and there are few cases in which changes in the source strength are likely to change the weather enough to affect the dilution-diffusion factors. An exception occurs for large plumes released from point sources, such as the emissions from a power plant stack. When the source emission rate is changed due to the rollback factor r, the temperature of the gases being emitted also may change, causing the "effective height" at which the plume is emitted to change. With the resulting change in plume rise, the dilution-diffusion variables can be affected. Once again, if r remains constant during the post-control state, and the nonlinearity caused by the micrometeorology is well understood, it may be possible to apply the rollback theory by substituting a new value for r for the one obtained for the linear case. Incorporating such nonlinearity into the theory is beyond the scope of this book but is a subject for future research.

Previous Rollback Concepts

These theorems and their mathematical basis constitute a Statistical Theory of Rollback (STR), and this theory is presented in this book for the first time. Similar equations have appeared in the environmental literature but without a theoretical basis. For example, Larsen[14,15] considers the deterministic case in which there is both a "background concentration" b in the pre-control state and adjusts the pre-control air quality C_{pre} due to a "growth factor" g in the source due to anticipated future increases in source emissions. His model seeks to achieve post-control concentrations that attain some "design value," such as an air quality standard, and the desired level of air quality becomes q. Thus the "effective" rollback factor in Larsen's model is given by

$$r = \frac{\textit{adjusted post-control air quality}}{\textit{adjusted pre-control air quality}} = \frac{q-b}{gC_{pre}-b}$$

Since the source "control factor" (proportion by which the source is reduced) is given by $s = 1 - r$, Larsen[14] suggests the following source control equation:

$$s = 1 - \frac{q-b}{gC_{pre}-b} = \frac{gC_{pre}-q}{gC_{pre}-b}$$

Georgopoulos and Seinfeld[16] suggest a probabilistic version of this equation: they replace the desired level of air quality q by a random variable Q and consider its expected value $E[Q]$, and they replace the pre-control concentration C_{pre} by its expected value $E[C_{pre}]$:

$$r = \frac{gE[C_{pre}] - E[Q]}{gE[C_{pre}] - b}$$

Their formulation is consistent with the Statistical Theory of Rollback developed in this book, although one also could imagine replacing the background concentration b by its expected value $E[b]$.

The following section illustrates why the product law, with which development of the STR begins, is applicable to air and water pollutants, and the last section of this chapter illustrates the statistical rollback concept by applying the theory to the lognormal distribution and to data from a field measurement study.

Environmental Transport Models in Air and Water

To understand why the product of two random variables is important in air pollution problems, it is important to examine the basic structure of the general Gaussian plume diffusion equation discussed on page 158 of Chapter 6:

$$X(d,w,h,u) = \frac{Q}{2\pi u \sigma_w(d)\sigma_h(d)} e^{-\frac{1}{2}\left[\frac{w^2}{\sigma_y^2(w)} + \frac{h^2}{\sigma_h^2(d)}\right]}$$

The notation for some of the variables has been changed from that used in Chapter 6 (section on "Plume Model," pages 154–159). Here, Q still denotes the source strength [mass/time]. However, d (formerly x in Chapter 6) denotes the distance from the source to the receptor point, w (formerly y in Chapter 6) denotes the horizontal distance from the plume centerline, h (formerly z in Chapter 6) denotes the vertical distance from the plume centerline, and u denotes the wind speed. Thus, $X(d,w,h,u)$ is the downwind pollutant concentration at a particular point for a particular wind speed. To simplify this equation, all the terms except Q can be combined together into one diffusion variable $D(d,w,h,u)$ reflecting the effect of downwind diffusion with dimensions of [time/length^3]. Thus, the concentration at a particular point can be written as:

$$X(d,w,h,u) = QD(d,w,h,u)$$

If the wind speed is a random variable U, then $D = D(d,w,h,U)$ above will be a random variable representing the effects of meteorological conditions. Since the source term Q will depend on the emissions from the source and will not depend on the meteorological conditions, the concentration X at the receptor point can be written as the product of two independent random variables:

$$X = QD$$

Because this equation is written in the same form as the product law analyzed above, the same conclusions will apply: If the source strength Q is multiplied by r, the expected value $E[X]$ also will be multiplied by r, and the coefficient of variation will remain fixed. For example, if the source strength Q is multiplied by $r = 0.30$, then the observed concentrations will be reduced to 30% of their prior values; the expected value will be reduced by 70%, and the coefficient of variation — which is attributed to meteorological conditions — will be unchanged. (As is the common practice for transport models, we assume that the pollutant from this source does not engage in chemical reactions that would introduce nonlinearity.)

As mentioned earlier in this chapter, a case exists in which the source and dilution-diffusion variables are not independent. The pollutant emission rate from a combustion source affects the gas exit temperature and velocity, which, in turn, affects the height of the plume rise. Here, the random variable D depends in some manner on r and Q. It is not clear whether such dependencies have a significant effect on predictions made by the rollback method, and additional research is needed on this phenomenon. In the meantime, it is assumed that this dependency exerts a negligible effect on the resulting concentrations. Usually, the standard Gaussian plume dispersion model (see page 158 of Chapter 6), when applied to air pollution sources, ignores such possible interactions.

The Theory of Successive Random Dilutions discussed in Chapter 8 (see pages 235–239 on "Water Quality") provides another example. According to that theory, pollutants released into the environment at some initial concentration c_o undergo a succession of dilutions that can be represented by a product of dimensionless dilution factors $D_1D_2...D_m$, yielding the final concentration C_m:

$$C_m = c_o D_1 D_2 \ldots D_m = c_o \prod_{i=1}^{m} D_i$$

Suppose we substitute the random variable C for the fixed value c_o and we lump all the dilution factors into a single random variable $D = D_1D_2...D_m$. If we denote the final concentration by X instead of C_m, then this equation can be written as the product of the independent random variables C and D:

$$X = CD$$

This equation for a water pollutant is identical in form to the air pollutant diffusion equation discussed above, except that the upstream concentration source C is substituted for Q, and D represents dilution in a stream rather than atmospheric dispersion.

For example, consider an upstream source discharging a pollutant at concentrations represented by the random variable C. The concentrations measured many miles downstream are represented by the random variable X. Suppose that the source operates for many years and that data on the downstream concentrations are collected daily over 10 years. A decision is made to control the effluent, reducing the expected value of the source concentrations by 70%, so the rollback factor $r = 0.30$. If the coefficient of variation of the source remains unchanged, then the mean of the downstream concentrations will decrease by 70%, and the coefficient of variation of the downstream concentrations will be unchanged.

An implicit assumption above is that the dilution-diffusion process (represented by the random variable D) is determined by nature and that its mean and variance are essentially the same in the pre-control and post-control states. Ideally, one needs to observe the pre-control state over a long enough time period to reflect conditions of nature that are also likely to occur in the post-control state. For example, meteorological conditions observed over one 10-year period are unlikely to differ greatly from meteorological conditions observed over any other 10-year period. In practice, shorter time periods may be used if reasonable steps are taken to ensure the data are "representative" of longer-term conditions.

What is the meaning of the rolled-back (post-control) frequency distribution of concentrations and how can it be used? Ordinarily, we would like to use the post-control frequency distribution to make predictions about quantiles in the post-control period. For example, the analyst may be interested in the quantile x_{80}, the value that at least 80% of the concentrations lie below and 20% exceed, or in the quantile x_{98}, the value that at least 98% of the concentrations lie below and 2% exceed. The validity of predictions based on the post-control frequency distribution will depend on the validity of the dilution-diffusion factors in the pre-control state, and the extent to which post-control dilution-diffusion conditions match pre-control conditions. Ideally, we would like to predict the *expected value* of each quantile of interest; for example, $E[X_{80}]$ and $E[X_{98}]$ in the post-control state. However, predicting an expected value from real data usually yields an approximation, because a true expected value requires asymptotic conditions. If the approximate nature of these predictions is understood, and if an effort is made to use a data set that is fairly representative of the conditions under study, then good predictions are possible. Application of the STR, like other predictive models, requires judgment by the analyst. Similar judgment is required, for example, when one applies an atmospheric diffusion model, because the analyst must choose meteorological data (wind speed and stability) that are appropriate for the post-control conditions in which the model is to be applied.

APPLICATION TO ENVIRONMENTAL PROBLEMS

As noted in Chapter 8, the lognormal distribution commonly appears in a great variety of environmental topic areas. Examples given there include the concentrations of pollutants observed in outdoor air, indoor air, streams, lakes, ground waters; radionuclides in soils, plants, fishes, birds, and mammals; additives in foods; radionuclides and trace metals in human tissue, blood, and feces; and precious metals in the earth. Indeed, a great variety of environmental data sets exists for which the two-parameter, three-parameter, and multi-parameter lognormal distributions seem well suited. The symmetrical normal distribution clearly is inappropriate for these environmental data sets. When we combine the lognormal distribution with the Statistical Theory of Rollback (STR) described above, the result is a powerful technique for predicting the effect of regulatory decisions on concentrations in the environment. As discussed above, the source control theory outlined earlier is sufficiently general that one does not need to show that a given distribution is lognormal, because the theory can be applied to any distribution.

Rollback of the Two-Parameter Lognormal Distribution

If we assume that a particular data set can be represented by the two-parameter lognormal distribution, then the equations given in Table 9.2 (page 262) apply. In the pre-control state, the arithmetic mean and arithmetic standard deviation are written as

$$\alpha_{pre} = e^{\left(\mu_{pre} + \frac{1}{2}\sigma^2_{pre}\right)}$$

$$\beta_{pre} = \sqrt{e^{(2\mu_{pre} + \sigma^2_{pre})}(e^{\sigma^2_{pre}} - 1)}$$

Now suppose we apply source control. By Theorem 2, the post-control expected value is given by the product of the rollback factor r and the pre-control expected value:

$$\alpha_{post} = r\alpha_{pre}$$

By Corollary 1 and the rules for variances (see Chapter 3, "Variances, Kurtosis, and Skewness," pages 41–42), the post-control standard deviation can be written as a function of the pre-control standard deviation:

$$\beta_{post} = r\beta_{pre}$$

If we now form the coefficient of variation in the post-control state, we see that r disappears from the result, and therefore the coefficient of variation in the pre-control state equals the coefficient of variation in the post-control state:

$$CV(X)_{post} = \frac{\beta_{post}}{\alpha_{post}} = \frac{r\beta_{pre}}{r\alpha_{pre}} = \frac{\beta_{pre}}{\alpha_{pre}} = CV(X)_{pre}$$

Thus, for the lognormal distribution, the coefficients of variation in the pre- and post-control states are the same, a property that holds for all distributions by Corollary 2. For the lognormal distribution, the coefficient of variation in both states is a function of just one parameter, σ:

$$CV(X)_{post} = CV(X)_{pre} = \frac{\beta_{pre}}{\alpha_{pre}} = \frac{e^{\mu+\frac{\sigma^2}{2}}\sqrt{e^{\sigma^2}-1}}{e^{\mu+\frac{\sigma^2}{2}}} = \sqrt{e^{\sigma^2}-1}$$

Examination of the other equations for the lognormal distribution in Table 9.2 (page 262) for the pre- and post-control states indicates that some of the parameters will be multiplied by r while others will remain constant. For example, the geometric mean (also the median) is scaled by r:

$$\mu_{g_{post}} = r\mu_{g_{pre}} \quad \text{and} \quad x_{50_{post}} = rx_{50_{pre}}$$

By comparison, the standard geometric deviation remains fixed:

$$\sigma_{g_{post}} = \sigma_{g_{pre}}$$

By Theorem 4, all the quantiles of the distribution are scaled geometrically, or multiplied by r. Thus, the concentration at the 10% frequency in the pre-control state, x_{10}, becomes rx_{10} in the post-control state.

An idealized example for nitrogen dioxide (NO_2) concentrations from Georgopoulos and Seinfeld[16] illustrates these concepts graphically (Figure 9.7). In the pre-control state, the geometric mean NO_2 concentration was $\mu_g = 0.10$ ppm, and the standard geometric deviation was $\sigma_g = 1.4$. Since it is assumed that the distribution follows a two-parameter lognormal distribution with median $x_{50} = \mu_g = 0.10$ ppm, the pre-control frequency distribution is plotted as a

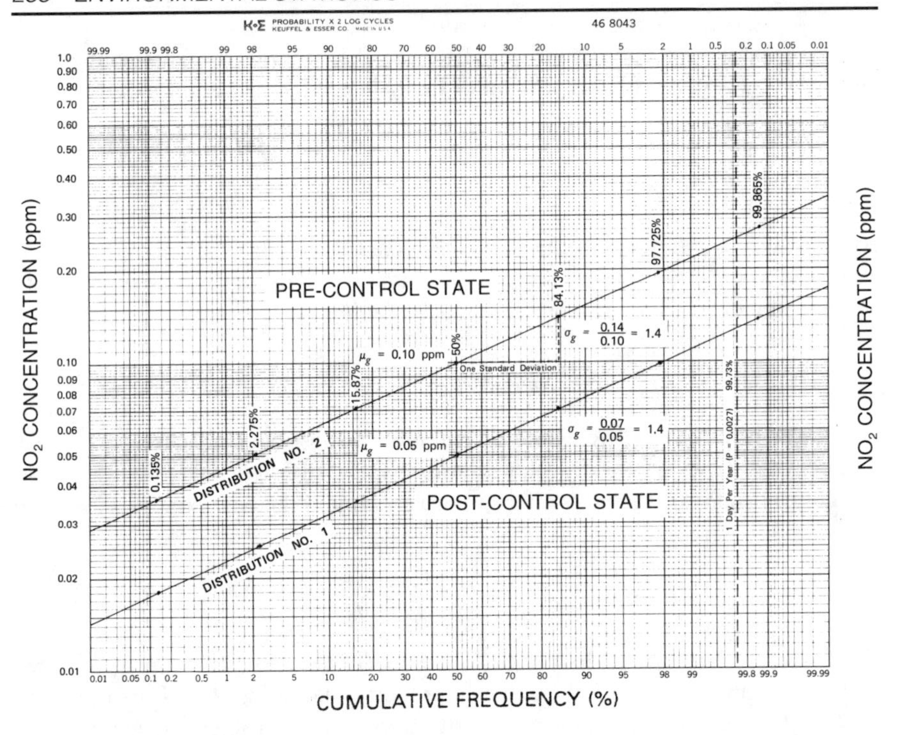

Figure 9.7. Frequency distributions of nitrogen dioxide (NO_2) concentrations with a two-parameter lognormal distribution and r = 0.5 (Based on Reference 16).

straight line on logarithmic-probability paper with its quantile for the 84.13% frequency given by $x_{84.13} = \sigma_g x_{50}$ = (1.4)(0.10) = 0.14 ppm. Suppose that the source is to be controlled by r = 0.50 with the consequence that all NO_2 concentrations are reduced by 50%. Then the new geometric mean in the post-control state is given by $\mu_g(\text{post}) = r\mu_g(\text{pre})$, or $\mu_g(\text{post})$ = (0.50)(0.10 ppm) = 0.05 ppm. By Corollary 2, the coefficient of variation is unchanged, so the post-control distribution will have the same standard geometric deviation, or σ_g = 1.4. The result is to shift the line representing the pre-control distribution on logarithmic probability paper downward everywhere by the same vertical distance. The result is a new line representing the post-control distribution that is exactly parallel to the uncontrolled distribution. All of the quantiles (every point along the line on the logarithmic scale) are multiplied by 50%. The arithmetic mean and arithmetic standard deviation also are multiplied by 50%. (See Problem 7 at the end of this chapter for a more detailed treatment of this example.)

Rollback of Other Distributions

The two-parameter lognormal distribution described above plots as a straight line on logarithmic-probability paper; however, other distributions and data collected in field measurements often show curvature when plotted on logarithmic-probability paper. The statistical theory developed above also can be applied to these cases. On logarithmic-probability paper, a fixed concentration ratio corresponds to a fixed vertical distance on the logarithmic scale. Therefore, if all the concentrations are multiplied by the fixed coefficient r, where $0 < r < 1$, then the post-control curve shifts a fixed vertical distance downward. On the logarithmic scale, the new post-control curve lies the same vertical distance everywhere below the pre-control curve, but the horizontal (frequency) intercepts do not change.

For example, suppose that SO_2 concentrations have been measured for a number of years near a single source (perhaps a power plant) and that the frequency distribution plots as a concave-downward curve (Figure 9.8). Suppose also that the background SO_2 concentration is zero and that we seek to reduce

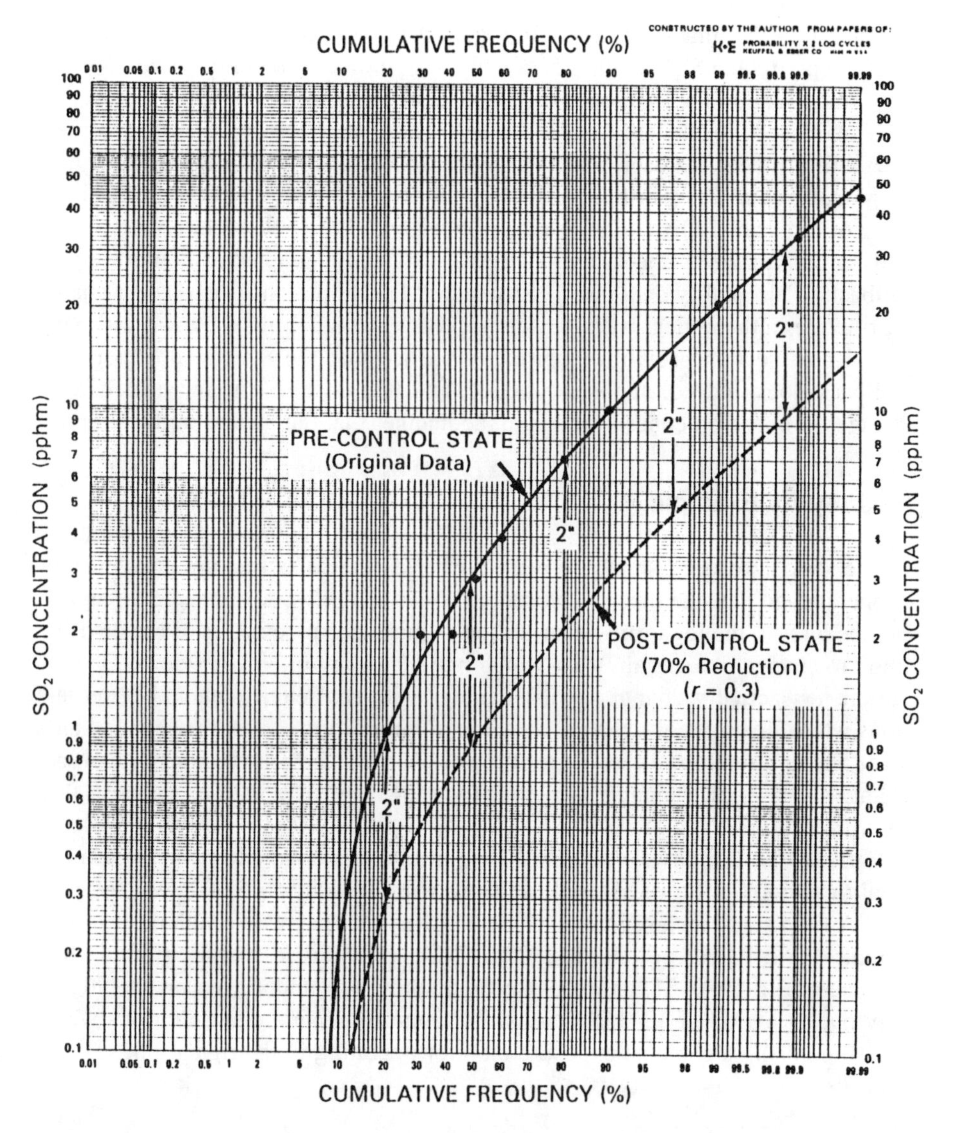

Figure 9.8. Logarithmic-probability plot of sulfur dioxide concentrations that do not follow a two-parameter lognormal distribution, showing the result of reducing concentrations by 70% ($r = 0.3$).

emissions from this source by 70%, giving a rollback factor of $r = 0.30$. After the analyst plots the data points on logarithmic-probability paper, a smooth curve is drawn through the data points. On the original graph paper used in this figure, the vertical distance between 3 pphm and 10 pphm — which corresponds to r = 3 pphm/10 pphm = 0.3 — was measured to be 2 inches (that is, 2″)*. To generate the curve of the frequency distribution predicted in the post-control state, the analyst draws a second curve that is exactly 2″ below the pre-control curve at every point as in Figure 9.8.

In Figure 9.8, plotting a smooth curve through the data permitted the STR to be applied with relative ease. The SO_2 data plotted in Figure 9.8 for the hypothetical power plant also happens to be the same data for SO_2 measured in Washington, DC, that was plotted in Figure 9.5. Thus, in Figure 9.8, we actually are applying the STR to the LN3C model. Use of a probability model to provide smoothing to the frequency distribution of observed data greatly facilitates the statistical rollback analysis. However, it is not necessary to specify a formal probability model for each data set, and one also could apply the STR to the empirical frequency distribution of the raw observations. Application of arbitrary or empirically-fit curves was suggested in 1992 by Larsen,[17] who originally proposed the rollback approach and provided examples of how to apply rollback to air quality data in a 1961 paper.[14] It is hoped that the present chapter provides a theoretical basis for this approach, which previously was lacking.

The next section provides empirical evidence that the Statistical Theory of Rollback does, indeed, apply in practice, and that changes in emissions affect changes in observed concentrations in the manner that the theory predicts.

Field Study Example

The serious reader is likely to want further evidence that nature does indeed cooperate and that the theories presented above really work in practice. Applying the STR to the lognormal distribution requires the parameters of the distribution to change in a certain way, which may not be intuitively obvious. Do the parameters actually change in that manner with real environmental observations when the emissions change by a known proportion? What is needed to answer these questions is an experiment in which the source emissions are changed in a known manner while all the other natural conditions remain the same.

Fortunately, a highway exposure field study was conducted on El Camino Real, a major California arterial highway with 30,000 to 45,000 vehicles per day. A field study on a segment of this highway that passes through three cities (Menlo Park, Palo Alto, and Los Altos, CA) provides a "testing ground" for this theory. In 1980–81, an exposure field study was undertaken on this highway over a period of 13–1/2 months.[18] Carbon monoxide concentrations were measured using a personal exposure monitor operating in the right-front seat of

*Figure 9.8 has been photographically reduced in size. The original vertical distance between the two curves plotted on Keuffel and Esser paper (No. 46–8043, "Probability by 2 Log Cycles") was found to be approximately 2 inches for a ratio of 0.3.

the passenger compartment of a moving test vehicle, and the vehicle always drove the exact same route, keeping up with traffic along the way. Each trip covered 11.8 miles (5.9 miles in each direction) and lasted between 31 and 61 minutes.

In 1991–92, the study was repeated on the same portion of the roadway.[19] Of 131 trips taken over 15 months in the second study, 96 were made on the same day of the week and the same time of day (closest to the original date) to match the original conditions 11 years before as much as possible. The window position and vehicle air exchange rate were similar in both field studies, and the monitoring instruments were equivalent. Emissions for the two time periods were calculated by combining information on traffic volumes, exhaust emissions by model year based on roadside tests, and vehicle registrations by model year in a mathematical model. Details of the emissions model are beyond the scope of this book and are described in a paper.[20] These calculations showed that emissions declined to 47% of their 1980–81 levels over the 11-year period. Thus, for this situation, $r = 0.47$.

If the frequency distribution of the average CO trip exposures in 1980–81 is plotted on logarithmic-probability paper, the resulting dots lie on a nearly straight line ("Observed" in Figure 9.9). A line fit through these points by visual estimation passes through the median at $x_{50} = 9.2$ ppm and the 84.13% frequency at $x_{84.13} = 13.8$ ppm. Thus, the standard geometric deviation σ_g is given by

$$\sigma_g = \frac{x_{84.13}}{x_{50}} = \frac{13.8}{9.2} = 1.5$$

To predict the exposures in 1991–92 using the STR developed in the previous sections of this chapter, we set $r = 0.42$ from the emission estimate. Then we compute the median for 1991–92 as $x_{50} = (0.42)(9.2 \text{ ppm}) = 3.9$ ppm. By Theorem 1, the standard geometric deviation will be constant at $\sigma_g = 1.5$. Thus the predicted lognormal distribution for 1991–92 should be a straight line passing through the median of $x_{50} = 3.9$ ppm and the 84.13% frequency of $x_{84.13} = (1.5)(3.9) = 5.8$ ppm. As Figure 9.9 shows, the dots plotted to represent the frequency distribution of observations in 1991–92 lie almost exactly on this line labeled "Predicted by Emissions Model." Thus, the STR works well in this situation.

In 1966, the National Air Pollution Control Administration (NAPCA), a predecessor to the Environmental Protection Agency, used a mobile sampling van to measure CO exposures on arterial highways in 11 U.S. cities.[21] Although the trips were taken during two-week periods in different parts of the year, it was possible to construct a composite frequency distribution combining data from all cities. Once again, the standard geometric deviation has a value of approximately $\sigma_g = 1.5$, but the median is much higher, reflecting the emissions of motor vehicles in the U.S. before emission control systems were in widespread use.

The good fit of the predicted distribution to the observed distribution provides a basis for predicting the distribution of exposures in future years. Applying the same emissions model[20] and using best-case estimates of vehicle emissions for the years 2000–2001, we obtain a rollback factor, compared with 1991–92, of $r = 0.40$. By the statistical theory, the predicted distribution for

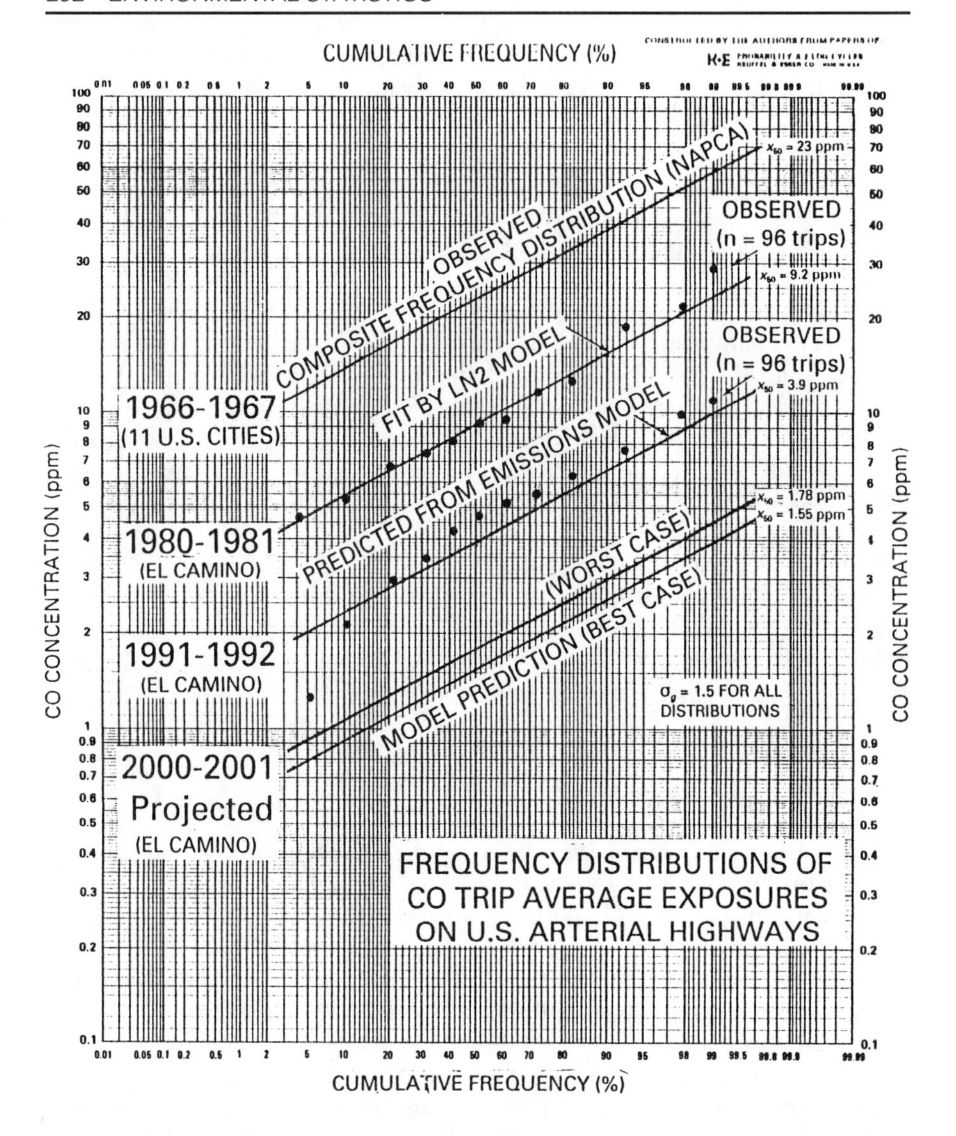

Figure 9.9. Logarithmic-probability plot of the CO exposures inside a moving vehicle measured on the El Camino Real urban arterial highway in California in 1980–81 and again in 1991–92.[19] The straight line in 1980–81 was fit visually to the data, and the statistical rollback technique predicts the straight line in 1991–92 from predicted emissions.[20] An earlier field study (1966–67) shows a similar standard geometric deviation.[21] The statistical rollback technique predicts the 2000–2001 distribution from predicted emissions.

2000–2001 will be a straight line passing through the median $x_{50} = (0.40)(3.9) = 1.55$ ppm with the same standard geometric deviation of $\sigma_g = 1.5$, so the quantile at the 84.13% frequency is given by $x_{84.13} = (1.55)(1.5) = 2.3$ ppm.

Because this approach predicts the entire frequency distribution, we can make estimates of any quantile, although one should always stay within the range of the observed data. Since there were 96 trips in the two periods that were matched by time of day, season, and day of the week,[19] the highest single concentration corresponds to a frequency of $f = i/(n + 1) = 96/97 = 99\%$ using the plotting formula given in Chapter 3 (page 56 in the section on "Histograms and Frequency Plots"). Thus, the expected value of the maximum concentration for 96 trips (99% frequency on the straight lines) in Figure 9.9 declines from 58 ppm in 1966–67 to 23 ppm in 1980–81 to 10 ppm in 1991–92 to only 4.0 ppm in 2000–2001. The worst-case projection gives a rollback factor of $r = 0.46$, so the predicted maximum may be as high as (0.46)(10 ppm) = 4.6 ppm in 2000–2001. Correspondingly, the median CO exposure drops from 23 ppm in 1966–67 to 9.2 ppm in 1980–81 to 3.9 ppm in 1991–92 to a range of 1.55–1.78 ppm in 2000–2001. This striking decline in vehicular exposures can be attributed to the success of emission control systems installed on new motor vehicles in the U.S., supplemented in California by a vehicle smog inspection program. The effect of emission control systems is gradual, because the strictest controls apply to model years in the early 1990's, and it takes many years for the new models to make their way into the entire fleet of vehicles. The highest 1% of the CO concentrations projected for the year 2000–2001 can be seen from the bottom lines of the graph to be only 4.0–4.6 ppm, a very low concentration range from a health standpoint. This concentration should be viewed as the expected highest value of one out of 100 trips taken on this highway during the first year of the next century. These predictions by the STR suggest, that, with no additional regulations, CO exposures experienced by drivers and passengers will decline to extremely low concentrations that are unlikely to constitute a significant health threat in future years on U.S. arterial highways .

CONCLUSIONS

The theory and principles presented in this chapter should require a minimum of assumptions and are intended to provide the analyst with easy-to-use, practical techniques for understanding and predicting environmental changes. The techniques have a strong empirical foundation, since they rely on observed data and distributions. Unlike traditional diffusion models, these statistical techniques permit the analyst to make predictions of expected concentrations over very short averaging times, such as 1-hour trips on arterial highways. It also is hoped the theoretical basis for these techniques, which is presented for the first time in this book, will provide data analysts, researchers, and regulatory officials with useful tools for analyzing environmental problems and making predictions about them. Because of the generality of the Statistical Theory of Rollback (STR), it is hoped that future investigators will evaluate its potential usefulness by comparing its predictions with field measurements in air pollution control, water pollution control, and other environmental fields.

PROBLEMS

1. Using the same approach that was used to derive the PDF of the two-parameter lognormal distribution, show that the PDF of the three-parameter lognormal distribution is given by

$$f_X(x) = \frac{1}{(x-a)\sigma\sqrt{2\pi}} e^{-\frac{1}{2}\left(\frac{x-a-\mu}{\sigma}\right)^2} \quad \text{for } x \geq a;\ a, \sigma > 0;\ -\infty < \mu < +\infty$$

Plot the PDF of this distribution for $a = 2$, $\mu = 3$, and $\sigma = 2$.

2. Derive the following equation for the mode of the 2-parameter lognormal probability model by beginning with the PDF and setting its first derivative equation to zero:

$$x_m = e^{\mu - \sigma^2}$$

3. For an example given in this chapter, $\mu = 2.2$ and $\sigma = 0.41$. What is the corresponding geometric mean and standard geometric deviation? [Answers: $\mu_g = 9.03$, $\sigma_g = 1.51$] What is the arithmetic mean and arithmetic standard deviation? [Answers: $\alpha = 9.82$ and $\beta = 4.20$] What is the variance of the distribution? [Answer: $Var(X) = 17.7$] What is the value of X at the 84.13% frequency? [Answer: 13.6]
4. Suppose the geometric mean of a lognormally distributed random variable is given as $\mu_g = 100.0$ and the standard geometric deviation is $\sigma_g = 2.500$. What are the values of its (a) normal parameters and (b) arithmetic parameters? [Answers: (a) $\mu = 4.605$, $\sigma = 0.916$; (b) $\alpha = 152.2$, $\beta = 174.5$] What is the quantile at the frequency of 84.13%? [Answer: 250.0] What is the value of the mode? [Answer: 43.2]
5. Calculate all the parameter values of the lognormal probability model for the two cases plotted in Figure 9.1: (a) $\mu_g = 3$, $\sigma_g = 1.5$, and (b) $\mu_g = 3$, $\sigma_g = 2$. [Answers: (a) $\mu = 1.099$, $\sigma = 0.4055$, $\alpha = 3.257$, $\beta = 1.377$; (b) $\mu = 1.099$, $\sigma = 0.6931$, $\alpha = 3.815$, $\beta = 2.996$] What are the values of the mode in (a) and (b)? [Answers: (a) 2.545; (b) 1.855] Although the medians for (a) and (b) are fixed at $x_{50} = 3$, both the arithmetic mean and standard deviation are larger in (b) than (a); then why is the mode smaller in (b) than in (a)? [Hint: Plot the PDF for both distributions.]
6. Suppose the arithmetic mean of a lognormal distribution is $\alpha = 20$ μg/m^3 and the variance is $\beta^2 = 100$ (μg/m^3)^2. What is the coefficient of variation of the distribution? [Answer: $v = \{100\}^{0.5}/20 = 0.5$] What is the normal standard deviation of this distribution? [Answer: $\sigma = \{\ln(v^2 + 1)\}^{0.5} = 0.472$] What is the normal mean of this distribution? [Answer: $\mu = \ln \alpha - 0.5\sigma^2 = 2.88$] What are the values of the geometric parameters of this distribution? [Answer: $\mu_g = e^{\mu} = 17.8$ μg/m^3; $\sigma_g = e^{\sigma} = 1.60$] What is the value of the mode? [Answer: 14.3 μg/m^3]
7. In the example describing statistical rollback of NO_2 concentrations in Figure 9.7, the reduction in the sources is 50%, and $r = 0.5$. Using the values for the geometric mean, $\mu_g = 0.1$ ppm, and the standard geometric deviation, $\sigma_g = 1.4$, given in Figure 9.7, calculate: (a) the pre-control normal parameters, and (b) the post-control normal parameters. [Answers: (a) $\mu = -2.303$, $\sigma = 0.3365$; (b) $\mu = -2.996$, $\sigma = 0.3365$] Using these results, calculate the arithmetic mean and standard deviation for (a) the pre-control and (b) post-control states. [Answers: (a) $\alpha = 0.1058$ ppm, $\beta = 0.03663$ ppm; (b) $\alpha = 0.05290$ ppm, $\beta = 0.01831$ ppm] Show that the arithmetic mean for

the post-control state is r times the arithmetic mean for the pre-control state. Explain why this happens for (a) the two-parameter lognormal distribution and (b) any distribution. Show that the arithmetic standard deviation for the post-control state is r times the arithmetic standard deviation for the pre-control state. Explain why this happens for (a) the two-parameter lognormal distribution and (b) any distribution. From these results, compute the coefficients of variation for the two states and show that they are the same. Explain why this happens for (a) the two-parameter lognormal distribution and (b) why it must happen for any distribution. Using the techniques for evaluating the CDF of the normal distribution described in Chapter 7, compute the values of the quantiles of the two-parameter lognormal distribution at the frequencies of 0.1%, 1%, 10% 20%, 30%, 40%, 50%, 60%, 70%, 80%, 90%, 99%, and 99.9% for the two states and make a table comparing them. Notice that Figure 9.7 shows a vertical line corresponding to a frequency of one day per year (99.73%); compute the pre- and post-control NO_2 concentrations for this frequency and explain the possible relevance of this value to air pollution control regulations.

REFERENCES

1. Kapteyn, J.C., *Skew Frequency Curves in Biology and Statistics* (Groningen, Noordhoff: Astronomical Laboratory, 1903).
2. Aitchison, J., and Brown, J.A.C., *The Lognormal Distribution*, (London, England: Cambridge University Press, 1973).
3. Kahn, Henry D., "Note on the Distribution of Air Pollutants," *J. Air Poll. Control Assoc.* 23(11):973 (November 1973).
4. Larsen, R.I., "Future Air Quality Standards and Industrial Control Requirements," *Proceedings of The Third National Conference on Air Pollution*, PHS Publication No. 1649, U.S. Government Printing Office, Washington, DC (1967).
5. Larsen, R.I., and H.W. Burke, "Ambient Carbon Monoxide Exposures," Paper No. 69–167 presented at the 62nd Annual Meeting of the Air Pollution Control Association (1969).
6. Larsen, R. I., "Relating Air Pollutant Effects to Concentration and Control," *J. Air Poll. Control Assoc.*, 20: 214–225 (1970).
7. Larsen, R.I., "A Mathematical Model for Relating Air Quality Measurements to Air Quality Standards," U.S. Environmental Protection Agency, Research Triangle Park, NC, Publication No. AP–89 (November 1971).
8. Mage, David T., and Wayne R. Ott, "An Evaluation of the Methods of Fractiles, Moments, and Maximum Likelihood for Estimating Parameters When Sampling Air Quality Data from a Stationary Lognormal Distribution," *Atmos. Environ.*, 18 (1): 163–171 (1984).
9. Nehls, G.J., and G.G. Akland, "Procedures for Handling Aerometric Data," *J. Air Poll. Control Assoc.*, 23:180 (1973).
10. Ott, W.R., D.T. Mage, and V.W. Randecker, "Testing the Validity of the Lognormal Probability Model," U.S. Environmental Protection Agency, Washington, DC, EPA-600/4–79–040, June 1979.

11. Mage, D. T., "On the Lognormal Distribution of Air Pollutants," Proceedings of the Fifth Meeting of the Expert Panel on Air Pollution Modeling, NATO/CCMS N.35, Roskilde, Denmark, June 4–6 (1974).
12. Ott, Wayne R., and David T. Mage, "A General Purpose Univariate Probability Model for Environmental Data Analysis," *Comput. & Ops. Res.*, 3:209–216 (1976).
13. Mage, David T., and Wayne R. Ott, "Refinements of the Lognormal Probability Model for Analysis of Aerometric Data," *J. Air Poll. Control Assoc.*, 28(8): 796–798 (1978).
14. Larsen, Ralph, "A Method for Determining Source Reduction Required to Meet Air Quality Standards," *J. Air Poll. Control Assoc.*, 11(2): 71-76 (February 1961).
15. Larsen, Ralph, "Determining Reduced-Emission Goals Needed to Achieve Air Quality Goals—A Hypothetical Case," *J. Air Poll. Control Assoc.* 17(2): 823–829 (December 1967).
16. Georgopoulos, Parros G., and John H. Seinfeld, "Statistical Distributions of Air Pollutant Concentrations," *Environmental Science and Technology*, 16(70):401A–416A (1982).
17. Larsen, Ralph I., personal communication, U.S. Environmental Protection Agency, Research Triangle Park, NC (1992).
18. Ott, Wayne, Paul Switzer, and Neil Willits, "Carbon Monoxide Exposures Inside an Automobile Traveling on an Urban Arterial Highway," *J. Air & Waste Manag. Assoc.*, 44:1010–1018 (August 1994).
19. Ott, W., P. Switzer, and N. Willits, "Trends of In-Vehicle CO Exposures on a California Arterial Highway Over One Decade," Paper No. 93-RP-116B.04 presented at the 86th Annual Meeting of the Air and Waste Management Association, Denver, CO, June 14–18, 1993.
20. Yu, Liya, Lynn Hildemann, and Wayne Ott, "A Mathematical Model for Predicting Trends in Carbon Monoxide Emissions and Exposures on Urban Arterial Highways" (in press).
21. Lynn, David, Elbert Tabor, Wayne Ott, and Raymond Smith, "Present and Future Commuter Exposure to Carbon Monoxide," Paper No. 67–5 presented at the 60th Annual Meeting of the Air Pollution Control Association, Cleveland, OH, June 1967.

Index

N